HADRON TRANSITIONS IN THE QUARK MODEL

HADRON TRANSITIONS IN THE QUARK MODEL

A. LE YAOUANC, LL. OLIVER, O. PÈNE
and J.-C. RAYNAL

Laboratoire de Physique Théorique et Hautes Energies, Orsay

GORDON AND BREACH SCIENCE PUBLISHERS
New York London Paris Montreux Tokyo Melbourne

Gordon and Breach Science Publishers

Post Office Box 786
Cooper Station
New York, New York 10276
United States of America

Post Office Box 197
London WC2E 9PX
England

58, rue Lhomond
75005 Paris
France

Post Office Box 161
1820 Montreux 2
Switzerland

3–14–9, Okubo
Shinjuku-ku, Tokyo
Japan

Private Bag 8
Camberwell, Victoria 3124
Australia

Library of Congress Cataloging-in-Publication Data

Hadron transitions in the quark model.
Bibliography: p.
Includes index.
1. Quark models. 2. Hadron spectroscopy.
I. Le Yaouanc, Alain, 1944–
QC793.5.Q2522H33 1987 539.7′216 87–8579
ISBN 2–88124–214–6

 Printed in Great Britain by Galliard (Printers) Ltd, Great Yarmouth.

Contents

Preface ix

1. INTRODUCTION 1

1.1 Outlook and Motivation 1

1.1.1 Quarks and hadrons 1

1.1.2 Transitions of hadrons 3

1.1.3 Fundamental theory versus the quark model 5

1.1.4 Basic features of the quark model 10

1.1.5 Main emphases of the book 14

1.1.6 Particular topics not covered by this book 16

1.2 Quarks, Interactions and Symmetries 18

1.2.1 Strong interactions of quarks 18

1.2.2 Electroweak interactions of quarks 20

1.2.3 Flavor symmetries 23

1.2.4 Chiral symmetry 26

1.3 The Hadron Spectrum 29

1.3.1 Flavor independence and color saturation 29

1.3.2 Meson spectrum 33

1.3.3 Baryon spectrum and baryon wave functions 39

1.3.4 The quark–quark potential 47

1.3.5 Spin-dependent effects and relativistic potentials 52

2. NONRELATIVISTIC QUARK MODEL OF TRANSITIONS 59

2.1 Electroweak Transitions 59

2.1.1 Radiative decays 60

2.1.1(a) The transition Hamiltonian 60

2.1.1(b) General presentation of the nonrelativistic approximation 61

2.1.1(c) Practical calculations 63

2.1.2 Semileptonic weak decays 70

2.1.3 Quark-antiquark annihilation into lepton pairs 74

2.1.4 Nonleptonic weak matrix elements 80

2.2 Strong Decays 87

2.2.1 Elementary-meson emission 88

2.2.2 OZI rule, OZI-forbidden decays 92

2.2.2(a) General discussion 92
2.2.2(b) Relation between cross-sections and annihilation widths 94
2.2.2(c) Calculation of η_c decay into hadrons 96
2.2.3 OZI-allowed decays through pair creation 98
2.2.3(a) General formulation of the quark-pair-creation (QPC) model 99
2.2.3(b) Comparison between the QPC and elementary-emission models 103
2.2.3(c) Some examples of explicit calculations in the QPC model 104
2.2.3(d) The Cornell model of strong decays 108
2.2.3(e) Brief comments on QCD string models and further comparison between the pair-creation models 110
2.2.4 Rearrangement models of diquonium decay 113
2.2.5 Computation of matrix elements in the QPC model 116
2.2.5(a) Baryon decay 116
2.2.5(b) Harmonic-oscillator wave functions, centrifugal barrier and anti-$SU(6)_W$ signs 119
2.2.5(c) Meson decay 121

3. THE QUARK MODEL BEYOND THE NONRELATIVISTIC APPROXIMATION 123

3.1 Introduction 123

3.2 Current-Matrix Elements in the Dirac Equation 126
3.2.1 Introduction 126
3.2.2 The bag model and the question of quark masses 127
3.2.3 Decays by quantum emission 128
3.2.4 General remarks on the v/c expansion 131
3.2.5 Systematic v/c expansion 135
3.2.6 Pion transitions 140
3.2.7 Radiative transitions 143
3.2.8 Two-quark interactions 147

3.3 Hamiltonian Methods 148

3.4 Bethe–Salpeter Amplitudes and the Mandelstam Formalism 154
3.4.1 Introduction 154
3.4.2 Meson annihilation into leptons 159
3.4.3 Transition-matrix elements of currents 160
3.4.4 The $\pi^0 \to \gamma\gamma$ decay 163
3.4.5 Strong-interaction decays 170
3.4.6 Recovering the naive quark model in the Mandelstam formalism 175

3.4.7 Center-of-mass motion and internal wave functions 179
3.4.8 QCD radiative corrections to current vertices 180

4. PHENOMENOLOGICAL APPLICATIONS 183
4.1 General Remarks 183
4.1.1 Basic processes and experimental measurements 183
4.1.2 Parameters of the nonrelativistic quark model 184
4.1.3 Ambiguities of the nonrelativistic quark model 189
4.1.4 Algebraic predictions and symmetry approaches 191
4.1.5 General status of the naive quark model for light quarks 193
4.2 Semileptonic Weak Decays 194
4.2.1 Baryon semileptonic decays 194
4.2.2 Meson semileptonic decays 199
4.3 Radiative Transitions 200
4.3.1 Radiative transitions of baryons 201
4.3.1(a) The nonrelativistic approximation 201
4.3.1(b) SU(6) analysis 207
4.3.1(c) The Roper-resonance radiative decay 209
4.3.2 Meson radiative transitions 210
4.4 Strong Decays Allowed by the Okubo–Zweig–Iizuka Rule 219
4.4.1 Introduction 219
4.4.2 Strong decays by pion emission. Direct and recoil terms 220
4.4.3 The decays of the Δ trajectory into Nπ 223
4.4.4 Examples of pion emission with two partial waves: $N^*\to\Delta\pi$, $B\to\omega\pi$ 225
4.4.5 $N^*\to\pi\pi N$ in the quark-pair-creation model 228
4.4.6 The decays of radial excitations 231
4.4.6(a) $\rho'(1600)$ decays 232
4.4.6(b) Charmonium decays above the $D\bar{D}$ threshold 233
4.4.6(c) Roper resonance decays 238
4.5 Meson $q\bar{q}$ Annihilation 242
4.5.1 Introduction 242
4.5.2 Meson annihilation into lepton pairs 244
4.5.3 Hadronic OZI-forbidden rates 248
4.6 Nonleptonic Weak Decays 250
4.6.1 General framework 250
4.6.2 The most naive model of hyperon decays 260
4.6.3 Short-distance QCD corrections 268
4.6.4 Relativistic corrections 275
4.6.5 Corrections due to excited baryon intermediate states 277

APPENDIX 281

LITERATURE SURVEY 287

REFERENCES 297

INDEX 307

Preface

A l'exemple d'icelui vous convient être sages, pour fleurer, sentir et estimer ces beaux livres de haute graisse, légers au pourchas et hardis à la rencontre. Puis, par curieuse leçon et méditation fréquente, rompre l'os et sucer la substantifique moelle, c'est à dire ce que j'entends par ces symboles pythagoriques, avec espoir certain d'être faits escors et preux à la dite lecture, car en icelle bien autre goût trouverez, et doctrine plus absconse, laquelle vous révèlera de très hauts sacrements et mystères horrifiques, tant en ce qui concerne votre religion que aussi l'état politique et vie économique.

RABELAIS, *Gargantua*, prologue

Hadronic physics in the quark model has become an area of investigation so wide and specialized that a study of a particular aspect such as hadron spectroscopy or hadronic transitions at low energy can be justified. As there are already good accounts of the spectroscopy of hadrons in the literature, we have considered it both important and interesting to deal with the topic of hadronic transitions. The aim of this book, then, is to study electromagnetic, weak and strong transitions between hadrons in the context of the quark model.

The level of the book corresponds approximately to that of postgraduate students with a good background in quantum mechanics and also some experience with quantum field theory and the standard SU(2) × U(1) × SU(3) model of electroweak and strong interactions. The book is intended to be of use to a wide audience of particle physicists, both experimentalists and theorists. The methods of the quark model developed here are applicable not only to the subjects that we treat in detail, but also to a number of other areas of investigation.

In our choice of the topics we have naturally often been motivated by our own research. Although this choice is thus necessarily somewhat subjective, we have tried to avoid too biased a view and to give a fair account of the main lines of investigation in a limited field, which is precisely defined in the Introduction

(Section 1.1). To complete the references in the text and to guide the reader in the subjects that have been too briefly treated, we provide a survey of the relevant literature at the end of the book. The choice of subjects to cover should not imply a judgement on their value, but is a result of our personal choices.

The book is organized in the following way. Chapter 1 is an introductory review of the basis of the quark model. We have tried first to make precise in Section 1.1 the present status of the naive quark model within the recent developments in particle physics. Then, in Section 1.2, there is an introduction to the quark interactions and their symmetries according to the standard $SU(2) \times U(1) \times SU(3)$ model. Finally, there is a brief but self-contained account of hadron spectroscopy in Section 1.3.

We have made a clear distinction between methods (Chapters 2 and 3) and phenomenological applications (Chapter 4). Chapter 4 is approximately self-contained, and the interested reader who is already familiar with the quark model can read it immediately.

In Chapter 2 we describe methods for calculating transition amplitudes—electromagnetic, weak and strong—within the nonrelativistic quark model. Although electroweak processes (Section 2.1) can be studied by standard methods, this is not the case for strong interactions (Section 2.2), and we discuss in this latter case a number of models that have been proposed.

Chapter 3 is devoted to methods beyond the nonrelativistic approximation. This is perhaps the more specialized contribution of the book. The various methods proposed in the literature for the study of relativistic corrections are described, with a critical account of their advantages and drawbacks: the Dirac equation in a central potential (Section 3.2), Hamiltonian methods (Section 3.3) and the Bethe–Salpeter amplitudes and Mandelstam formalism (Section 3.4). We emphasize which types of corrections to the nonrelativistic quark model of Chapter 2 are obtained. Also, processes like $\pi^0 \to \gamma\gamma$, linked to special concepts in quantum field theory, could not be treated in Chapter 2 and are included here.

Finally, Chapter 4 is an account of phenomenological applications. We have treated different phenomena with a unified set of parameters to make consistent the comparison of the quark model with a great variety of data. We have made a selection of those applications that can give a fair survey of the wide success of the quark model, and also of the open problems in the area. We have also emphasized the distinctive features of the more specific and predictive quark model in comparison with general algebraic schemes. After general remarks on the parameters and the practical application of the model (Section 4.1), we study semileptonic transitions, radiative transitions, strong processes, annihilation processes and finally nonleptonic weak decays, in

Sections 4.2–4.6 respectively. The data and particle symbols are taken from the Particle Data Group (PDG) tables (1984 edition).

The main components of this book were first intended as a contribution to a second volume of the book by Flamm and Schöberl (1982). The reader who wants more detailed descriptions of the fundamentals of the quark model and of spectroscopy than are given in Chapter 1 can profitably consult the latter book. We should like to thank Dieter Flamm for having stimulated us to write this book.

We should also like to thank Mireille Calvet for having typed the manuscript, and Jacqueline Bellone and Patricia Flad for their help in its preparation.

ALAIN LE YAOUANC
LLUIS OLIVER
OLIVIER PÈNE
JEAN-CLAUDE RAYNAL

CHAPTER 1

Introduction

Hunc igitur terrorem animi tenebrasque necessest non radii solis neque lucida tela diei discutiant, sed naturae species ratioque.†

T. LUCRETI, *Cari De Rerum Natura*, II

Can we actually "know" the universe ? My God, it's hard enough finding your way around in Chinatown.

WOODY ALLEN, *Getting Even*

I tax not you, you elements, with unkindness.

W. SHAKESPEARE, *King Lear*, III.2

1.1 OUTLOOK AND MOTIVATION

1.1.1 Quarks and Hadrons

In the past 20 years, decisive progress in our understanding of particle physics has been achieved. Matter and its interactions are now understood in terms of a few building blocks: the quarks and the leptons, which are fermions, and the *gauge* bosons — the photon, W and Z bosons, and gluons, associated with the gauge invariance group. The fundamental interactions are trilinear interactions between the fermions and the gauge bosons, analogous to that of quantum electrodynamics (QED), plus interactions between the gauge bosons themselves, associated with the nonabelian character of the gauge groups.

This well-defined picture constitutes a whole program that has been widely tested and is already very convincing. Electromagnetic and weak interactions are unified in the Glashow–Salam–Weinberg model based on the $SU(2) \times U(1)$ gauge group: it has been beautifully confirmed by the observation of the W and Z bosons. The point that remains obscure is the status of the additional Higgs particles. We shall frequently call these interactions *electroweak*. However, we

† This dread and darkness of the mind cannot be dispelled by the sunbeams, the shining shafts of day, but only by an understanding of the outward form and inner workings of nature.

are particularly interested in the new picture that has emerged for the strong interactions. Just as electromagnetic interactions are induced by the exchange of virtual photons between charges, so strong interactions are related to the exchange of *gluons* between *color* sources. Color is a set of quantum numbers characterizing the behavior of the source under the gauge group $SU(3)_c$, just as the electric charge characterizes the behavior of sources under the electromagnetic gauge group U(1). A specific feature of strong interactions is that owing to the nonabelian character of $SU(3)_c$ the gluons themselves are colored, whereas photons are not charged. Among the building blocks enumerated above, only quarks and gluons are colored and are affected by the strong interaction. The quantum field theory of strong interactions is known as quantum chromodynamics (QCD). The present theory of the electroweak and strong interactions based on the gauge group $SU(3)_c \times SU(2) \times U(1)$ is called the *Standard Model.*

Hadrons, which are the subject of this book, have a conceptually simple definition. They are composite particles made out of the color constituents, quarks and gluons, but which are themselves *color-neutral.* Just as electrically neutral atoms undergo electromagnetic interactions because of the charge of their constituents, so hadrons, in spite of being colorless, still undergo strong interactions. Of course, these interactions are weaker than those between the colored elementary constituents. In fact, there is strong evidence that the interaction energy between colored sources tends to infinity with separation; this is attributed to the self-interaction of gluons—a distinctive feature of QCD in contrast with QED. The net effect of the very strong repulsions and attractions that are generated by color is the combination of quarks and gluons into color-neutral hadrons. Within hadrons, the forces between the constituents are attractive—so strongly attractive in fact that the constituents cannot be separated: this property is termed *confinement* of quarks and gluons within color-neutral states.

Strictly speaking, the number of quarks and gluons within a given hadron is not defined in quantum field theory. However, within the approximation of the nonrelativistic quark model, one can associate a particular hadron with a definite number of quarks. The usual hadrons are then composed of quarks only: mesons of a quark and an antiquark, and baryons of three quarks. Other hadrons can be constructed in this way, but have not yet been observed. These include *glueballs*, containing only gluons, *hybrids*, containing both quarks and gluons, as well as hadrons containing more than three quarks—*diquonia*, with two quark–antiquark pairs and *dibaryons*, with six quarks. Our discussion here, however, will concentrate on the usual hadrons: mesons and baryons.

1.1.2 Transitions of Hadrons

Consider the various types of particles distinguished by different values of the conserved quantum numbers N_B (baryon number), N_ℓ (lepton number): gauge bosons and mesons, leptons and baryons. Only the lowest-mass states in each of these categories are stable: the photon, the electron and the neutrinos, and the proton. Quarks and gluons, being colored, cannot exist separately because of confinement. Certain particles decay only through electroweak interactions, and since these are relatively weak, the corresponding lifetimes are still sizable, and they are classified as *stable* particles by the Particle Data Group (1984).

Most hadrons, however, are much less stable because they decay strongly, i.e. they are not even stationary states under the action of the QCD Hamiltonian H_{QCD}. Such states have very short lifetimes, not directly observable, and constitute a very large part of the hadron spectrum. These hadron levels are characterized by their degeneracy with a continuum of multihadron states to which they can decay by QCD interactions. For example, $\Delta(1236)$ is degenerate with $N\pi$ states, and N(1520) with $N\pi$ and $N\pi\pi$ states. Certain states avoid decay because of the conservation of *flavor* quantum numbers by the QCD interaction. Indeed, each flavor or species of quark u, d, s, c, b or t can appear or disappear only by creation or annihilation with an antiquark of the same species, or, in terms of quark diagrams, any quark line has a definite species. This principle is only violated by the charged weak interactions. Let us then associate with each flavor an additive quantum number N_i, which is the number of quarks minus the number of antiquarks of flavor i. N_i must be conserved in a strong or electromagnetic process, and each hadron has a definite set of quantum numbers $\{N_i\}$. In each sector of hadrons having the same set of quantum numbers $\{N_i\}$, the lowest-mass state can decay only through the weak interaction if at least one $N_i \neq 0$; it can also decay electromagnetically if all $N_i = 0$. Examples of the first case are the decays of π^+ ($N_u = +1, N_d = -1$), n, Λ, D, ..., which decay weakly, or of the p, which cannot decay within the standard model $SU(2) \times U(1) \times SU(3)_c$. An example of the second case is the π^0 ($N_u = N_d = 0$), which decays electromagnetically through $\pi^0 \to 2\gamma$. The higher-mass hadrons in each flavor sector will in general decay strongly to the lowest-mass state plus a π. This may be still forbidden by energy–momentum conservation, as in the decay of Σ^0, which proceeds electromagnetically, $\Sigma^0 \to \Lambda\gamma$, or by additional conservation laws, P and C parity, and isospin conservation. Otherwise, the high-mass states are strongly unstable.

The above discussion on unstable hadrons has been rather imprecise. The concept of the mass of the unstable particle has been used, although such a state is, by definition, not an eigenstate of energy–momentum. The rigorous treatment of an unstable state is a standard question. The most direct way is to separate the Hamiltonian into a part H_0 for which the decaying state as well as the multihadron continuum are eigenstates, and a part H_I that is responsible for the transition to the continuum and which one hopes can be treated using perturbation theory. This perturbation H_I will also shift the energy levels. The separation $H = H_0 + H_I$ is trivial when the particle decays only through the electroweak interactions: $H_0 = H_{QCD}$, $H_I = H_{EW}$ (the electroweak Hamiltonian). On the other hand, if the particle is not stationary under H_{QCD}, i.e. if it decays strongly, there is no known systematic procedure for making such a separation. In fact, as we will see later, there are only phenomenological approaches based on the quark model, which try to define H_I through the creation of a $q\bar{q}$ pair for example.

It happens that some strongly unstable particles have a rather small width for an important reason. These are the hadrons that undergo only strong decays forbidden by the so-called Okubo–Zweig–Iizuka (OZI) rule (Okubo, 1963; Zweig, 1964; Iizuka, 1966), which applies when for example the quark–antiquark species of a meson annihilate and do not appear in the final state (see Chapter 2). This rule is reinforced in the case of heavy flavors like the $c\bar{c}$ mesons ψ below the $D\bar{D}$ threshold. In this case the decay amplitude itself is small and can be calculated by perturbation theory, because the strength of the QCD interaction decreases—a property of QCD called *Asymptotic Freedom.* However, even in this case, one cannot define such states as eigenstates of some part H_0 of H_{QCD}. A powerful method that circumvents the problem of an explicit separation of the Hamiltonian is to define the unstable state as a resonance in a scattering amplitude of stable particles. The unstable particle appears as a complex pole in the complex energy plane, or a bump in real energy. For instance, the $\Delta(1236)$ manifests itself as a bump in πN scattering. This is how strongly unstable particles are observed, as their lifetime is too short to observe a track. The definition of the unstable state as a complex pole in some energy variable applies not only to the scattering amplitude, but also to other quantities, which are easier to calculate—in particular to the correlation functions of local operators, a method that is used in lattice QCD calculations and QCD sum rules.

The idea of unstable hadrons that decay strongly but are nevertheless somewhat comparable to the stable ones, is fundamental in the analysis of low- and medium-energy processes. Strong decays of excited hadron states form a large component of the available experimental data. The decay of resonances to multihadron states can often be viewed as a decay sequence through intermediate resonances. For example, $N(1520) \to N\pi\pi$ can be viewed as $N(1520) \to \Delta(1236)\pi$, $\Delta(1236) \to N\pi$. The strong decays can thereby be reduced to a chain of two-body decays. In the same way, electroweak decays where the final state includes a photon or a lepton pair and a number of stable hadrons can quite often be described by a transition between the initial hadron and a resonance, followed by the strong decay of the resonance; for example $D \to K\pi e\nu$ can be viewed as $D \to K^* e\nu$, $K^* \to K\pi$. By such methods, the structure of decays is greatly simplified and more easily related with theoretical schemes. This procedure, on the other hand, allows the measurement of new quantities. For example, the transition $N(1520) \to \Delta\pi$ would not be otherwise observable. Since strong decays are much more rapid than electroweak ones, it might be feared that electroweak transitions of unstable hadrons such as $\Delta \to N\gamma$ would be impossible to measure, because they would be completely masked by the strong ones. But they are indirectly observable, because the transition $\Delta \to N\gamma$ partial width is also measured in the process $\gamma N \to N\pi$ in the vicinity of the $\Delta(1236)$ resonance. In the same way, the weak semileptonic transition $\Delta^{++} \to p\mu^+\nu_\mu$ is measurable through the neutrino-production reaction $\nu_\mu p \to \mu^- p\pi^+$. Finally, certain quantities are very akin to transition amplitudes, although they do not correspond to actual decays allowed by phase space. For example, the magnetic moment of the proton μ_p has essentially the same structure as the $\Delta \to N\gamma$ amplitude. The same applies to the axial couplings of the nucleon that can be measured by the reaction $\nu_\mu n \to \mu^- p$ at low momentum transfer or to the pion–nucleon coupling. The set of experimental data for particles decaying by the electroweak interactions is thus greatly enlarged by the hadron levels that decay strongly and by the various couplings of the stable states.

1.1.3 Fundamental Theory Versus the Quark Model

It is relatively easy to treat the effects of electroweak interactions by perturbation theory. However, for any type of transition under study, we are confronted with the problem of dealing with QCD. Hadrons, either stable or unstable, involve QCD essentially and there is no known general solution for this complicated field theory. On the other hand, one cannot expect much from perturbation theory in a situation that is fundamentally not one of weak coupling.

The usual way to treat local interactions is through perturbation theory, by expanding various quantitites in powers of the coupling constant. More precisely, taking as an example bound states in QED, one must first sum the effect of multiple Coulomb interactions, which generate the bound states, by solving the Schrödinger equation with a Coulomb potential. Any number of Coulomb exchanges within a bound state contribute with the same order of magnitude, in spite of the additional powers of $\alpha = e^2/4\pi$ in each Coulomb interaction. Once the Schrödinger equation has been solved, any matrix element can be approximated by a matrix element between Schrödinger wave functions, to be corrected by terms of order α, α^2, ... These corrective terms correspond to relativistic kinetic energy effects, to transverse photon exchanges, etc.

This procedure is not possible for QCD. The difficulties are the well-known difficulties of strong interaction theory. Since the strong interactions are characterized by large coupling constants, perturbation theory is not useful. However, this old idea is made much more precise by QCD. First, all the strong interactions are expressed in terms of a dimensionless coupling constant $\alpha_s = g_s^2/4\pi$. However, owing to the asymptotic freedom of QCD, this coupling constant is not always large. Indeed, it depends on a renormalization scale μ connected with the typical momentum transfers or gluon momenta in the process. At large μ, $\alpha_s(\mu^2)$ becomes small, which allows perturbation theory to be used in the so-called *hard processes.* Hard processes occur for example in electron– or neutrino–nucleon deep inelastic scattering, and also in some decays. This is the case, as we have already indicated, of the OZI-forbidden strong decays of heavy quarkonia (bound states of a quark and an antiquark of the same flavor) and also of some weak decays of heavy flavors. On the other hand, for a small renormalization scale, α_s is large. This small scale $\mu < 1\ \mathrm{fm}^{-1}$ is the one relevant for the strong decays of hadrons composed of light quarks, and in general for the OZI-allowed strong decays, for electromagnetic transitions (except when connected with OZI-rule violation, as in $J/\psi \to \eta\gamma$), and for weak decays of light flavors like hyperon or kaon decays. In these latter cases, which are the main part of our subject, perturbation theory in α_s is hopeless. In any case, even in processes where one can use perturbative QCD, there are in general parts where the relevant α_s is not small. For instance, in hard scattering processes only the parton process (e.g. the scattering of a quark by a photon, a gluon or another quark) is perturbatively calculable. However, the relation between the partons (quarks and gluons) and the observed hadrons is described by structure and fragmentation functions, which are not calculable in perturbation theory. Analogously, in the annihilation of J/ψ into light hadrons, one calculates by perturbative QCD the annihilation of the $c\bar{c}$ pair into hard gluons. However, the result has to be multiplied by the wave function of the $c\bar{c}$ system at the origin, a quantity that is not calculable by perturbation theory, i.e. with a Coulomb potential.

The fact that α_s is large is not the only point preventing the use of perturbative QCD. There are in QCD two main phenomena that are essentially nonperturbative and cannot be obtained even by summing the whole perturbation series. The first one is *confinement*, described above. The second is *dynamical breaking of chiral symmetry*. Both are connected with nonzero vacuum expectation values, which would vanish at any order in perturbation theory: the so-called *gluon condensate* and the *quark condensate*. The first seems related, as we have already mentioned, to an interaction energy that increases with distance (Leutwyler, 1979; Baker, Ball and Zachariasen,

1985), in contrast with the Coulomb energy. The second one gives a dynamical mass of the order of several hundred MeV to the light quarks u and d, which would have otherwise a mass of only a few MeV. These very important phenomena are essentially nonperturbative. The dynamics of light quarks really does not look like the case of very small masses bound by a Coulomb potential, but rather like massive particles in a potential growing with distance up to infinity, as will be illustrated by the nonrelativistic quark model. Therefore the whole perturbative scheme seems to be useless. In the last ten years, two alternatives have been proposed on the basis of field theory, in order to deal quantitatively with this new situation. One is the method of *QCD duality sum rules* (Shifman, Vainshtein and Zakharov, 1979); the other is *lattice QCD*, either in a strong-coupling expansion or in a numerical approach (Wilson, 1974; Kogut and Susskind, 1975; Susskind, 1977).

The *QCD duality sum rule* approach uses perturbation theory while accounting at the same time in a semiempirical manner for the above mentioned nonperturbative phenomena. Using the short-distance expansion of products of local operators (Wilson, 1969), certain quantities can be calculated in terms of the gluon and quark vacuum condensates (to be specified empirically) and of perturbatively calculable coefficients. On the other hand, these quantities are also calculable in terms of hadron properties: masses, decay widths, branching ratios of the various possible states. This *duality* between the above quark and gluon QCD description and the one in terms of hadron properties therefore allows calculation of these properties, in particular the decay widths and branching ratios, in terms of perturbative QCD and the vacuum condensates.

The *lattice* approach is much more ambitious, because it aims at completely calculating hadronic properties from first principles, i.e. from the QCD Lagrangian on a lattice. This method does not need such empirical ingredients as vacuum condensates, but in fact calculates them. The physical idea is that the long-distance properties of QCD are presumably the most important. A good picture can be already obtained with a large lattice spacing, and one hopes to improve the accuracy by going to small lattice spacing, i.e. approaching the continuum. These long-distance properties of QCD that are crucial are again confinement and dynamical breaking of chiral symmetry. They are indeed present even in very crude approximations on the lattice, while it is impossible to get them even by using the whole perturbation series. The lattice method allows nonperturbative calculations in two ways. First, it permits *strong-coupling* calculations, based on the assumption that the unrenormalized coupling constant is very large. Secondly, it permits Monte Carlo numerical calculations, based on the Feynman path-integral formulation of field theory. It becomes possible, on finite lattices, to make direct calculations of the path integral by numerical methods, and the approximation of finite lattices turns out to be surprisingly good.

In spite of these two attempts at a theoretical treatment of hadrons on the basis of QCD, this book will exclusively present the older *quark-model* approach (Gell-Mann, 1964a,b; Zweig, 1964), and especially the *nonrelativistic quark model* (Dalitz, 1965, 1966, 1967a,b; Dalitz and Sutherland, 1966; Morpurgo, 1965, 1969), which we discuss in the next chapter. The nonrelativistic quark model relies only for some very general features on QCD. Why do we make this very selective and rather conservative choice?

It must first be said that, in spite of the progress achieved by the above two methods, these are far from covering such a large domain of facts as the naive quark model—all the more so in the particular area of hadron transitions. Although there is no difficulty in principle in calculating any hadronic property with these methods, they have practical limitations. These are especially severe for lattice QCD, at least with the present possibilities of numerical calculation. The prospect of estimating decay amplitudes is limited to a few cases. QCD sum rules are less powerful in principle, but they are much easier to handle and have already yielded a much greater number of results, in particular for strong and radiative decays. However, it would still be a very arduous and maybe impossible task to describe by this method the decays of the huge number of baryonic excitations that are successfully described by the nonrelativistic quark model. Even for the simpler meson radial excitations, the proponents of sum rules themselves recognize the superiority of the nonrelativistic approach. The exclusive two-body decays are a complicated issue because they involve a three-current Green function, while the simpler calculations concern correlation functions of two currents. Moreover, the results are reliable only for the lowest-mass state in the channels defined by the currents.

On the whole, the first objection against the sophisticated methods is that of *technical complexity*, which makes concrete calculations difficult or even impossible in practice. Another objection is the difficulty in obtaining *physical insight.* In contrast with the nonrelativistic quark model, the above sophisticated methods hardly give an idea of the mechanism and of the various factors leading to the final result. This is especially true for the lattice approach—most of all for the numerical calculations. There is at present no physical picture underlying the calculation. This is to be contrasted with the old QED perturbation theory, which is able to describe the various contributions very clearly through Feynman graphs. Certainly, the QCD sum rules do better in this respect than lattice numerical calculations, in particular because they put forward the important physical distinction between short- and long-distance contributions. They emphasize the two physical facts of vacuum condensation of gluons and quarks, respectively related to confinement and chiral symmetry breaking. However, the way they consider hadron transitions remains rather indirect, since *a priori* it is only an average of several different hadron states that can be pictured through their semiperturbative QCD approach. The outstanding advantage of the quark model is that it allows the direct calculations of the relevant matrix element for each definite decay of each definite hadron, each hadron being represented by its wave function. It has an especially transparent direct relation to experimental data.

Of course, there is a price to pay for the simplicity of the nonrelativistic picture. Not only is it an often crude approximation, but, more basically, it lacks a theoretical foundation in QCD. It is not a well-defined approximation of QCD, like perturbation theory, but a model whose approximations are theoretically uncontrolled. The relativistic corrections by which one can try to improve it are also at present out of theoretical control, especially for light quarks. The naive nonrelativistic quark model certainly has some features of QCD, and indeed *it has suggested many basic features of QCD*. However, the main argument in its favor is its impressive empirical success.

A balanced point of view should consider the three methods as complementary rather than contradictory. The most ambitious method, the lattice approach, allows in principle the calculation of everything from first principles, i.e. from the QCD Lagrangian. However, because of its technical complexity, it should rather be chosen to elucidate a limited number of theoretically crucial issues, namely fundamental phenomena such as gluon or quark condensates or non-leptonic weak decays. The QCD sum rules, although closer to hadronic phenomenology, are useful in particular for those phenomenological issues that may depend crucially on detailed QCD field-theoretic features. On the other hand, in everyday phenomenology, the quark model will probably retain its present predominance (which has even increased in recent years). However, future efforts should certainly be devoted to theoretical understanding of its not yet understood basic approximations.

Confronted by this rather ambiguous theoretical situation, one might be tempted to mix together the results of the various approaches. This would indeed be necessary if our aim was completeness and the accumulation of the maximum of theoretical information. Our choice is different, however. In the present book we should like to show how a huge amount of experimental data on hadron transitions can be described in a simple, coherent, single framework, which is the quark model, mainly in its naive nonrelativistic formulation. It is already a big enough task to deal with it adequately on its own, and it is better not to bring in the discussion of quite different methods, which should be considered separately.

For the same reasons of coherence and homogeneity, we have not tried to account for the very important contribution of *perturbative* QCD concerning the category of *hard* decays, which was commented upon earlier in this discussion of field-theoretic methods. We recall once more that we are dealing here with OZI-forbidden strong decays of heavy flavors, for example $J/\psi \to$light hadrons. It must be emphasized that in this question the use of perturbative QCD instead of the naive quark model is actually connected with a different *physical* situation in these hard decays. This reinforces the motivation for considering them separately or explaining them in connection with other hard processes. We shall, however, study *total* decays like $J/\psi \to$all, as they involve the hadron wave function in a simple way, and the QCD aspect is very simple to treat in analogy with QED. We shall have also to borrow from perturbative QCD some *radiative corrections* to quark-model calculations, because these radiative corrections happen to be important. This is the case especially for e^+e^- annihilation.

1.1.4 Basic Features of the Quark Model

The quark model historically preceded QCD and the idea of gluons, but one can present it now in the context of QCD. The first step is to integrate out the gluon fields and the second is to formulate simplifying assumptions for the resulting equations of motion of the quark fields. The most drastic simplification is that adopted in the nonrelativistic quark model: the equation of motion is assumed to be the second-quantization form of the Schrödinger equation for particles interacting through a two-body potential. This does not imply that the situation is actually nonrelativistic and that the Schrödinger equation is justified as in a weak-binding approximation. If this were so then the potential and the masses would be those implied by the Lagrangian, with account taken of perturbative renormalization effects. There would then be a Coulomb-like gluon-exchange potential, and a mass given by the mass term in the Lagrangian. In fact, on the one hand, one works with a *confining* potential that does not bear any resemblance to the Coulomb potential. On the other hand, one considers quark *constituent masses* that are also quite different, especially for the light quarks u, d, and s, from the masses generated by the Lagrangian mass term renormalized on the usual scale $\mu \sim 1$ GeV.

This is due to the fact that the potential and the constituent masses must include the two essential nonperturbative effects described in the previous sections: confinement and dynamical breaking of chiral symmetry with fermion mass generation. In the presence of such effects, the potential and the quark masses have no simple connection with the QCD Lagrangian, although they should ultimately be deduced from it. A significant drawback of the nonrelativistic quark model in this respect, however, is that it does not reflect one of the consequences of the dynamical breaking of chiral symmetry: the presence of a Goldstone boson (see Section 1.2.4), which should correspond to the pion (Nambu and Jona-Lasinio, 1961a,b). The pion does not, as it should, play the role of a Goldstone boson. Chiral symmetry, even when dynamically broken, implies that the divergence of the axial current for light quarks should be very small, of the order of the *current masses* (some MeV). This is the famous *PCAC* property (partial conservation of axial current). On the contrary, in the nonrelativistic quark model there is a large *explicit* breaking of chiral symmetry, and the divergence is of the order of the *constituent masses* (hundreds of MeV). It can therefore be concluded that, while the nonrelativistic quark model reflects the dynamical mass generation of fermions, it does so in a much too naive way. This defect is not always of practical consequence. It can be expected to be important whenever the pion or the kaon, which are approximate Goldstone bosons, are involved. Fortunately, it is found that the consequences of treating the pion in the nonrelativistic quark model — a bold procedure indeed — are frequently quite satisfactory.

An essential feature of the Schrödinger equation is that the number of particles is fixed, and hadron states can be constructed with a definite number of quarks: mesons with $q\bar{q}$, and baryons with three quarks. This construction will be detailed in the following sections. Such a model with a definite number of constituents can be justifiably called the *constituent quark model.* The corresponding quanta are called *constituent quarks*; they are not simply related to the quark fields in the QCD Lagrangian.

The paradox presented by the quark model is that, on the one hand, with the great simplifications afforded by the nonrelativistic equations, it is able to successfully describe a large domain of data. However, on the other hand, the velocity of quarks in the Schrödinger equation — the velocity of the constituent quarks — is by no means small for the lightest u, d and s quarks. As we will show in Chapter 4, in discussing the basic parameters of the model one finds numerically $v/c \sim 1$. This happens in spite of the fact that the nonrelativistic constituent-quark masses are by far larger than the so-called current masses of u, d and s. As a consequence of this large velocity, the relativistic effects that correct the Schrödinger equation, are expected to be large for the light quarks.

In the face of such a situation, it is possible to propose various directly relativistic schemes as alternatives to the nonrelativistic quark model. These methods retain the essential features of eliminating the gluon degrees of freedom and keeping a two-body long-range potential between quarks. The covariant models have several drawbacks: besides their technical complexities and their theoretical problems (such as ghosts), they have a wide variety of possible choices (in fact restricted by the necessity to reproduce the main successes of the nonrelativistic model), and their final agreement with experiment is no better than the nonrelativistic model, indeed it is often worse. We prefer methods where the interaction is described by an instantaneous potential. These methods are only partially relativistic. Their main relativistic features are a relativistic expression for kinetic energy, and quark Dirac spinors instead of nonrelativistic Pauli spinors. If one wants to know whether or not these models lead to an improvement, the best and often the only practicable method is to use a systematic v/c expansion, which has the two advantages of starting from the rather successful nonrelativistic model and of yielding an unambiguous algorithm to implement corrections order by order. In the case of heavy quarks (charmonium) the corrections arising from the models that we develop in Chapter 3 give a significative improvement. In the case of light quarks, even if a v/c expansion seems theoretically unjustified, v/c not being small, the corrections do seem to go in the right direction.

For all these reasons, we have kept the nonrelativistic quark model as the starting point of our development on light quarks (u, d, s), as well as on heavy quarks (c, b). This is the point of view strongly advocated by Morpurgo for a very long time (for a recent reference see Morpurgo, 1983). The relativistic schemes will be presented with reference to it rather than vice versa, and moreover we will select models that reproduce the nonrelativistic one to lowest order.

Some specific comments must now be made on the *heavy* quarks (c, b and possibly t). These quarks, with masses $\gtrsim 1.5$ GeV, are much heavier than u, d and s. For such quarks one is approaching a nonrelativistic situation as v/c becomes really small. For example, for the $b\bar{b}$ mesons Υ (bottonium), the nonrelativistic predictions concerning the radiative decays become accurate to within a few per cent. Moreover, since the constituent- and the current-quark masses are both large, they become close to one another, and the nonrelativistic quark model approaches to the Coulomb-like approximation. Indeed, although the potential remains confining, independently of the quark mass, it has a Coulomb part that at higher quark mass becomes more and more important in determining the wave function of low-lying states. However, this limit is only really approached for very large masses (>100 GeV). As to the charmonium system $c\bar{c}$, it represents an intermediate situation between the light mesons and the $b\bar{b}$ system. Quark velocities are still sizable and lead to sizable corrections to the nonrelativistic quark model. On the other hand, these corrections are sufficiently small to make sense. Finally, systems like the D and the B mesons made out of light and heavy quarks remain relativistic.

Because of the place it occupies in the specialized literature, we feel obliged to explain why we do not consider the *bag model* (Chodos *et al.*, 1974; Degrand *et al.*, 1975). The motivation of coherence and homogeneity is one of our reasons, but a comparison must be made to justify our choice. Certainly, this is a very active approach, which has covered a wide variety of phenomenological applications. The model shares the same idea of separating the quark from the gluon degrees of freedom, at least in the first approximation, since the quarks obey a free Dirac equation inside a deep square-well potential (the bag), with zero fields outside. The main advantages of the bag model seem to be the following. First, it is an independent-particle model, so that multiquark states can easily be constructed. Secondly, it uses the Dirac equation, which is the simplest relativistic equation, and which can be solved exactly in a spherical cavity. The price to pay for an independent-particle model is well known, however: the center-of-mass motion is not that of a free particle. This is a problem when one wants to calculate hadron transitions, except in the case of elementary quantum emission (see Chapter 3), and leads, in the spectrum, to spurious states. Restoring the expected behavior of the center-of-mass is possible, but this technical complication somewhat spoils the initial simplicity of the model. The bag model thus accounts for the internal relativistic quark motion in a simple way, but leads to a complicated scheme when it describes the center-of-mass motion of hadrons.

This is one of the main reasons why we prefer the nonrelativistic quark model, as it describes on equal footing both the quark internal and the hadron center-of-mass motions. Although the nonrelativistic approximation on its own is not justified, we prefer this simple starting point, which we can further correct by a systematic v/c expansion. Moreover, although in the bag model certain relativistic effects are treated exactly through the use of the Dirac equation, other corrections, which are *a priori* also important, are omitted or treated perturbatively. It appears to us to be more coherent to start from the simplest nonrelativistic approximation, and to consider all the corrections in a systematic, order-by-order, expansion, as will be presented in Chapter 3.

Finally, as to the concrete description of the spectrum, two remarks are in order: first, for heavy quarks, the final picture obtained in the bag model after consideration of bag deformation (Hasenfratz and Kuti, 1978) is very close to the nonrelativistic potential model: the $q\bar{q}$ potential is derived from the energy of the classical configuration of the gluon field, with bag boundary conditions, created by the quarks as classical point charges; one obtains a string-like configuration and a linear potential at large distance, and a Coulomb force at short distance. On the other hand, for light quarks, even after departing from the spherical-cavity approximation, one is not able to obtain a satisfactory description of the baryon excitation spectrum except by introducing new freedom (De Grand, 1976). On the whole, the nonrelativistic potential model appears as a simpler and more satisfactory approach to the hadron spectroscopy.

1.1.5 Main Emphases of the Book

This book is devoted mainly to the nonrelativistic quark model, and to the discussion of corrections that can improve it. The reasons underlying this choice have been given above. In addition, within this general framework, we have emphasized several particular points, which we shall try to justify here.

We must first apologize for not having spent much effort in justifying the quark model from QCD. We do not go beyond the very general ideas given in this introduction. This is once more justified by our desire to present a coherent development from well-defined hypotheses rather than to be complete. Many things can be said, but only on a rather speculative level, and this could confuse the reader. Nor do we want to be superficial and justify the Schrödinger equation or the harmonic oscillator by QCD. We prefer to emphasize the domains where QCD has really achieved remarkable breakthroughs. Examples are the proposal in spectroscopy to explain spin-dependent forces by gluon exchange (De Rújula, Georgi and Glashow, 1975) or the perturbative QCD calculation of heavy quarkonium decay to light hadrons (Appelquist and Politzer, 1975). It may well be that an analogous breakthrough will take place in the domain of soft decays—especially strong decays. Meanwhile, we prefer to insist on the mainly *empirical* foundation of the quark model.

On the other hand, we have tried to clarify the status of the quark model in two other aspects. First we have tried to start whenever possible from a second-quantized formulation and to reformulate the basic assumptions in terms of *fields*. Although this may seem unnecessary and formalistic, it is actually important in many cases to get unambiguous answers: to relate quarks and antiquarks, to relate baryons and mesons, to take correctly into account the Fermi statistics, and to enumerate and weight the various contributions to matrix elements. Secondly, we have tried to clarify the meaning of the nonrelativistic approximation, which is in fact often ambiguous. In Chapter 3 we have developed a systematic v/c expansion, which clarifies the lowest-order approximation in v/c in each particular case, and the order in v/c of the various corrections. In Chapter 4 we discuss the *phenomenological* side of the ambiguities of the use of the nonrelativistic approximation, and we try to formulate what seem empirically the most efficient rules of the game.

In the phenomenological applications, we have emphasized *light-quark* physics much more than is fashionable today. There are several reasons for this. First, contrary to what is often heard, the (mainly nonrelativistic) quark model works not only for heavy quarks, but also for light quarks. It is certainly easier to justify the model for heavy quarks, and it works more accurately in this area. However, it has as many successful applications for light quarks, and, in fact, it began historically with them. A second reason is that, even with the advent of heavy-flavor physics, light hadrons still represent the main body of experimental data. Finally, since heavy-flavor physics has very frequently been reviewed, in particular in various international conferences, it is useful to provide a more balanced presentation where light quarks are also accounted for in a systematic way.

Another important emphasis in the comparison with experiment is on *explicit* matrix-element evaluation, on *absolute rate* calculations and not only, as it is often done, on ratios, selection rules and symmetry arguments. The prediction of ratios and selection rules was very important in the early stages of development of the quark model, and the development of the $SU(6)_w$ analysis of decays in the 1970s (Petersen and Rosner, 1972; Faiman and Rosner, 1973) was very important in confirming general features of the quark model and in excluding too-naive decay models. We think, however, that it is high time to be more ambitious and that matrix elements can and must be predicted quantitatively. Indeed, in our opinion, an important advance consists in the fact that we are able to fix the basic parameters of the quark model, which allow us to make such quantitative predictions. These basic parameters are the masses of the quarks and the strength of the confining potential, or equivalently some radius characterizing the ground-state wave function. They can be fixed coherently through hadronic spectroscopy and a few magnetic moments. The latter are necessary because, in the case of light quarks, the determination of masses through spectroscopy is too uncertain.

Once this is done, many electroweak transitions can be *predicted* within the ambiguities of the nonrelativistic approximation. Therefore it is misleading to put forward ratios or algebraic relations between decay amplitudes instead of looking at absolute rates. It is even worse to adjust parameters, which should be universal, in order to get empirical agreement in a particular study. We have tried to emphasize this relatively rigid character of the model, sometimes at the cost of having only crude agreement. This may be disappointing for the reader, but we feel on the contrary that this crude agreement is in fact already rewarding for an approach where all the necessary parameters have been fixed once and for all by spectroscopy or magnetic moments.

Such a relatively unambiguous treatment is not available for most *strong* decays, because the picture is much more phenomenological, and not simply imposed either by QCD or by the general approach of the quark model. In addition, experiment itself often gives only too crude results to test the models. We have not tried to conceal this situation, which is reflected in particular in the multiplicity of possible models of OZI-allowed decays. Our presentation is therefore rather different from the one on electroweak transitions, and the discussion of various models of strong decays in Chapter 2 has a somewhat inconclusive character. Analogously, in the confrontation of models with experiment, we have given a long discussion of the various qualitative achievements, especially on signs of amplitudes and the $SU(6)_w$ algebraic structure, because this is often the best that can be said. This account may seem discouraging to some readers, but it is the only one that is faithful to the present state of research. On the other hand, we remain consistent with our general attitude, as we do not try to improve the agreement with experiment by fitting quantities that can be actually predicted. An additional phenomenological parameter is present in OZI-allowed strong decays, which can be termed generically the *quark-pair-creation* constant. However, once it is fixed, the decay amplitudes are quantitatively calculable in each model. This is the reason why we have not followed the algebraic $SU(6)_w$ analysis, which fits reduced decay amplitudes, or the intermediate line of thought followed by Koniuk and Isgur (1980) in their extensive analysis of strong decays through elementary pion emission.

1.1.6 Particular Topics not Covered by This Book

Even when restricted to the potential quark model, the material to be explained is still considerable in comparison with the book of Kokkedee (1969). It is not possible to treat all the aspects of the model, and we do not want to just have a review, which would be reduced to simply quoting results and references. We have rather chosen to discuss carefully a restricted number of topics that seem to us to be crucial for evaluating the status of the quark model, and which illustrate the main types of concrete processes. We try to compensate for our omissions by an annotated bibliography.

For the sake of simplicity, we have not tried to account for the effects of unequal quark masses. In particular we have disregarded flavor $SU(3)_F$ breaking. We have avoided strange hadrons, except in weak decays. We do not treat in detail the D mesons, and we just borrow certain results concerning them to account for the charmonium OZI-allowed decays, which exhibit features of the wave functions and the decay models rather spectacularly. Explicit consideration of unequal quark masses would have greatly complicated the formulae, although we are conscious that discarding this leaves aside some interesting topics.

On the other hand, the striking effects discovered by Isgur and Karl (1978) and Isgur, Karl and Koniuk (1978) in the strange-baryon strong decays as a consequence of the s and u, d quark mass difference are in fact more of spectroscopical origin than due to some special decay mechanism. These effects, as well as the many SU(6) mixing effects, have been excluded from our study, although they should be taken into account in many concrete calculations.

We have not discussed the question of *form factors*, which represents the extension to virtual exchanges of certain hadronic transitions considered in the book. For example the Δ–N electromagnetic form factors studied in Δ electroproduction represent an extension of the $\Delta \to N\gamma$ amplitude to a virtual exchanged photon. On the one hand, they may seem to be very closely related to our subject and to enlarge the applications of the quark model. On the other hand, they involve new questions: the treatment of large hadron velocities, the quark structure, or the analytical properties as a function of q^2, the squared mass of the virtual quantum. These questions are interesting, but discussing them would have weakened the structure of the book as their status is controversial.

In the phenomenological applications (Chapter 4) we have discarded a large number of important decay processes for various reasons. For example, for the reasons explained above, exclusive *OZI-forbidden decays* have not been treated, and we have only considered the inclusive annihilation rate, which serves as an illustration of the combination of quark model and perturbative QCD methods necessary to handle such processes.

Some familiar decays have not yet acquired a standard interpretation. This is especially the case of the nonleptonic decay $K \to 2\pi$. A great amount of work has been done on these processes, but they remain very controversial. This seems to be a serious specific problem, but one that does not tell us much about the general methodology that this book tries to develop. Moreover, the $K \to 2\pi$ decays are probably not tractable through the naive quark models simply because chiral symmetry, which is not included in the simplest framework, becomes crucial. For heavy-flavor nonleptonic weak decays the difficulties are certainly much less considerable than for $K \to 2\pi$. Nevertheless, we do not yet have a clear understanding of D decays. We have left aside all the meson nonleptonic weak decays. Finally, in the case of *exotic* states, not composed of $q\bar{q}$ or qqq, such as diquonia $qq\bar{q}\bar{q}$, dibaryons qqqqqq, hybrids (composed of quarks and gluons) and glueballs (composed of gluons only), it is the very existence or identification of these states that remains controversial. The study of such states is for the moment largely speculative. Therefore, since the book aims primarily at presenting what can be considered on a clear basis, we do not develop this matter, although it is of primordial theoretical interest, and at the center of current research. We have decided to limit the discussion to the multiquark states, which follow the same Schrödinger equation as the mesons and baryons, and to the simplest of them, the so-called diquonium states $qq\bar{q}\bar{q}$.

Before entering into the subject of this book, the hadron transitions, we shall briefly review the quark quantum numbers, the quark interactions and their symmetries (Section 1.2), and the hadron spectrum according to the quark model (Section 1.3).

1.2 QUARKS, INTERACTIONS AND SYMMETRIES

We leave aside the gravitational interaction and restrict ourselves to the *standard model* of electroweak and strong interactions, $SU(2) \times U(1) \times SU(3)_c$. The *gauge bosons* mediate the different interactions: the photon, the $W^{\pm}$, Z^0 weak bosons, and the gluons that are responsible for the strong interaction. The *matter* particles interacting through the gauge bosons are spin-$\frac{1}{2}$ fermions, the *leptons* (e^-, μ^-, τ^- and the corresponding neutrinos), which have only electroweak interaction, and the *quarks*, which also interact through the strong interaction $SU(3)_c$. There are also Higgs scalar particles, which feel only electroweak interactions; they do not play an appreciable role in the phenomenology of low and medium energies, except that they are responsible for spontaneous generation of masses. Quarks and gluons do not appear as free particles, and are confined, bound within the *hadrons* (the nucleon, the pion, ...) by the quark–gluon interactions described by the $SU(3)_c$ gauge group of the theory of strong interactions, quantum chromodynamics (QCD).

There exist at least 6 different types of quarks, called *flavors*, the u, d, s, c, b quarks and the predicted but still to be confirmed t quark. These different flavors have widely different masses. The quarks u, c, t have charge $+\frac{2}{3}$ and d, s, b charge $-\frac{1}{3}$.

1.2.1 Strong Interactions of Quarks

With respect to the strong interaction, each flavor possesses three different colors. In the language of group theory one says that each quark flavor transforms as the fundamental representation **3** of the color gauge group $SU(3)_c$. Antiquarks form the conjugate representation $\bar{\mathbf{3}}$.

The gluons are spin-1 massless bosons that belong to the adjoint representation **8** of the gauge group $SU(3)_c$. They are the gauge bosons of the strong interaction.

The QCD Lagrangian describes the quark–gluon interactions, and also gluon self interactions because the gluons also carry color. The QCD Lagrangian is

$$L(x) = \mathrm{i}\sum_i \bar{q}_i(x)[\partial_\mu - ig_s \sum_a \tfrac{1}{2}\lambda^a A^a_\mu(x)]\gamma^\mu q_i(x)$$
$$-\tfrac{1}{4}\sum_a F^a_{\mu\nu}(x)F^{\mu\nu a}(x) - \sum_i \bar{q}_i(x) m_i q_i(x), \qquad (1.1)$$

where g_s is the dimensionless strong coupling constant; i runs over all the flavors. For each i, $q_i(x)$ is a Dirac spinor field and a 3-dimensional vector in color space. The 3×3 color Gell-Mann matrices λ^a ($a = 1, \ldots, 8$), traceless and Hermitian, satisfy the SU(3) commutation relations

$$[\lambda^a, \lambda^b] = 2\mathrm{i} f^{abc}\lambda^c, \tag{1.2}$$

where f^{abc} are the structure constants of the SU(3) Lie algebra. $A^a_\mu(x)(a=1, \ldots, 8)$ are the gluon fields, and the field tensor $F^a_{\mu\nu}(x)$ is defined by

$$F^a_{\mu\nu}(x) = \partial_\mu A^a_\nu(x) - \partial_\nu A^a_\mu(x) + g_s f^{abc} A^b_\mu(x) A^c_\nu(x). \tag{1.3}$$

Writing the gluon fields in color matrix form,

$$A_\mu = \sum_a A^a_\mu \lambda^a, \tag{1.4}$$

the QCD Lagrangian (1.1) is invariant under color gauge transformations

$$\left.\begin{aligned} &A_\mu(x) \to g(x) A_\mu(x) g^{-1}(x) + \frac{\mathrm{i}}{g_s} g(x) \partial_\mu g^{-1}(x), \\ &q(x) \to g(x) q(x), \end{aligned}\right\} \tag{1.5}$$

where $g(x)$, an $SU(3)_c$ matrix, is a smooth function of space–time.

QCD has two remarkable properties: *asymptotic freedom* and *confinement*. Although the coupling constant g_s is dimensionless, renormalization introduces a scale parameter Λ fitted by the data to be around 100–200 MeV, the true fundamental quantity in QCD. The coupling constant $\alpha_s = g_s^2/4\pi$ will then depend on the scale μ at which a given process occurs, and more precisely on the ratio μ/Λ. Because of the non-Abelian character of $SU(3)_c$, the gluons have self-interactions according to (1.1) and (1.3), and the corresponding vacuum polarization effects give an effective coupling $\alpha_s(\mu^2)$ that decreases logarithmically as μ increases, or, equivalently, at short distances. This is the property of asymptotic freedom, $\alpha_s(\mu^2) \to 0$ as $\mu \to \infty$, that explains the point-like structure of the hadron constituents discovered in deep inelastic (large-Q^2, large-invariant-mass) lepton–nucleon scattering.

On the contrary, $\alpha_s(\mu^2)$ grows as μ^2 decreases (long distances), leading (it is hoped) to a power-like potential $V(r) \sim r$ between the quarks and thus to the property of confinement: quarks and gluons are confined within the hadrons. This property cannot be dealt with in perturbation theory, it is a nonperturbative phenomenon which belongs to the theory of phase transitions. The color electric flux between hadrons is confined in very much the same way as the magnetic flux is quantized and confined within class II superconductors. Quarks do not appear as free particles but are confined within color-neutral, i.e. color-singlet, states, the hadrons. The simplest color-singlet states are quark–antiquark bound states (*mesons*) and three-quark bound states (*baryons*). The color wave functions of mesons and baryons are respectively,

$$\left.\begin{aligned} |\mathrm{M}\rangle &= \frac{1}{\sqrt{3}}\sum_{\alpha=1}^{3} |\mathrm{q}_\alpha \bar{\mathrm{q}}_\alpha\rangle, \\ |\mathrm{B}\rangle &= \frac{1}{\sqrt{6}}\sum_{\alpha,\beta,\gamma} \varepsilon^{\alpha\beta\gamma} |\mathrm{q}_\alpha \mathrm{q}_\beta \mathrm{q}_\gamma\rangle, \end{aligned}\right\} \tag{1.6}$$

where the indices α, β, γ run from 1 to 3 for the three quark colors; q_α, $\bar{q}_\alpha$ represent quarks and antiquarks of any flavor and spin projection, and $\varepsilon^{\alpha\beta\gamma}$ is the totally antisymmetric tensor. The quarks have baryon number $\frac{1}{3}$, which gives baryon number 0 for mesons and 1 for baryons. Also, since the quarks have an electric charge $\frac{2}{3}$ or $-\frac{1}{3}$, the charges of mesons and baryons are integers. It is striking that the quark model is able to describe the observed hadrons with their quantum numbers and the approximate spectrum pattern in terms of such a simple scheme.

The basic hypothesis of the quark model is that, once the gluon fields have been integrated out from the QCD action, one reaches a satisfactory approximate description of hadrons by using an effective Hamiltonian in terms of quarks only. We shall make this idea more precise in Section 1.3.

These quarks will of course differ from the quark fields in the QCD Lagrangian (1.1), and will be called *constituent quarks*. They have the same quantum numbers as the fundamental quarks of QCD, but differ in some dynamical properties, such as their masses.

1.2.2 Electroweak Interactions of Quarks

We now consider the theory of unified electroweak interactions, the Glashow–Salam–Weinberg model based on the gauge group SU(2) × U(1). The total Lagrangian has the form

$$L_{\mathrm{EW}} = L_{\mathrm{G}} + L_{\mathrm{H}} + L_{\mathrm{F}} + L_{\mathrm{GF}} + L_{\mathrm{HG}} + L_{\mathrm{HF}}. \tag{1.7}$$

L_{G} is the part containing only the gauge bosons. L_{H} contains the so-called Higgs bosons that are responsble for the spontaneous breaking of the gauge symmetry through a vacuum expectation value $\langle \mathrm{H} \rangle$ that gives masses to the $\mathrm{W}^\pm$ and Z^0 weak gauge bosons through the term L_{HG}. In the minimal model the Higgs bosons will also give masses to the fermions through the Yukawa couplings contained in L_{HF}. L_{F} is the fermion kinetic term.

Finally, L_{GF} is for coupling of gauge bosons to the fermions, quarks and leptons. This term is the one we will use extensively in this book, and we shall give it explicitly. To construct this term, one needs the representations of SU(2) × U(1) assigned to the fermions. Left-handed and right-handed fermions

$$\left.\begin{array}{l} f_L = \tfrac{1}{2}(1-\gamma_5)f, \\ f_R = \tfrac{1}{2}(1+\gamma_5)f \end{array}\right\} \tag{1.8}$$

are assigned to different representations of SU(2). Let us consider the first two *families*: $\{u, d, e^-, \nu_e\}$ and $\{c, s, \mu^-, \nu_\mu\}$. It happens that, under SU(2) × U(1), the linear combinations

$$\left.\begin{array}{l} \{u,\ d_C = d\cos\theta_C + s\sin\theta_C,\ e^-,\ \nu_e\}, \\ \{c,\ s_C = s\cos\theta_C - d\sin\theta_C,\ \mu^-,\ \nu_\mu\}, \end{array}\right\} \tag{1.9}$$

rather than the mass eigenstates, have a simple behavior. θ_C is the Cabibbo angle, $\sin\theta_C \approx 0.22$. We shall not consider in this book the weak interactions of heavier flavors such as b, t, nor will we examine *CP* violation. For the sake of completeness, we shall introduce below the Kobayashi–Maskawa matrix, which involves mixing angles among all the flavors in charge-changing currents. For the moment, for the sake of simplicity, we shall stick to the two-family scheme (1.9). Under SU(2) we have the behavior of *doublets* for the left-handed fermions, and of *singlets* for the right-handed fermions:

$$\left.\begin{array}{l} \begin{pmatrix} u \\ d_C \end{pmatrix}_L, \quad \begin{pmatrix} c \\ s_C \end{pmatrix}_L, \quad \begin{pmatrix} \nu_e \\ e^- \end{pmatrix}_L, \quad \begin{pmatrix} \nu_\mu \\ \mu^- \end{pmatrix}_L, \\ \\ (u)_R, \quad (c)_R, \quad (d_C)_R, \quad (s_C)_R, \quad (e^-)_R, \quad (\mu^-)_R, \end{array}\right\} \tag{1.10}$$

so that the weak isospin T_3 (generator of SU(2)) will be

$$T_3 = \begin{cases} +\tfrac{1}{2} & \text{for } u_L,\ c_L,\ \nu_e,\ \nu_\mu, \\ -\tfrac{1}{2} & \text{for } d_{CL},\ s_{CL},\ e_L^-,\ \mu_L^-, \\ 0 & \text{for } u_R,\ c_R,\ d_{CR},\ s_{CR},\ c_R,\ e_L^-,\ \mu_L^-. \end{cases}$$

One assumes here massless purely left-handed neutrinos.

Under U(1), we have the additive quantum number (*hypercharge*)

$$Y = \begin{cases} \tfrac{1}{3} & \text{for } u_L,\ c_L,\ d_{CL},\ s_{CL}, \\ -1 & \text{for } \nu_e,\ \nu_\mu,\ e_L^-,\ \mu_L^-, \\ \tfrac{4}{3} & \text{for } u_R,\ c_R, \\ -\tfrac{2}{3} & \text{for } d_{CR},\ s_{CR}, \\ -2 & \text{for } e_R^-,\ \mu_R^-, \end{cases}$$

so that the electric charge satisfies

$$Q = T_3 + \tfrac{1}{2}Y. \tag{1.11}$$

Let us call $W^\pm$, W^0 and B^0 the gauge bosons associated with the gauge group SU(2) × U(1). The Higgs H is a *doublet* $\begin{pmatrix} H^0 \\ H^- \end{pmatrix}$ under SU(2), H^0 being a neutral complex field, and has $Y=1$ under U(1). The spontaneous breaking induced by $\langle H \rangle \neq 0$ gives masses to $W^\pm$ and to a linear combination of W^0 and B^0 called Z^0, and leaves massless the orthogonal combination, the photon,

$$\left.\begin{aligned} Z^0_\mu &= W^0_\mu \cos\theta_w - B_\mu \sin\theta_w, \\ A_\mu &= W^0_\mu \sin\theta_w + B_\mu \cos\theta_w, \end{aligned}\right\} \tag{1.12}$$

where θ_w is the weak mixing angle.

The part of L_{EW} coupling the fermions to the gauge bosons is then

$$\begin{aligned} L_{GF} = eA^\mu J^{em}_\mu &+ \frac{e}{\sqrt{2}\sin\theta_w}(W^{+\mu} J^{(-)}_\mu + \text{h.c.}) \\ &+ \frac{e}{\sin\theta_w \cos\theta_w} Z^\mu J^{(0)}_\mu, \end{aligned} \tag{1.13}$$

where

$$\cos^2\theta_w = \frac{M^2_W}{M^2_Z} \tag{1.14}$$

gives the weak mixing angle θ_w. The Fermi coupling G_F, the electromagnetic coupling e and θ_w are related to the couplings g, g' of SU(2), U(1) respectively by

$$e = \frac{gg'}{(g^2+g'^2)^{1/2}}, \qquad \tan\theta_w = \frac{g'}{g}, \qquad G_F = \frac{g^2}{4\sqrt{2}M^2_W}. \tag{1.15}$$

The magnitude of the couplings and gauge boson masses is $\alpha = e^2/4\pi = \frac{1}{137}$, $\sin^2\theta_w = 0.23$, $G_F = 10^{-5} m_p^{-2}$, $M_W \approx 82$ GeV, $M_Z \approx 93$ GeV. The electromagnetic current in (1.13) is given by

$$J^{em}_\mu = \tfrac{2}{3}(\bar{u}\gamma_\mu u + \bar{c}\gamma_\mu c) - \tfrac{1}{3}(\bar{d}\gamma_\mu d + \bar{s}\gamma_\mu s) - (\bar{e}\gamma_\mu e + \bar{\mu}\gamma_\mu \mu), \tag{1.16}$$

the charged weak current by

$$\begin{aligned} J^{(+)}_\mu = &(\bar{u} \quad \bar{c})\gamma_\mu \tfrac{1}{2}(1-\gamma_5)\begin{pmatrix} \cos\theta_C & \sin\theta_C \\ -\sin\theta_C & \cos\theta_C \end{pmatrix}\begin{pmatrix} d \\ s \end{pmatrix} \\ &+ (\bar{\nu}_e \quad \bar{\nu}_\mu)\gamma_\mu \tfrac{1}{2}(1-\gamma_5)\begin{pmatrix} e^- \\ \mu^- \end{pmatrix}, \end{aligned} \tag{1.17}$$

and the neutral weak current by

$$J^{(0)}_\mu = J^{(3)}_\mu - \sin^2\theta_w J^{em}_\mu, \tag{1.18}$$

where $J_\mu^{\rm em}$ is given by (1.16) and $J_\mu^{(3)}$ by

$$J_\mu^{(3)} = \tfrac{1}{2}(\bar{u} \quad \bar{c})\gamma_\mu \tfrac{1}{2}(1-\gamma_5)\begin{pmatrix} u \\ c \end{pmatrix} - \tfrac{1}{2}(\bar{d} \quad \bar{s})\gamma_\mu \tfrac{1}{2}(1-\gamma_5)\begin{pmatrix} d \\ s \end{pmatrix}$$

$$+ \tfrac{1}{2}(\bar{\nu}_{\rm e} \quad \bar{\nu}_\mu)\gamma_\mu \tfrac{1}{2}(1-\gamma_5)\begin{pmatrix} \nu_{\rm e} \\ \nu_\mu \end{pmatrix} - \tfrac{1}{2}(\bar{e} \quad \bar{\mu})\gamma_\mu \tfrac{1}{2}(1-\gamma_5)\begin{pmatrix} e \\ \mu \end{pmatrix}, \qquad (1.19)$$

as it corresponds to the weak SU(2) isospin assignments quoted above.

In the case of three families (u, c, t; d, s, b) the charged-current Cabibbo matrix (1.17) is replaced by the Kobayashi–Maskawa matrix (Kobayashi and Maskawa, 1972)

$$(\bar{u} \quad \bar{c} \quad \bar{t}) \begin{pmatrix} c_1 & -s_1c_3 & -s_1s_3 \\ s_1c_2 & c_1c_2c_3 - s_2s_3e^{i\delta} & c_1c_2s_3 + s_2c_3e^{i\delta} \\ s_1s_2 & c_1s_2c_3 + c_2s_3e^{i\delta} & c_1s_2s_3 - c_2c_3e^{i\delta} \end{pmatrix} \begin{pmatrix} d \\ s \\ b \end{pmatrix}, \qquad (1.20)$$

where s_i and c_i are the sines and cosines of three angles ($s_1c_2 \sim s_1c_3 \sim -\sin\theta_{\rm C}$), and the fourth angle δ is responsible for CP violation.

Concerning the weak interactions, we shall mostly study the charged-current processes, mediated by W exchange. As we consider only low-energy processes, we can approximate the W propagator by

$$D_{\mu\nu}(M_{\rm W};\, x-y) \approx -\frac{g_{\mu\nu}}{M_{\rm W}^2}\delta(x-y), \qquad (1.21)$$

which gives the current–current interaction

$$H_{\rm W} = \frac{G_{\rm F}}{\sqrt{2}}\int {\rm d}\boldsymbol{x}\, \tfrac{1}{2}\{J_\mu^{(+)}(\boldsymbol{x}),\ J^{(-)\mu}(\boldsymbol{x})\}, \qquad (1.22)$$

where the charged current is given by (1.17), and $G_{\rm F}$ is the Fermi coupling defined above.

1.2.3 Flavor Symmetries

We have just seen that the electroweak *charged* interactions change flavor. Electromagnetic or neutral weak interactions, on the contrary, conserve flavor, but the quarks have neutral couplings that depend on the flavor through the electric charge or the third component of the weak isospin.

We have seen that not only does the QCD quark–quark–gluon interaction (1.1) conserve flavor, but also the strength is actually flavor-independent. The only dependence on flavor in the QCD Lagrangian is through the quark-mass terms in (1.1), which in principle are generated by the vacuum expectation value $\langle {\rm H} \rangle$ through $L_{\rm HF}$ in (1.17). Since at least for light quarks the mass differences are not very large, $m_d - m_u = 3$ MeV, $m_s - m = 150$ MeV ($m = m_u, m_d$), the strong interactions have an *approximate* global flavor symmetry $U(3)_F$ that is locally isomorphic to $U(1) \times SU(3)_F$, U(1) being just associated with the baryonic number. The approximate symmetry $SU(3)_F$ between the u, d, s quarks and its consequences for the hadron spectrum is indeed well supported by the data.

Quarks are then assigned to the **3** representation of $SU(3)_F$, and antiquarks to the conjugate $\bar{\mathbf{3}}$ representation. In the limit $m_u = m_d = m_s$ the QCD Lagrangian and the QCD vacuum are invariant under $SU(3)_F$ global rotations (α^a space–time independent):

$$q \to \exp\left(i \sum_a \alpha^a \lambda^a\right) q, \qquad q^\dagger \to q^\dagger \exp\left(-i \sum_a \alpha^a \lambda^a\right), \tag{1.23}$$

where q is now a vector $(u \quad d \quad s)$, and λ^a are the Gell-Mann SU(3) matrices acting on this 3-dimensional flavor space. It follows that the physical states, mesons and barons, will be gathered into approximately degenerate multiplets corresponding to irreducible representations of $SU(3)_F$. Let us first consider mesons. From (1.6) we have to combine one quark and one antiquark. We have therefore to decompose the product $\mathbf{3} \otimes \bar{\mathbf{3}}$ into irreducible representations:

$$\mathbf{3} \otimes \bar{\mathbf{3}} = \mathbf{1} + \mathbf{8} \tag{1.24}$$

Indeed, looking at the meson spectrum, we find 8 pseudoscalars with masses below 550 MeV and one pseudoscalar with a mass of the order of 1 GeV, the η'. Their charges agree with the decomposition (1.24). The flavor wave functions of this pseudoscalar meson nonet are

$$\pi^+ = u\bar{d}, \qquad \pi^0 = -\frac{1}{\sqrt{2}}(u\bar{u} - d\bar{d}), \qquad \pi^- = -d\bar{u}, \tag{1.25}$$

$$K^+ = u\bar{s}, \qquad K^0 = d\bar{s}, \qquad \bar{K}^0 = s\bar{d}, \qquad K^- = -s\bar{u}, \tag{1.26}$$

$$\eta \approx -\frac{1}{\sqrt{6}}(u\bar{u} + d\bar{d} - 2s\bar{s}), \qquad \eta' \approx -\frac{1}{\sqrt{3}}(u\bar{u} + d\bar{d} + s\bar{s}). \tag{1.27}$$

The approximate equality $\approx$ for η, η' comes from the fact that there is a small octet–singlet mixing for these states.

Analogously, a nonet of vector mesons is found with masses between 0.77 and 1.02 GeV, the ρ^+, ρ^-, ρ^0 mesons with the same flavor wave functions as the pions, the strange vector mesons K^{*+}, K^{*0}, $\bar{K}^{*0}$, K^{*-} with the same flavor wave function as the pseudoscalar K (1.26), and two neutral states ω, ϕ that are mixtures of the octet and singlet wave functions (1.27):

$$\omega = -\frac{1}{\sqrt{2}}(u\bar{u} + d\bar{d}), \qquad \phi = -s\bar{s}. \tag{1.28}$$

For baryons (1.6) we have to decompose the product $\mathbf{3}\otimes\mathbf{3}\otimes\mathbf{3}$:

$$\mathbf{3}\otimes\mathbf{3}\otimes\mathbf{3} = \mathbf{1} + \mathbf{8}' + \mathbf{8}'' + \mathbf{10}. \tag{1.29}$$

For the lowest-mass baryons we do indeed find an octet of spin–parity $J^P = \frac{1}{2}^+$, P(uud), N(udd), Σ^+(uus), Σ^0(uds), Σ^-(dds), Λ(uds), Ξ^0(uss), Ξ^-(dss), with masses between 0.94 and 1.31 GeV and a decuplet of $J^P = \frac{3}{2}^+$, Δ^{++}(uuu), Δ^+(uud), Δ^0(udd), Δ^-(ddd), Σ^+(uus), Σ^0(uds), Σ^-(dds), Ξ^0(uss), Ξ^-(dss), Ω^-(sss), with masses between 1.24 and 1.67 GeV. In the next section we shall give in detail the flavor wave functions of all these states and of excited baryon states as well. The reason why there is no $\frac{1}{2}^+$ SU(3) singlet baryon of low mass will be clear when considering the spatial and spin degrees of freedom. Let us recall that $SU(3)_F$ is an approximate symmetry, broken by the $m_s - m = 150$ MeV mass difference. The term in the QCD Lagrangian breaking $SU(3)_F$ is of the form (neglecting the $m_u - m_d$ mass difference),

$$(m_s - m)\bar{s}s \tag{1.30}$$

Therefore to first order in this $SU(3)_F$ breaking term, we expect that the hadrons within an $SU(3)_F$ multiplet will differ in mass linearly according to their strange-quark s or $\bar{s}$ content. This is indeed the case to a good approximation, for the $\frac{3}{2}^+$ and 1^- states,

$$\begin{aligned} &\Omega^-(1670) - \Xi(1530) \approx \Xi(1530) - \Sigma(1385) \approx \Sigma(1385) - \Delta(1232) \\ &\phi(1020) - K^*(890) \approx K^*(890) - \rho(770), \end{aligned} \tag{1.31}$$

where the particle symbols represent their masses. More generally, one can look at the $SU(3)_F$ properties of the breaking (1.30), which can be written:

$$\bar{s}s = \bar{q}\frac{1 - \sqrt{3}\,\lambda^8}{3}q, \tag{1.32}$$

where $q = (u \quad d \quad s)$, and realize that the breaking transforms as λ^8. The application of the Wigner–Eckart theorem to $SU(3)_F$ gives the famous Gell-Mann–Okubo mass formula (Gell-Mann, 1961; Okubo, 1962) for the octet:

$$\Sigma + 3\Lambda = 2(N + \Xi). \tag{1.33}$$

Since the masses of the u and d quarks differ only by a few MeV, there is an SU(2) subgroup of $SU(3)_F$, called the *isospin*, a flavor symmetry conserved to great accuracy. In the limit $m_u = m_d$, the QCD Lagrangian (1.1) and the vacuum of the theory are invariant under the isospin global rotations.

$$q \rightarrow e^{i\tau\cdot\alpha} q, \qquad \bar{q} \rightarrow \bar{q} e^{-i\tau\cdot\alpha} \tag{1.34}$$

where q and $\bar{q}$ are two-dimensional vectors in isospin space, $q = (u \quad d)$ and $\bar{q} = (\bar{d} \quad -\bar{u})$, and τ are the Pauli matrices. The strange quarks s, $\bar{s}$ are isosinglets, and in the limit of exact isospin, all hadrons with the same s, $\bar{s}$ content and differing only in u, d (or $\bar{u}$, $\bar{d}$) will be degenerate. For example, triplets of isospin are π^+, π^0, π^-, or ρ^+, ρ^0, ρ^-, or $\Sigma^+, \Sigma^0, \Sigma^-$. Doublets of isospin are K^+, K^0 or p, n. The isospin symmetry is broken by the small $m_d - m_u$ mass difference, and by electromagnetism, which distinguishes the d and u quarks as they have different charges ($-\frac{1}{3}$ and $+\frac{2}{3}$ respectively).

By inspection of the Lagrangian (1.1), we see that QCD conserves flavor. We can therefore define, as we have done implicitly for the strangeness, flavor quantum numbers corresponding to the charges

$$Q_i = \int dx\, q_i^\dagger(x) q_i(x), \tag{1.35}$$

where i labels the different flavors (i = u, d, s, c, b, t). These flavor charges are conserved and explain observed selection rules that are only violated at the level of the weak interactions. For example, strange or charmed quarks are always created in pairs s$\bar{s}$, c$\bar{c}$ by the strong interactions.

1.2.4 Chiral Symmetry

The pseudoscalar mesons are relatively light. The isospin triplet π^+, π^0, π^-, have masses $\sim$140 MeV, the isosinglet η has a mass $\sim$550 MeV and the corresponding strange mesons (kaons K^+, K^0, $\bar{K}^0$, K^-) have masses $\sim$495 MeV. These relative small masses on the hadron scale indicate the existence of an approximate global symmetry of L_{QCD} called chiral symmetry. This symmetry breaks down dynamically, giving an octet of so-called Nambu–Goldstone pseudoscalar bosons that would be massless if the symmetry were exact.

This approximate symmetry would become exact if the quark masses in L_{QCD} were exactly zero. The flavor symmetry $SU(3)_F$ of the preceding section is an approximate symmetry because the *mass differences* $m_u - m_d$, $m_s - m_u$, $m_s - m_d$, are small. Chiral symmetry demands something more: the absolute magnitudes of m_u, m_d, m_s also have to be small. This is very well verified in particular for the u, d quarks, which have Lagrangian masses, called, after renormalization, *current masses*, of the order of a few MeV.

Let us assume first that the current masses in L_{QCD} vanish for the u, d, s quarks. Then the QCD Lagrangian (1.1) restricted to these flavors is invariant under the following global transformations of the quark fields:

$$\left.\begin{aligned} q(x) &\rightarrow \exp\left[\mathrm{i}\gamma_5\left(\sum_{a=1}^{8}\frac{\lambda^a}{2}\alpha^a+\alpha^0\right)\right]q(x), \\ q^\dagger(x) &\rightarrow q^\dagger(x)\exp\left[-\mathrm{i}\gamma_5\left(\sum_{a=1}^{8}\frac{\lambda^a}{2}\alpha^a+\alpha^0\right)\right], \end{aligned}\right\} \tag{1.36}$$

where λ^a are the Gell-Mann matrices of the $SU(3)_F$ flavour symmetry, $\gamma_5 = \mathrm{i}\gamma^0\gamma^1\gamma^2\gamma^3$, and $\alpha^a\,(a=0,\ldots,8)$ are real numbers. These transformations, combined with the flavor U(3) transformations (1.23),

$$\left.\begin{aligned} q(x) &\rightarrow \exp\left[\mathrm{i}\left(\sum_{a=1}^{8}\frac{\lambda^a}{2}\beta^a+\beta^0\right)\right]q(x), \\ q^\dagger(x) &\rightarrow q^\dagger(x)\exp\left[-\mathrm{i}\left(\sum_{a=1}^{8}\frac{\lambda^a}{2}\beta^a+\beta^0\right)\right], \end{aligned}\right\} \tag{1.37}$$

form a U(3)×U(3) group called chiral U(3)×U(3). This group factorizes into $U(1)_V \times U(1)_A \times SU(3)_L \times SU(3)_R$. $U(1)_V$ has as generator the identity matrix 1 in flavor and Dirac space. $U(1)_A$ has 1 in flavor and γ_5 in Dirac space. $SU(3)_R$, $SU(3)_L$ have respectively

$$\tfrac{1}{2}(1\pm\gamma_5)\frac{\lambda^a}{2} \quad (a=1,\ldots,8), \tag{1.38}$$

where $\tfrac{1}{2}(1\pm\gamma_5)$ project the quark fields into right-handed and left-handed fields:

$$q_R = \tfrac{1}{2}(1+\gamma_5)q, \qquad q_L = \tfrac{1}{2}(1-\gamma_5)q. \tag{1.39}$$

As mentioned above, the invariance $U(1)_V$ is simply related to baryon-number conservation and is an exact symmetry. $U(1)_A$ has a very special status, as it is a symmetry that, although exact if $m=0$ at the Lagrangian level, is broken by quantum effects. This is the Adler–Bell–Jackiw axial anomaly (Adler, 1969; Bell and Jackiw, 1969). In the present context we shall only remark that this phenomenon is at the origin of the high mass of the $SU(3)_F$ singlet pseudoscalar, the η′(958) ('t Hooft, 1976). We shall for the moment ignore this flavor-singlet axial symmetry and shall concentrate on $SU(3)_L \times SU(3)_R$. The realization of this symmetry has two possible modes, or phases. If the QCD *vacuum* were *symmetric* under this group of transformations then the spectrum would be one of *parity doublets*, i.e. hadrons of the same mass and spin that appear degenerate in couples of opposite parity, owing to the $1 \leftrightarrow \gamma_5$ symmetry. For example, to 1^- mesons would correspond 1^+ mesons of the same mass. This pattern is not actually realized in nature, but we have rather a spectrum characterized by light 0^- mesons followed by heavier 1^- mesons, without even an approximate signal of parity doubling. It is the other mode of chiral symmetry that is approximately realized. What happens is that, although the QCD Lagrangian is approximately $SU(3)_L \times SU(3)_R$ invariant, the vacuum does not possess this invariance, because it is a condensate of $q\bar{q}$ pairs that breaks chirality. We have, in the vacuum, by a nonperturbative dynamical mechanism (Nambu and Jona-Lasinio, 1961a,b; Gell-Mann, Oakes and Renner, 1968)

$$\begin{aligned} \langle 0|\bar{u}u|0\rangle = \langle 0|\bar{d}d|0\rangle &\approx \langle 0|\bar{s}s|0\rangle \neq 0 \\ &= -(220\,\text{MeV})^3. \end{aligned} \tag{1.40}$$

The vacuum is only $SU(3)_F$-invariant, and $SU(3)_L \times SU(3)_R$ is said to be spontaneously or dynamically broken down to $SU(3)_F$, the flavor symmetry of the preceding section. Then the symmetry of the Lagrangian, broken by the vacuum, manifests itself through the existence of massless pseudoscalar bosons corresponding to the $SU(3)_L \times SU(3)_R$ generators which do not belong to $SU(3)_F$. This is Goldstone's theorem (Goldstone, 1961). In the case of chiral symmetry there is an octet of pseudoscalar Goldstone bosons, corresponding to the generators of $SU(3)_L \times SU(3)_R$ that do not leave the vacuum invariant. These massless bosons are the equivalent of the spin waves in a ferromagnet. In this latter case, it is rotational invariance that is spontaneously broken. The Goldstone bosons of chiral symmetry are the π^+, π^0, π^-, η and K^+, K^0, $\bar{K}^0$, K^-. Of course $SU(3)_L \times SU(3)_R$ is only an approximate symmetry because of the mass terms, and it is better realized for the subgroup $SU(2)_L \times SU(2)_R$. Because of the mass terms in L_{QCD} that explicitly break $SU(3)_L \times SU(3)_R$, the pseudoscalar mesons are not massless and are said to be quasi-Goldstone bosons. Their masses squared turn out to be proportional to $\langle \bar{q}q\rangle$ and to the current masses, as $m_\pi^2 \propto m_d + m_u$ and $m_K^2 \propto m_s + m_u$. An analysis of the pseudoscalar meson spectrum along these lines gives, for the current masses (Fritzsch, Gell-Mann and Leutwyler, 1973; Leutwyler, 1974; Weinberg, 1977)

$$m_u \approx 4\,\text{MeV}, \qquad m_d \approx 7\,\text{MeV}, \qquad m_s \approx 150\,\text{MeV}. \tag{1.41}$$

Another effect of the vacuum condensate is to dynamically generate a mass for the quarks that is much larger than the current mass. Let us consider the condensate of $q\bar{q}$ pairs that has nonzero chirality. A quark moving in this nontrivial vacuum will feel the action of the condensed pairs exactly as a particle with spin in a ferromagnet feels the magnetic field created by all the aligned spins. This interaction has the effect of creating an energy gap between the zero-momentum quark and the vacuum. This energy gap is an effective dynamical mass, which is present even in the limit of zero Lagrangian or current mass. It is reasonable to assume that this is precisely the constituent-quark mass. From the nucleon mass and the nucleon magnetic moment, as we shall see later, the effective masses turn out to be of the order of

$$m_u \approx m_d \approx 300\,\text{MeV}, \qquad m_s \approx 450\,\text{MeV}, \tag{1.42}$$

which differ widely from the current masses (1.41).

From the special role played by the pion as a Goldstone boson one can deduce several relations between amplitudes implying low-energy pions and matrix elements of axial currents $\bar{q}\gamma^\mu\gamma_5(\lambda^a/2)q$. These soft-pion theorems will not be studied in this book, and we refer the reader to the literature dealing with Current Algebra (Adler and Dashen, 1968). There are recent developments on the dynamical breaking of chiral symmetry within QCD in the Coulomb gauge or for confining potentials (Finger, Horn and Mandula, 1979; Amer *et al.*, 1983; Le Yaouanc *et al.*, 1984, 1985a,b; Adler and Davis, 1984): the $\langle\bar{q}q\rangle$ condensate, the quark-dynamical mass and the meson spectrum are deduced from a chiral-invariant interaction at the quark level.

Chiral symmetry dynamical breaking has taught us something about constituent quarks: they are massive. This may encourage us to use as a first step a nonrelativistic approximation to describe the interactions between quarks. Up to now we have deduced approximate symmetries from QCD. Now we will proceed with a model whose basic hypotheses cannot be derived from QCD and find their justification only *a posteriori* in their empirical successes. But, while oversimplified, this model will give us a more concrete grasp of the structure of hadrons.

1.3 THE HADRON SPECTRUM

We shall now describe the spectrum of hadrons. The most economical way of doing it is to use the nonrelativistic quark model, very successful in predicting the flavor, J^{PC} (spin, parity, charge-conjugation) quantum numbers of hadrons and the approximate masses of the different excited states.

1.3.1 Flavor Independence and Color Saturation

The basic hypothesis of the nonrelativistic quark model is that the constituent quarks obey the nonrelativistic Schrödinger equation to a reasonably good approximation. As we shall discuss in Chapters 3 and 4, this is a questionable statement, and we can only argue that, at least for light quarks, the best defence of the nonrelativistic quark model is its impressive empirical success.

The Hamiltonian in terms of quark fields is

$$H = \int d\boldsymbol{x} \sum_i q_i^\dagger(\boldsymbol{x}) \beta \left(m_i - \frac{\Delta}{2m_i} \right) q_i(\boldsymbol{x})$$
$$+ \tfrac{1}{2} \int d\boldsymbol{x}\, d\boldsymbol{y}\, V_0(\boldsymbol{x}-\boldsymbol{y}) \sum_{ija} q_i^\dagger(\boldsymbol{x}) \frac{\lambda^a}{2} q_i(\boldsymbol{x})\, q_j^\dagger(\boldsymbol{y}) \frac{\lambda^a}{2} q_j(\boldsymbol{y}), \tag{1.43}$$

where i, j are *flavor* indices, and the matrices λ^a act on color space. This corresponds to assuming a spin-independent interaction, and $V_0(\boldsymbol{x}-\boldsymbol{y})$ is a central potential, dependent only on $|\boldsymbol{x}-\boldsymbol{y}|$. This expression has the important feature that the interaction term is independent of flavor: the potential is the same between each pair of flavors, and all the flavor dependence comes through the masses in the kinetic term of (1.43). Note that this flavor independence of the potential is just a transposition to the quark potential model of this general feature of the QCD Lagrangian (1.1). Note also that the particular color structure of the interaction (1.43) can be inferred from the one-gluon-exchange Coulomb approximation to L_{QCD}, with $V_0(\boldsymbol{x}) = \alpha_s/|\boldsymbol{x}|$. We shall, however, assume this particular structure for other potentials, in particular for potentials that confine the quarks and grow with distance.

In our nonrelativistic approximation (1.43), the expansion of the quark field on the eigenstates of the bilinear part of the Hamiltonian is

$$q_i(\boldsymbol{x}) = \frac{1}{(2\pi)^{3/2}} \int d\boldsymbol{p}\, e^{i\boldsymbol{p}\cdot\boldsymbol{x}} \sum_s \left[b_{is}(\boldsymbol{p}) \begin{pmatrix} \chi_s \\ 0 \end{pmatrix} + d_{is}^\dagger(-\boldsymbol{p}) \begin{pmatrix} 0 \\ \chi_s^{\text{C}} \end{pmatrix} \right], \tag{1.44}$$

where χ is the quark Pauli spinor, while χ_s^{C} is related to χ_s so as to give the desired spin properties to the four-spinor. b_i annihilates a quark, while $d_i^\dagger$ creates an antiquark. The interaction term in (1.43) conserves separately the number of quarks and of antiquarks, and can therefore be diagonalized in sectors with a fixed number of quarks and antiquarks.

Let us now consider how the Hamiltonian (1.43) acts on a quark–antiquark state. The state will be written, in Fock space,

$$|q_{is}(\boldsymbol{x})\bar{q}_{js'}(\boldsymbol{y})\rangle = \frac{1}{(2\pi)^3} \left(\int d\boldsymbol{p}\, e^{-i\boldsymbol{p}\cdot\boldsymbol{x}} b_{is}^\dagger(\boldsymbol{p}) \right) \left(\int d\boldsymbol{p}'\, e^{-i\boldsymbol{p}'\cdot\boldsymbol{y}} d_{js'}^\dagger(\boldsymbol{p}') \right) |0\rangle, \tag{1.45}$$

where $|0\rangle$ is the vacuum state. Let us call the second term in (1.43) $\mathscr{V}$; it contains the potential. This term has the following mean value in the $q\bar{q}$ state (1.45):

$$\langle q(\boldsymbol{x}')\bar{q}(\boldsymbol{y}')|\, \mathscr{V}\, |q(\boldsymbol{x})\bar{q}(\boldsymbol{y})\rangle = -\sum_a \frac{\lambda_q^a}{2} \frac{\tilde{\lambda}_{\bar{q}}^a}{2} V_0(\boldsymbol{x}-\boldsymbol{y})\, \delta(\boldsymbol{x}-\boldsymbol{x}')\, \delta(\boldsymbol{y}-\boldsymbol{y}'), \tag{1.46}$$

where $\tilde{\lambda}^a$ is the transpose of λ^a, the subscripts q, $\bar{q}$ imply action on the quark or the antiquark, and the minus sign comes from the anticommutation of fermion fields.

If we consider a qq state, analogous to the q$\bar{q}$ one (1.45), we have

$$\langle q(x')q(y')|\mathscr{V}|q(x)q(y)\rangle = \sum_a \frac{\lambda_q^a}{2}\frac{\lambda_q^a}{2} V_0(x-y)\delta(x-x')\delta(y-y'), \qquad (1.47)$$

The qq and q$\bar{q}$ states will decompose, from the point of view of color, into the irreducible representations

$$\left.\begin{aligned} \mathbf{3}\otimes\bar{\mathbf{3}} &= \mathbf{1}+\mathbf{8}, \\ \mathbf{3}\otimes\mathbf{3} &= \bar{\mathbf{3}}+\mathbf{6}. \end{aligned}\right\} \qquad (1.48)$$

To know the interaction within a qq or a q$\bar{q}$ pair in a color representation R given in (1.48) we need to compute

$$\langle R|\sum_a \frac{\lambda_1^a}{2}\frac{\lambda_2^a}{2}|R\rangle, \qquad (1.49)$$

where 1 means a quark and 2 a quark or an antiquark. In this last case λ^a will be $-\bar{\lambda}^a$ for particle 2. It is useful to decompose (1.49) into

$$\begin{aligned} \langle R|\sum_a \frac{\lambda_1^a}{2}\frac{\lambda_2^a}{2}|R\rangle &= \tfrac{1}{2}\Big\{\langle R|\sum_a \left(\frac{\lambda_1^a}{2}+\frac{\lambda_2^a}{2}\right)^2|R\rangle \\ &\quad -\langle R|\sum_a\left(\frac{\lambda_1^a}{2}\right)^2+\sum_a\left(\frac{\lambda_2^a}{2}\right)^2|R\rangle\Big\} \qquad (1.50) \\ &= \tfrac{1}{2}[C(R)-2C(\mathbf{3})], \end{aligned}$$

where $C(R)$ is the color Casimir of the qq or q$\bar{q}$ representation R, and $C(\mathbf{3})$ is the color Casimir of the quark fundamental representation **3**. From

$$\left.\begin{aligned} C(\mathbf{1}) &= 0, \\ C(\mathbf{8}) &= 3, \\ C(\mathbf{3}) &= C(\bar{\mathbf{3}}) = \tfrac{4}{3}, \\ C(\mathbf{6}) &= \tfrac{10}{3}, \end{aligned}\right\} \qquad (1.51)$$

we obtain the potentials in the various color states:

$$\left.\begin{aligned} \langle(q\bar{q})_1|\mathscr{V}|(q\bar{q})_1\rangle &= -\tfrac{4}{3}V_0, \\ \langle(q\bar{q})_8|\mathscr{V}|(q\bar{q})_8\rangle &= +\tfrac{1}{6}V_0, \\ \langle(qq)_{\bar{3}}|\mathscr{V}|(qq)_{\bar{3}}\rangle &= -\tfrac{2}{3}V_0, \\ \langle(qq)_6|\mathscr{V}|(qq)_6\rangle &= +\tfrac{1}{3}V_0, \end{aligned}\right\} \qquad (1.52)$$

where the color representation is given by the indices. We see that to obtain an attractive confining force within a meson, which is a color singlet according to

(1.6), we need to assume, with the sign convention (1.43), that $V_0(\boldsymbol{x}) \to -\infty$ for $|\boldsymbol{x}| \to \infty$. We then obtain a repulsive force for a color octet. For example, $V_0(\boldsymbol{x}) = -|\boldsymbol{x}|$ will give a potential $+\frac{4}{3}|\boldsymbol{x}|$ for a q$\bar{\text{q}}$ color singlet and a repulsive potential $-\frac{1}{6}|\boldsymbol{x}|$ for a q$\bar{\text{q}}$ color octet.

Moreover, for a qq in $\bar{\mathbf{3}}$ state, the force will also be attractive, and for a color **6** state it will be repulsive. Let us now see the consequences of the color structure of (1.43) for the baryons, which have the color-singlet wave function (1.6). From the point of view of color, a qqq state can be decomposed into the irreducible representations

$$\mathbf{3} \otimes \mathbf{3} \otimes \mathbf{3} = (\bar{\mathbf{3}} + \mathbf{6}) \otimes \mathbf{3} = (\mathbf{1} + \mathbf{8}') + (\mathbf{8}'' + \mathbf{10}). \tag{1.53}$$

Baryons are color singlets. In (1.53) only the $\bar{\mathbf{3}} \otimes \mathbf{3}$ contributes to the right-hand-side singlet. It follows that within a baryon each qq pair is in a $\bar{\mathbf{3}}$ state, making the qq force attractive, with attractive strength equal to one half of the attraction within a meson (a q$\bar{\text{q}}$ color singlet), as we see from (1.52). Note that this relation $V_{qq} = \frac{1}{2} V_{q\bar{q}}$ is well verified phenomenologically (Richard, 1981). Furthermore, color octets or decuplets qqq and color octets q$\bar{\text{q}}$ are unstable because of the repulsive forces between quarks. This phenomenon is called *color saturation*: hadrons appear as color-neutral or singlet states. Moreover, on the hadron energy scale, mesons and baryons appear naturaly as approximately degenerate. Indeed, a bound diquark $\bar{\mathbf{3}}$ is attracted or repelled by a quark according to its combined color state (singlet or octet respectively) with almost the same potential as between the quark (a **3** state) and the antiquark (a $\bar{\mathbf{3}}$ state) within a meson. In this picture we consider a configuration in which the interquark distance in the diquark is negligible compared with the quark–diquark distance. The resulting state is a qqq color singlet with the same energy scale as the q$\bar{\text{q}}$ color singlets. Any other color configuration is broken apart by the potential.

Other color singlets, such as qq$\bar{\text{q}}\bar{\text{q}}$ or 6q, can also be considered, although there is not yet any experimental evidence for these states, except for some still controversial candidates.

Thus, the very simple model (1.43), inspired from Coulomb one-gluon exchange, describes at the same time the confinement of quarks within mesons and baryons. We should note that if one takes, instead of a four-component Lorentz-vector potential as in (1.43), a Lorentz-scalar color potential, i.e. an interaction term of the form

$$\mathscr{V}_s = \tfrac{1}{2} \int d\boldsymbol{x}\, d\boldsymbol{y}\, V_0(\boldsymbol{x}-\boldsymbol{y}) \sum_{ija} \bar{q}_i(\boldsymbol{x}) \frac{\lambda^a}{2} q_i(\boldsymbol{x})\, \bar{q}_j(\boldsymbol{y}) \frac{\lambda^a}{2} q_j(\boldsymbol{y}), \tag{1.54}$$

then one gets an additional minus sign for the q$\bar{\text{q}}$ coupling as compared with (1.52). Therefore one does not obtain in this case simultaneous confinement of baryons and mesons. A four-component Lorentz-vector confining potential possesses, on the contrary, the property of color saturation. However, detailed analyses of the meson spectrum (L–S splittings) do not support a pure Lorentz-vector potential, and some admixture of Lorentz scalar is needed, as we shall discuss in Section 1.3.4.

1.3.2 Meson Spectrum

From (1.46) and (1.52) we obtain the Hamiltonian for color-singlet $q_i\bar{q}_j$ systems:

$$H = \frac{\boldsymbol{p}_x^2}{2m_i} + \frac{\boldsymbol{p}_y^2}{2m_j} - \tfrac{4}{3}V_0(\boldsymbol{x}-\boldsymbol{y}), \tag{1.55}$$

where $\boldsymbol{x}$ ($\boldsymbol{y}$) is the position of the quark of flavor i (antiquark of flavor j). From now on we shall omit color and call

$$V(\boldsymbol{r}) = -\tfrac{4}{3}V_0(\boldsymbol{r}) \tag{1.56}$$

Let us consider for the moment a single flavor, $i=j=$charm for example. The Hamiltonian (1.55) will then be

$$H = \frac{\boldsymbol{P}^2}{4m_c} + \frac{\boldsymbol{p}^2}{m_c} + V(\boldsymbol{r}), \tag{1.57}$$

where $\boldsymbol{P}=\boldsymbol{p}_x+\boldsymbol{p}_y$ is the center-of-mass momentum of the $c\bar{c}$ system, $\boldsymbol{r}=\boldsymbol{x}-\boldsymbol{y}$ is the relative distance, $\boldsymbol{p}=\frac{1}{2}(\boldsymbol{p}_x-\boldsymbol{p}_y)$ is the corresponding relative momentum. As we assume $V(\boldsymbol{r})$ to be a spin-independent central potential, (1.57) is invariant under separate orbital and spin rotations. The eigenstates of the Schrödinger equation

$$\left(-\frac{\Delta}{m_c} + V(\boldsymbol{r})\right)\psi(\boldsymbol{r}) = E\psi(\boldsymbol{r}) \tag{1.58}$$

will be of the form

$$\psi_L^m(\boldsymbol{r}) = Y_L^m(\hat{\boldsymbol{r}})\frac{v_L(r)}{r}, \tag{1.59}$$

where $v_L(r)$ is the radial wave function.

The different levels of the charmonium $c\bar{c}$ system will be classified according to $N\,^{2S+1}L_J$, where N is the radial quantum number, $\boldsymbol{L}$ the $q\bar{q}$ orbital angular momentum, $\boldsymbol{S}$ the total $q\bar{q}$ spin 0 or 1, and $\boldsymbol{J}$ the total angular momentum $\boldsymbol{J}=\boldsymbol{L}+\boldsymbol{S}$, the meson spin. For each N,L there will be $J=L$ or $J=|L-1|$, L, $L+1$ states. For example, the ground state will correspond to $N=1$, $S=0$ or 1, which gives a spin triplet $1\,^3S_1$ and a spin singlet $1\,^1S_0$: the $J/\psi(3096)$, $\eta_c(2980)$ respectively.

The $q\bar{q}$ states have definite discrete parity P and charge-conjugation C (if the quark and the antiquark have matching flavor). Consider first a $q_i\bar{q}_j$ color-singlet state (the flavors i and j can be different):

$$|q_{is}, \bar{q}_{js'}\rangle = N \int \mathrm{d}\boldsymbol{p}\, \psi(\boldsymbol{p}) \sum_a b^\dagger_{ais}(\boldsymbol{p}) d^\dagger_{ajs'}(-\boldsymbol{p})|0\rangle, \tag{1.60}$$

where $\psi(\boldsymbol{p})$ is the spatial wave function in momentum space, and the indices $(a\,i\,s)$, $(a\,j\,s')$ refer to color, flavor and spin. The parity transformation of the fermion field is given by

$$q(\boldsymbol{x}) \to \gamma^0 q(-\boldsymbol{x}). \tag{1.61}$$

Using (1.44), we obtain

$$b^\dagger(\boldsymbol{p}) \to b^\dagger(-\boldsymbol{p}), \qquad d^\dagger(\boldsymbol{p}) \to -d^\dagger(-\boldsymbol{p}), \tag{1.62}$$

and we therefore find, on applying the parity operator to the state (1.60),

$$\mathrm{P}|q_{is}, \bar{q}_{js'}\rangle = (-1)^{L+1}|q_{is}, \bar{q}_{js'}\rangle, \tag{1.63}$$

where we have used $\psi(-\boldsymbol{p}) = (-1)^L \psi(\boldsymbol{p})$.

Let us now see how charge conjugation applies to a state of the type (1.60). Starting from

$$\mathrm{C} b^\dagger_{ais}(\boldsymbol{p}) = d^\dagger_{ais}(\boldsymbol{p}), \qquad \mathrm{C} d^\dagger_{ais}(\boldsymbol{p}) = b^\dagger_{ais}(\boldsymbol{p}), \tag{1.64}$$

we obtain

$$\begin{aligned}
\mathrm{C}|q_{is}, \bar{q}_{js'}\rangle &= N \int \mathrm{d}\boldsymbol{p}\, \psi(\boldsymbol{p}) \sum_a d^\dagger_{ais}(\boldsymbol{p}) b^\dagger_{ajs'}(-\boldsymbol{p})|0\rangle \\
&= -N \int \mathrm{d}\boldsymbol{p}\, \psi(-\boldsymbol{p}) \sum_a b^\dagger_{ajs'}(\boldsymbol{p}) d^\dagger_{ais}(-\boldsymbol{p})|0\rangle \qquad (1.65) \\
&= (-1)^{L+1} N \int \mathrm{d}\boldsymbol{p}\, \psi(\boldsymbol{p}) \sum_a b^\dagger_{ajs'}(\boldsymbol{p}) d^\dagger_{ais}(-\boldsymbol{p})|0\rangle.
\end{aligned}$$

Note that when the quark and the antiquark have the same flavor (e.g. $c\bar{c}$), up to the phase $(-1)^{L+1}$, C amounts to the exchange of spin indices. The antiquark's spin states are described by the standard Pauli spinors χ_s, related to χ_s^{C} in (1.44) by

$$\chi_s^{\mathrm{C}} = -\mathrm{i}\sigma^2 \chi_s^*. \tag{1.66}$$

Since the spin $S=1$ and 0 mesons have respectively symmetric and antisymmetric wave functions,

$$\left.\begin{aligned}&\chi_1^{+1}=\uparrow\uparrow, \qquad \chi_1^0=\frac{1}{\sqrt{2}}(\uparrow\downarrow+\downarrow\uparrow), \qquad \chi_1^{-1}=\downarrow\downarrow,\\&\chi_0=\frac{1}{\sqrt{2}}(\uparrow\downarrow-\downarrow\uparrow),\end{aligned}\right\} \tag{1.67}$$

we conclude that $C=(-1)^{L+S}$.

We can now tabulate the charmonium ($c\bar{c}$) states with their assigned $N\,^{2S+1}L_J$ and P, C quantum numbers (Figure 1.1). The ground state has the quantum numbers 1^3S_1, $J^{PC}=1^{--}(J/\psi)$ and 1^1S_0, $J^{PC}=0^{-+}(\eta_c)$. The first orbital excitation, the χ_c states, corresponds to the quantum numbers $1^3P_J(J^{PC}=0^{++},\ 1^{++},\ 2^{++})$ and $1^1P_1(J^{PC}=1^{+-})$. We have also tabulated the bottonium $b\bar{b}$ states.

Let us now consider the mesons made out of light quarks. We have seen that, because of the spin-independence and the central character of the potential adopted as a first approximation, the spectrum is invariant under independent spin and spatial rotations. The spectrum is SU(2) × O(3)-invariant. SU(2) corresponds to the total quark spin $S=0$, 1 and O(3) to the orbital angular momentum L. In the case of light quarks we have moreover three almost degenerate flavors u, d and s that form the fundamental representation **3** of the $SU(3)_F$ symmetry introduced in the preceding section. Including the quark spin degrees of freedom, the spin–flavor–space symmetry is then SU(6) × O(3), SU(6) containing $SU(2)\times SU(3)_F$. The Hamiltonian (1.55) for three degenerate flavors is indeed invariant under flavor, spin and simultaneous flavor and spin rotations, corresponding to the SU(6) group. The fundamental representation **6** of SU(6) will be formed by nonrelativistic Pauli spinors

$$u\uparrow,\quad u\downarrow,\quad d\uparrow,\quad d\downarrow,\quad s\uparrow,\quad s\downarrow, \tag{1.68}$$

where the arrows ↑ and ↓ represent spin up and down along some quantization axis. We shall find the meson content in terms of SU(6) multiplets by forming the product

$$\mathbf{6}\otimes\bar{\mathbf{6}}=\mathbf{1}+\mathbf{35}, \tag{1.69}$$

and it is useful to decompose these representations according to their SU(2)-spin and $SU(3)_F$ content, $SU(2)\times SU(3)_F\subset SU(6)$:

$$\left.\begin{aligned}\mathbf{1}\ &=(1,\mathbf{1}),\\ \mathbf{35}&=(1,\mathbf{8})+(3,\mathbf{1})+(3,\mathbf{8}),\end{aligned}\right\} \tag{1.70}$$

where the notation $(2S+1,\ \boldsymbol{\alpha})$ means a representation of $SU(2)\times SU(3)_F$. We see that the singlet **1** representation corresponds to an $S=0$ meson, singlet of SU(3), and that the **35** contains an $S=0$ octet, an $S=1$ octet and an $S=1$ singlet. These states correspond, for the ground state $L=0$, $N=1$, respectively

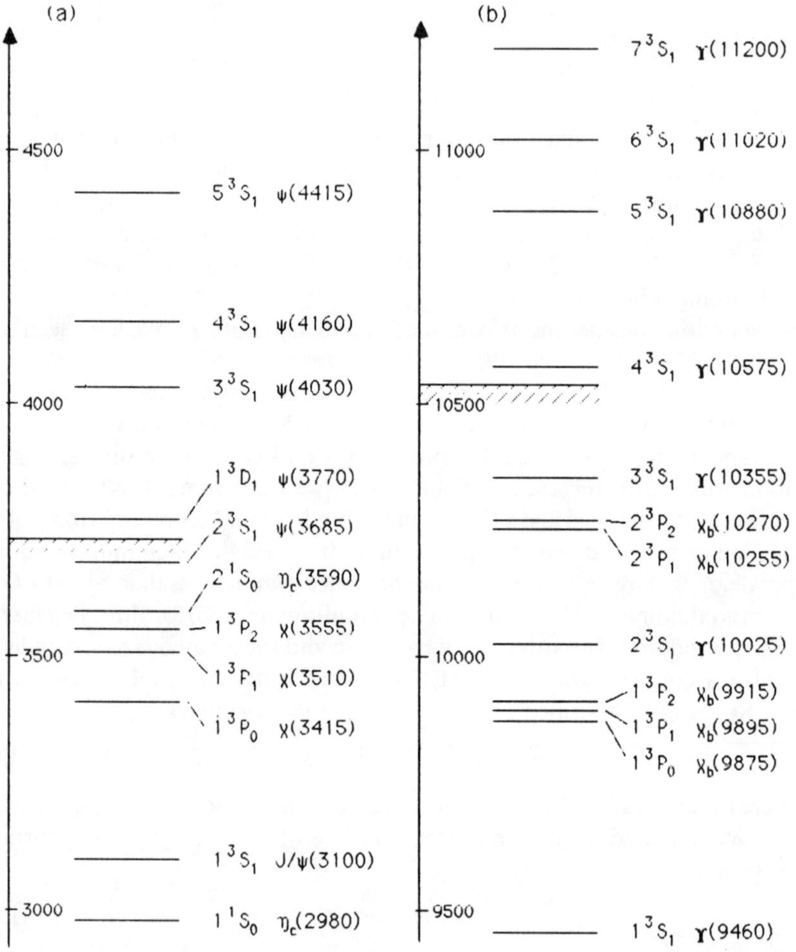

Figure 1.1 The charmonium $c\bar{c}$ (a) and bottonium $b\bar{b}$ (b) spectra in MeV. The notation of the levels is $N^{2S+1}L_J$, where N is the principal quantum number, S the total $q\bar{q}$ spin, L the relative orbital angular momentum, and J the spin of the bound state. The $D\bar{D}$ and $B\bar{B}$ thresholds are indicated by dashed lines.

to the $0^-\eta'$, the pseudoscalar 0^- octet (π, η, K), the vector 1^- octet (ρ, K^*, a linear combination of ω, ϕ), and the vector 1^- singlet (the orthogonal combination of ω, ϕ). From the point of view of C–parity ($C=(-1)^{L+S}$), the states that have $q\bar{q}$ with equal flavor, π^0, η and η', have $J^{PC}=0^{-+}$, and ρ^0, ω and ϕ have $J^{PC}=1^{--}$.

If we now go to the orbitally excited states, we find first the $N=1$, $L=1$ mesons. We must combine $L=1$ of O(3) with $S=0$ or 1 of the SU(6) flavor–spin multiplets **1**+**35**. In each multiplet we shall only quote the corresponding observed isovector. According to (1.63), we obtain an **8**+**1** of flavor with $J^P=1^+$, $J^{PC}=1^{+-}$ for the $B^0(1235)$; an **8**+**1** with $J^P=0^+$, $J^{PC}=0^{++}$ for $\delta^0(970)$; an **8**+**1** with $J^P=1^+$, $J^{PC}=1^{++}$ for $A_1^0(1270)$; and an **8**+**1** with $J^P=2^+$, $J^{PC}=2^{++}$ for $A_2^0(1320)$.

Also, some radially excited states with the same J^{PC} quantum numbers as the ground state have been identified, $\rho'(1600)$, $\phi'(1680)$, ..., and also some $L=2$, $S=1$ states like g(1690), $J^{PC}=3^{--}$.

We give in Figure 1.2 the main states of the light-meson spectrum with their attributed $N\,^{2S+1}L_J$ and P,C quantum numbers.

Finally, in Figure 1.3 we give the spectrum of *charmed* mesons, mesons made out of a charmed quark c and a light quark, like the $J^P=0^-$ mesons D (D^+: $c\bar{d}$; D^0: $c\bar{u}$), the F^+ meson $c\bar{s}$, and the corresponding $J^P=1^-$ states D^* and F^*.

It is clear from Figures 1.1 and 1.2 that the large symmetry for light flavors SU(6) × O(3) or the smaller symmetry that we have for the $c\bar{c}$ or $b\bar{b}$ systems is strongly broken by spin-dependent interactions. For example, there are large mass splittings within the $L=1$, **35**+**1** states, and even larger ones within the **35**+**1** ground state. In this latter case, however, a more involved phenomenon appears, namely the dynamical breaking of chiral symmetry discussed in Section 1.2.4.

After this brief description of the spectrum, we can write down the wave function of any meson in the form

$$\Psi_J^M(\mathbf{r}) = N\varphi_c\varphi_F \sum_m \langle L, S; M-m, m \mid J, M\rangle \chi_S^m \psi_L^{M-m}(\mathbf{r}) \tag{1.71}$$

where φ_c denotes the color wave function (1.6), φ_F the flavor wave function (for example $-(u\bar{u}-d\bar{d})/\sqrt{2}$ for π^0 and ρ^0, $-s\bar{s}$ for ϕ, $c\bar{c}$ for J/ψ, $c\bar{d}$ for D^+). The Clebsch–Gordan coefficient couples the spin wave functions (1.67) (spin-triplet $S=1$ or spin-singlet $S=0$) with the orbital angular momentum. $\psi_L^m(\mathbf{r})$ is the spatial wave function (1.59), the solution of the Schrödinger equation (1.58).

We shall make in Sections 1.3.4 and 1.3.5 a quick survey of the quark–quark potential—its spin-independent as well as its spin-dependent part. But before this discussion we shall first review the baryon spectrum and the simplest baryon wave functions according to the harmonic-oscillator potential.

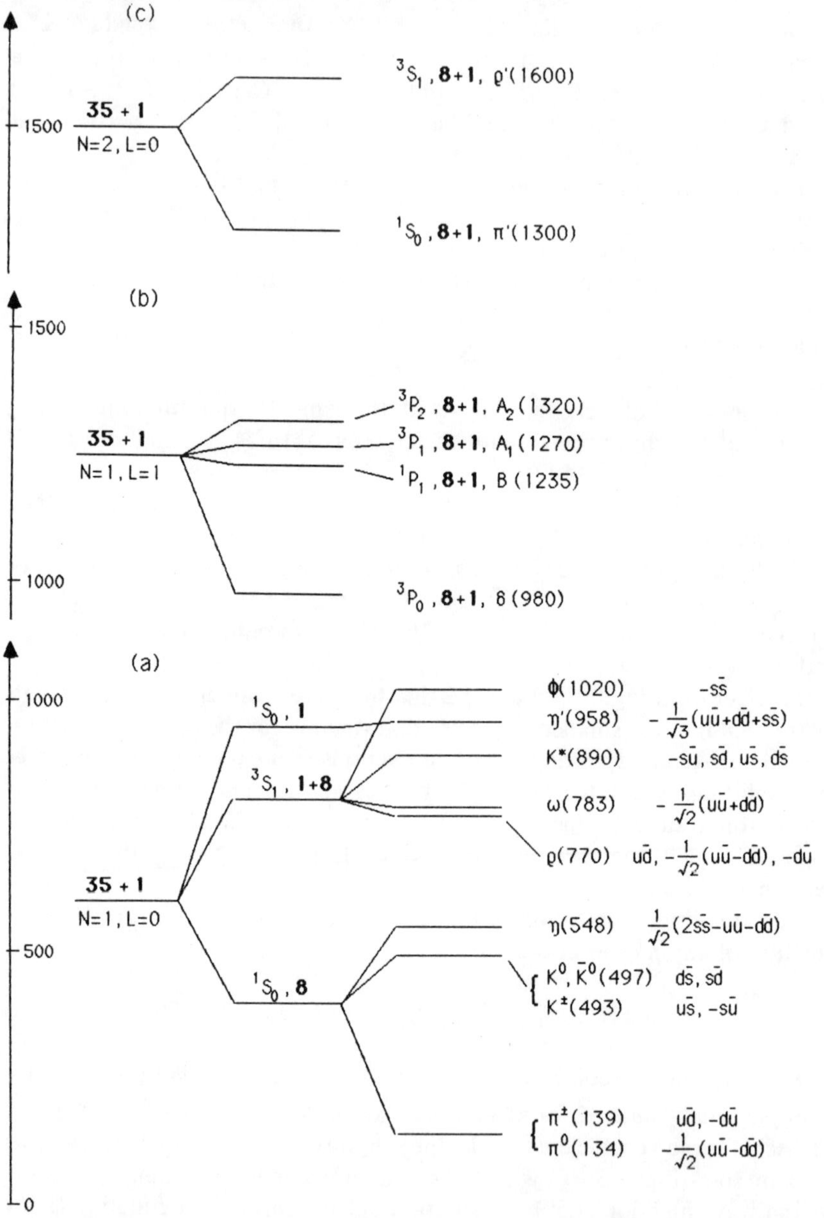

Figure 1.2 The spectrum of mesons made up of light quarks u, d and s, according to the SU(6)×O(3) classification. The SU(6)×O(3) multiplets (**35**+**1**, L) split into ${}^{2S+1}L_J$–SU(3) (**8** or **1**) multiplets. In (a) the quark content of all ground-state mesons is made explicit. In (b) and (c) only the I=1 meson of each nonet **8**+**1** is quoted.

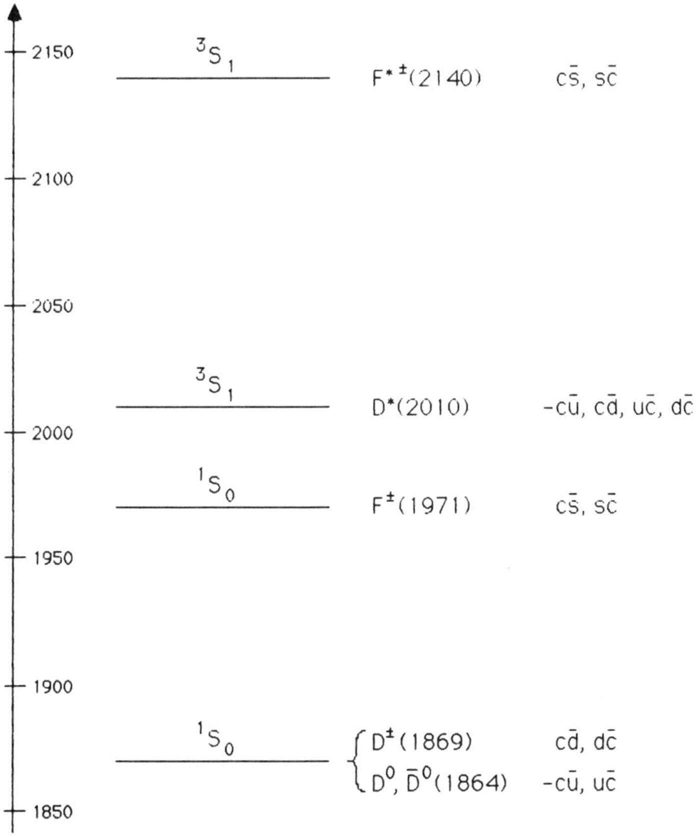

Figure 1.3 The ground state of the explicitly charmed mesons.

1.3.3 Baryon Spectrum and Baryon Wave Functions

The baryons are color-singlet qqq states with a wave function totally antisymmetric with respect to the color, flavor, spin and space degrees of freedom. We shall concentrate on the baryons made up of light quarks u, d and s that span the large SU(6) × O(3) symmetry. It is possible to extend these considerations to baryons made out of heavier quarks (like the Λ_c composed of cud) by taking the large flavor mass differences properly into account.

To build the wave functions of baryons, one uses the representations of the group of permutations of 3 objects. Besides the symmetric (S) and the antisymmetric (A) one-dimensional representations, indicated by ^{s}, and ^{a}, there is the so-called mixed (M) representation of dimension 2, whose basis vectors will be indicated by $'$ and $''$. Denoting them by e' and e'', one chooses e' antisymmetric and e'' symmetric under the exchange $(1\leftrightarrow 2)$, and their relative sign is fixed by specifying their transformation under the exchange $(1\leftrightarrow 3)$:

$$\left.\begin{aligned} e' &\rightarrow \frac{1}{2}e' - \frac{\sqrt{3}}{2}e'', \\ e'' &\rightarrow -\frac{\sqrt{3}}{2}e' - \frac{1}{2}e''. \end{aligned}\right\} \tag{1.72}$$

One has to combine the various degrees of freedom (spin, flavor, space and color) into an antisymmetric total wave function. The combination of two degrees of freedom goes as follows (we denote by e and f the vectors to be combined, and by g the result):

$$\left.\begin{aligned} &\mathrm{S}\otimes\mathrm{S} \quad (\text{or } \otimes\mathrm{A} \text{ or } \otimes\mathrm{M}) = \mathrm{S} \quad (\text{or A or M}) \\ &\qquad g = e^{s}f; \\ &\mathrm{A}\otimes\mathrm{A} = \mathrm{S} \\ &\qquad g^{s} = e^{a}f^{a}; \\ &\mathrm{A}\otimes\mathrm{M} = \mathrm{M} \\ &\qquad g' = e^{a}f'', \qquad g'' = -e^{a}f'; \\ &\mathrm{M}\otimes\mathrm{M} = \mathrm{S}+\mathrm{A}+\mathrm{M} \\ &\qquad g^{s} = \frac{1}{\sqrt{2}}(e'f' + e''f''), \\ &\qquad g^{a} = \frac{1}{\sqrt{2}}(e'f'' - e''f'), \\ &\qquad g' = \frac{1}{\sqrt{2}}(e'f'' + e''f'), \qquad g'' = \frac{1}{\sqrt{2}}(e'f' - e''f''). \end{aligned}\right\} \tag{1.73}$$

The link between SU(N) and the permutations is that the tensors of given symmetry, here S, A, M$'$ or M$''$, form an irreducible representation of SU(N), with equivalent representations for M′ and M″. The following table gives the dimensions of these representations for N=2, 3 and 6.

$$\text{SU(2)} \quad \mathbf{2}\otimes\mathbf{2}\otimes\mathbf{2}=\mathbf{4}_S+\mathbf{2}_{M'}+\mathbf{2}_{M''}$$

$$\text{SU(3)} \quad \mathbf{3}\otimes\mathbf{3}\otimes\mathbf{3}=\mathbf{10}_S+\mathbf{8}_{M'}+\mathbf{8}_{M''}+\mathbf{1}_A$$

$$\text{SU(6)} \quad \mathbf{6}\otimes\mathbf{6}\otimes\mathbf{6}=\mathbf{56}_S+\mathbf{70}_{M'}+\mathbf{70}_{M''}+\mathbf{20}_A$$

For a given symmetry, inside a representation of SU(N), the wave functions can be chosen according to the conventions in use.

Spin

The SU(2) spin wave functions are denoted by χ; χ^s has a spin $S=\frac{3}{2}$, χ' and χ'' have a spin $S=\frac{1}{2}$; explicitly, one takes

$$\left.\begin{aligned} &\chi^s(S_z=+\tfrac{3}{2})=\uparrow\uparrow\uparrow,\\ &\chi'(S_z=+\tfrac{1}{2})=\frac{1}{\sqrt{2}}(\uparrow\downarrow\uparrow-\downarrow\uparrow\uparrow),\\ &\chi''(S_z=+\tfrac{1}{2})=\frac{1}{\sqrt{6}}(2\uparrow\uparrow\downarrow-\downarrow\uparrow\uparrow-\uparrow\downarrow\uparrow). \end{aligned}\right\} \tag{1.74}$$

Other values of S_z are obtained by applying $\sigma_-(1)+\sigma_-(2)+\sigma_-(3)$ and normalizing to one. In terms of the standard coupling with Clebsch–Gordan coefficients, these wave functions are

$$\left.\begin{aligned} \chi^s&=[[\tfrac{1}{2},\tfrac{1}{2}]_1,\tfrac{1}{2}]_{3/2},\\ \chi'&=[[\tfrac{1}{2},\tfrac{1}{2}]_0,\tfrac{1}{2}]_{1/2},\\ \chi''&=[[\tfrac{1}{2},\tfrac{1}{2}]_1,\tfrac{1}{2}]_{1/2}. \end{aligned}\right\} \tag{1.75}$$

Flavor

The SU(3) flavor symmetry was described in Section 1.2.3. We give here the wave functions in detail. The SU(3) flavor wave functions are denoted by φ: φ^s corresponds to the decuplet, φ' and φ'' correspond to the octet, and φ^a to the singlet of SU(3). Explicitly, one has

decuplet

	φ^s (to be symmetrized)	
$\Delta^{++}, \Delta^+, \Delta^0, \Delta^-$	*uuu*, *uud*, *udd*, *ddd*	
$\Sigma^+, \Sigma^0, \Sigma^-$	*uus*, *uds*, *dds*	(1.76)
Ξ^0, Ξ^-	*uss*, *dss*	
Ω^-	*sss*	

octet

$$
\left.
\begin{array}{lll}
 & \varphi' & \varphi'' \\
\mathrm{p} & \frac{1}{\sqrt{2}}(udu-duu) & \frac{1}{\sqrt{6}}(2uud-duu-udu) \\
\mathrm{n} & \frac{1}{\sqrt{2}}(udd-dud) & \frac{1}{\sqrt{6}}(dud+udd-2ddu) \\
\Sigma^+ & \frac{1}{\sqrt{2}}(suu-usu) & \frac{1}{\sqrt{6}}(suu+usu-2uus) \\
\Sigma^0 & \frac{1}{2}(sud+sdu-usd-dsu) & \frac{1}{2\sqrt{3}}(sdu+sud+usd+dsu-2uds-2dus) \\
\Sigma^- & \frac{1}{\sqrt{2}}(sdd-dsd) & \frac{1}{\sqrt{6}}(sdd+dsd-2dds) \\
\Lambda & \frac{1}{2\sqrt{3}}(usd+sdu-sud-dsu-2dus+2uds) & \frac{1}{2}(sud+usd-sdu-dsu) \\
\Xi^0 & \frac{1}{\sqrt{2}}(sus-uss) & \frac{1}{\sqrt{6}}(2ssu-sus-uss) \\
\Xi^- & \frac{1}{\sqrt{2}}(sds-dss) & \frac{1}{\sqrt{6}}(2ssd-sds-dss)
\end{array}
\right\} \qquad (1.77)
$$

singlet

$$\varphi^a = \frac{1}{\sqrt{6}}(uds+dsu+sud-dus-usd-sdu) \qquad (1.78)$$

Spin–flavor

By combining the wave functions of spin, χ^s, χ' and χ'', and those of $SU(3)_F$ φ^s, φ', φ'' and φ^a, one gets those of SU(6), built with the states (1.68), and their decomposition under spin–SU(3):

symmetric

$$
\left.
\begin{array}{l}
|\mathbf{56}, S=\tfrac{1}{2}, \mathbf{8}\rangle = \frac{1}{\sqrt{2}}(\varphi'\chi' + \varphi''\chi''), \\
|\mathbf{56}, S=\tfrac{3}{2}, \mathbf{10}\rangle = \varphi^s\chi^s;
\end{array}
\right\} \qquad (1.79)
$$

mixed symmetric

$$\left.\begin{aligned}
&|\mathbf{70},\,S=\tfrac{1}{2},\,\mathbf{8}\rangle'=\frac{1}{\sqrt{2}}(\varphi'\chi''+\varphi''\chi'), && |\mathbf{70},\,S=\tfrac{1}{2},\,\mathbf{8}\rangle''=\frac{1}{\sqrt{2}}(\varphi'\chi'-\varphi''\chi''),\\
&|\mathbf{70},\,S=\tfrac{3}{2},\,\mathbf{8}\rangle'=\varphi'\chi^{s}, && |\mathbf{70},\,S=\tfrac{3}{2},\,\mathbf{8}\rangle''=\varphi''\chi^{s},\\
&|\mathbf{70},\,S=\tfrac{1}{2},\,\mathbf{10}\rangle'=\varphi^{s}\chi', && |\mathbf{70},\,S=\tfrac{1}{2},\,\mathbf{10}\rangle''=\varphi^{s}\chi'',\\
&|\mathbf{70},\,S=\tfrac{1}{2},\,\mathbf{1}\rangle'=\varphi^{a}\chi'', && |\mathbf{70},\,S=\tfrac{1}{2},\,\mathbf{1}\rangle''=-\varphi^{a}\chi';
\end{aligned}\right\}\tag{1.80}$$

antisymmetric

$$\begin{aligned}
&|\mathbf{20},\,S=\tfrac{1}{2},\,\mathbf{8}\rangle=\frac{1}{\sqrt{2}}(\varphi'\chi''-\varphi''\chi'),\\
&|\mathbf{20},\,S=\tfrac{3}{2},\,\mathbf{1}\rangle=\varphi^{a}\chi^{s}.
\end{aligned}\tag{1.81}$$

Color

One sees that the **56** multiplet gives a good description of the low-lying baryons, as it contains exactly a spin-$\frac{1}{2}$ octet and a spin-$\frac{3}{2}$ decuplet. Without color, these SU(6) wave functions would have to be combined with an *antisymmetric* spatial wave function — very unlikely for the ground state. This paradox was the first indication of a new degree of freedom, called color. It is solved by assuming that quarks have 3 colors, and that hadrons always belong to the singlet representation of color SU(3), (1.6).

It then follows that for a baryon, a three-quark state, the color wave function is antisymmetric; therefore, to satisfy Fermi statistics, the wave function must now be *symmetric* with respect to the other degrees of freedom. For the ground state this implies a symmetric spatial wave function.

Space

The spatial wave functions are denoted by ψ. In the approximation of spin-independent forces, the ψ are solutions of a Hamiltonian that is invariant under permutation of the quark coordinates. Each level will have definite symmetry and will be associated with the SU(6) multiplet of the same symmetry, S, M or A, to build a symmetric function. Besides, the orbital angular momentum $\boldsymbol{L}$ is added to the quark spin $\boldsymbol{S}$ to give the spin $\boldsymbol{J}$ of the baryon.

The levels and their symmetries are dependent on the potential, but, quite generally, the ground-state level is symmetric with $L^P=0^+$, and the next level is of mixed symmetry with $L^P=1^-$. We give in the following the harmonic-oscillator wave functions, which may be used as a first approximation, up to level $N=2$.

The harmonic-oscillator Hamiltonian is

$$H = \sum_{i=1}^{3} \frac{p_i^2}{2m} + \frac{1}{6} m\omega^2 \sum_{i<j} (r_j - r_j)^2. \tag{1.82}$$

It is convenient to use the following relative coordinates, which form a doublet M′, M″ of permutation symmetry,

$$\left.\begin{aligned} \rho &= \frac{1}{\sqrt{2}}(r_1 - r_2), & p_\rho &= \frac{1}{\sqrt{2}}(p_1 - p_2), \\ \lambda &= \frac{1}{\sqrt{6}}(r_1 + r_2 - 2r_3), & p_\lambda &= \frac{1}{\sqrt{6}}(p_1 + p_2 - 2p_3), \end{aligned}\right\} \tag{1.83}$$

and the usual center-of-mass position and momentum

$$R = \tfrac{1}{3}(r_1 + r_2 + r_3), \qquad P = p_1 + p_2 + p_3. \tag{1.84}$$

In terms of these new variables, (1.82) is

$$H = \frac{P^2}{6m} + \frac{p_\rho^2}{2m} + \frac{p_\lambda^2}{2m} + \tfrac{1}{2} m\omega^2(\rho^2 + \lambda^2), \tag{1.85}$$

where the first term describes the center-of-mass kinetic energy, and the rest the internal dynamics. The level spacing is ω.

We shall write the internal wave functions normalized with respect to the measure:

$$\int \prod_i \mathrm{d}r_i\, \delta\left(\frac{1}{3}\sum_i r_i\right) = 3\sqrt{3} \int \mathrm{d}\rho\, \mathrm{d}\lambda. \tag{1.86}$$

Let us define

$$\psi_0(\rho, \lambda) = (3\sqrt{3}\, R^6 \pi^2)^{-1/2} \exp\left(-\frac{\rho^2 + \lambda^2}{2R^2}\right), \tag{1.87}$$

where $R^2 = 1/m\omega$. We then have the following spatial wave functions (Faiman and Hendry, 1968):

1 state $N = 0$ (ground state)

$$\psi^s(\mathbf{56}, 0^+) = \psi_0(\rho, \lambda); \tag{1.88}$$

6 states $N = 1$

$$\left.\begin{aligned}\psi'(\mathbf{70},1^-)&=\left(\frac{8}{3}\pi\right)^{1/2}R^{-1}\mathcal{Y}_1^M(\boldsymbol{\rho})\psi_0(\boldsymbol{\rho},\boldsymbol{\lambda}),\\ \psi''(\mathbf{70},1^-)&=\left(\frac{8}{3}\pi\right)^{1/2}R^{-1}\mathcal{Y}_1^M(\boldsymbol{\lambda})\psi_0(\boldsymbol{\rho},\boldsymbol{\lambda});\end{aligned}\right\}\tag{1.89}$$

21 states $N=2$

$$\left.\begin{aligned}\psi^s(\mathbf{56},0^+)&=\left(\frac{1}{3}\right)^{1/2}R^{-2}[3R^2-(\boldsymbol{\rho}^2+\boldsymbol{\lambda}^2)]\psi_0(\boldsymbol{\rho},\boldsymbol{\lambda}),\\ \psi^s(\mathbf{56},2^+)&=\left(\frac{8}{15}\pi\right)^{1/2}R^{-2}[\mathcal{Y}_2^M(\boldsymbol{\rho})+\mathcal{Y}_2^M(\boldsymbol{\lambda})]\psi_0(\boldsymbol{\rho},\boldsymbol{\lambda}),\\ \psi'(\mathbf{70},0^+)&=\left(\frac{1}{3}\right)^{1/2}R^{-2}[2\boldsymbol{\lambda}\cdot\boldsymbol{\rho}]\psi_0(\boldsymbol{\rho},\boldsymbol{\lambda})\\ \psi''(\mathbf{70},0^+)&=\left(\frac{1}{3}\right)^{1/2}R^{-2}[\boldsymbol{\rho}^2-\boldsymbol{\lambda}^2]\psi_0(\boldsymbol{\rho},\boldsymbol{\lambda})\\ \psi'(\mathbf{70},2^+)&=\left(\frac{8}{3}\pi\right)^{1/2}R^{-2}\sum_m\langle 1,1;\,M-m,m\,|\,2,\,M\rangle\\ &\quad\mathcal{Y}_1^m(\boldsymbol{\rho})\mathcal{Y}_1^{M-m}(\boldsymbol{\lambda})\psi_0(\boldsymbol{\rho},\boldsymbol{\lambda}),\\ \psi''(\mathbf{70},2^+)&=\left(\frac{8}{3}\pi\right)^{1/2}R^{-2}[\mathcal{Y}_2^M(\boldsymbol{\rho})-\mathcal{Y}_2^M(\boldsymbol{\lambda})]\psi_0(\boldsymbol{\rho},\boldsymbol{\lambda}),\\ \psi^a(\mathbf{20},1^+)&=\left(\frac{8}{3}\right)^{1/2}R^{-2}[\boldsymbol{\rho}\times\boldsymbol{\lambda}]\psi_0(\boldsymbol{\rho},\boldsymbol{\lambda})\end{aligned}\right\}\tag{1.90}$$

Spin–flavor–space

It is enough now to combine SU(6) and space in a symmetric way to get the total wave functions (the color is just an overall factor that we do not write).

Combining the **56** of SU(6) with the spatial wave function $L^P=0^+$, one gets the low-lying baryons:

$$\left.\begin{aligned}\mathbf{8},\,J=\frac{1}{2}\qquad&\frac{1}{\sqrt{2}}(\varphi'\chi'+\varphi''\chi'')\psi_0,\\ \mathbf{10},\,J=\frac{3}{2}\qquad&\varphi^s\chi^s\psi_0.\end{aligned}\right\}\tag{1.91}$$

Combining the **70** of SU(6) with the sptial wave function $L^P=1^-$, one gets the wave functions of the negative-parity baryons:

$$\left.\begin{array}{ll}
\mathbf{8},\ J=\frac{1}{2},\frac{3}{2} & \frac{1}{2}[(\varphi'\chi''+\varphi''\chi')\psi'+(\varphi'\chi'-\varphi''\chi'')\psi'']_J,\\
\mathbf{8},\ J=\frac{1}{2},\frac{3}{2},\frac{5}{2} & \frac{1}{\sqrt{2}}[\varphi'\chi^s\psi'+\varphi''\chi^s\psi'']_J,\\
\mathbf{10},\ J=\frac{1}{2},\frac{3}{2} & \frac{1}{\sqrt{2}}[\varphi^s\chi'\psi'+\varphi^s\chi''\psi'']_J,\\
\mathbf{1},\ J=\frac{1}{2},\frac{3}{2} & \frac{1}{\sqrt{2}}[\varphi^a\chi''\psi'-\varphi^a\chi'\psi'']_J,
\end{array}\right. \tag{1.92}$$

where the index J under the bracket means that the spatial and spin wave functions are coupled by Clebsch–Gordan coefficients to get a total angular momentum J.

As a final example, we write the wave functions of the level $N=2$, which have the same SU(3) and J^P quantum numbers as the low-lying baryons, and may mix with their wave functions when SU(6)-breaking effects are taken into account.

For the octet, $J^P=\frac{1}{2}^+$, there are 4 wave functions:

$$\left.\begin{array}{lll}
\mathbf{56},\ S=\frac{1}{2}, & L=0 & \frac{1}{\sqrt{2}}(\varphi'\chi'+\varphi''\chi'')\psi^s(0^+),\\
\mathbf{70},\ S=\frac{1}{2}, & L=0 & \frac{1}{2}[(\varphi'\chi''+\varphi''\chi')\psi'(0^+)\\
 & & \quad +(\varphi'\chi'-\varphi''\chi'')\psi''(0^+)],\\
\mathbf{70},\ S=\frac{3}{2}, & L=2 & \frac{1}{\sqrt{2}}[\varphi'\chi^s\psi'(2^+)+\varphi''\chi^s\psi''(2^+)]_{J=1/2},\\
\mathbf{20},\ S=\frac{1}{2}, & L=1 & \frac{1}{\sqrt{2}}[(\varphi'\chi''-\varphi''\chi')\psi^a(1^+)]_{J=1/2}.
\end{array}\right\} \tag{1.93}$$

For the decuplet, $J^P=\frac{3}{2}^+$, there are 3 wave functions:

$$\left.\begin{array}{lll}
\mathbf{56},\ S=\frac{3}{2}, & L=0 & \varphi^s\chi^s\psi^s(0^+),\\
\mathbf{56},\ S=\frac{3}{2}, & L=2 & [\varphi^s\chi^s\psi^s(2^+)]_{J=3/2},\\
\mathbf{70},\ S=\frac{1}{2}, & L=2 & \frac{1}{\sqrt{2}}[\varphi^s\chi'\psi'(2^+)+\varphi^s\chi''\psi''(2^+)]_{J=3/2}.
\end{array}\right\} \tag{1.94}$$

In Figure 1.4 we give a simplified table with some of the observed baryon states and their masses, quantum numbers, and tentative classification under $SU(6) \times O(3)$.

1.3.4 The Quark–Quark Potential

We want here to be more detailed—from both the theoretical and phenomenological points of view—about the quark–quark potential. We shall not give a complete review of this wide subject, but we will be a little more precise in order to give a background for the subsequent treatment of transitions.

In Section 1.3.2 we have assumed the simplest Ansatz for the $q\bar{q}$ potential—a central spin-independent form. For baryons in Section 1.3.3 we have been still more specific, as we have assumed the harmonic oscillator, a confining potential that moreover allows separation of the motion of the center-of-mass in this three-body problem. These forms have been useful in enumerating and classifying the spectra of mesons ($q\bar{q}$) and baryons (qqq).

If we start from the QCD Lagrangian (1.1), and deduce the Hamiltonian in the Coulomb gauge—just to find the analogy with electromagnetism—we have

$$\begin{aligned} H(\boldsymbol{x}) = {} & \bar{q}(\boldsymbol{x})\boldsymbol{\gamma}\cdot\left(-\mathrm{i}\boldsymbol{\nabla} + g_s \sum_a \frac{\lambda^a}{2}\boldsymbol{A}^a\right) q(\boldsymbol{x}) \\ & - \frac{1}{2} g_s^2 \sum_a \rho^a(\boldsymbol{x}) \left(\frac{1}{\boldsymbol{\nabla}^2} + \ldots\right) \rho^a(\boldsymbol{x}) + \frac{1}{2}\sum_a [(\boldsymbol{E}^{a\mathrm{T}})^2 + (\boldsymbol{B}^a)^2] + \ldots, \end{aligned} \tag{1.95}$$

where higher order omitted terms contain only powers of the gluon fields, including at the end terms which are absent in the classical Hamiltonian (Christ and Lee, 1980) and where the charge density receives contributions from the quark and gluon fields:

$$\rho^a(\boldsymbol{x}) = \bar{q}(\boldsymbol{x})\gamma^0 \frac{\lambda^a}{2} q(\boldsymbol{x}) - f^{abc} \boldsymbol{A}^b \cdot \boldsymbol{E}^c. \tag{1.96}$$

In these equations $\boldsymbol{E}^a$ and $\boldsymbol{B}^a$ are the color electric and magnetic fields. The longitudinal part of $\boldsymbol{E}^a$ is implicitly defined by Gauss' law:

$$\boldsymbol{\nabla} \cdot \boldsymbol{E}^a = -g_s \rho^a. \tag{1.97}$$

To lowest order in α_s, and neglecting for the moment the spin-dependent part, we obtain from (1.95) a colored quark–quark potential of Coulomb form, attractive in both the $q\bar{q}$ color singlet and in the qq color $\bar{\mathbf{3}}$ channels, as we have seen in Section 1.3.1:

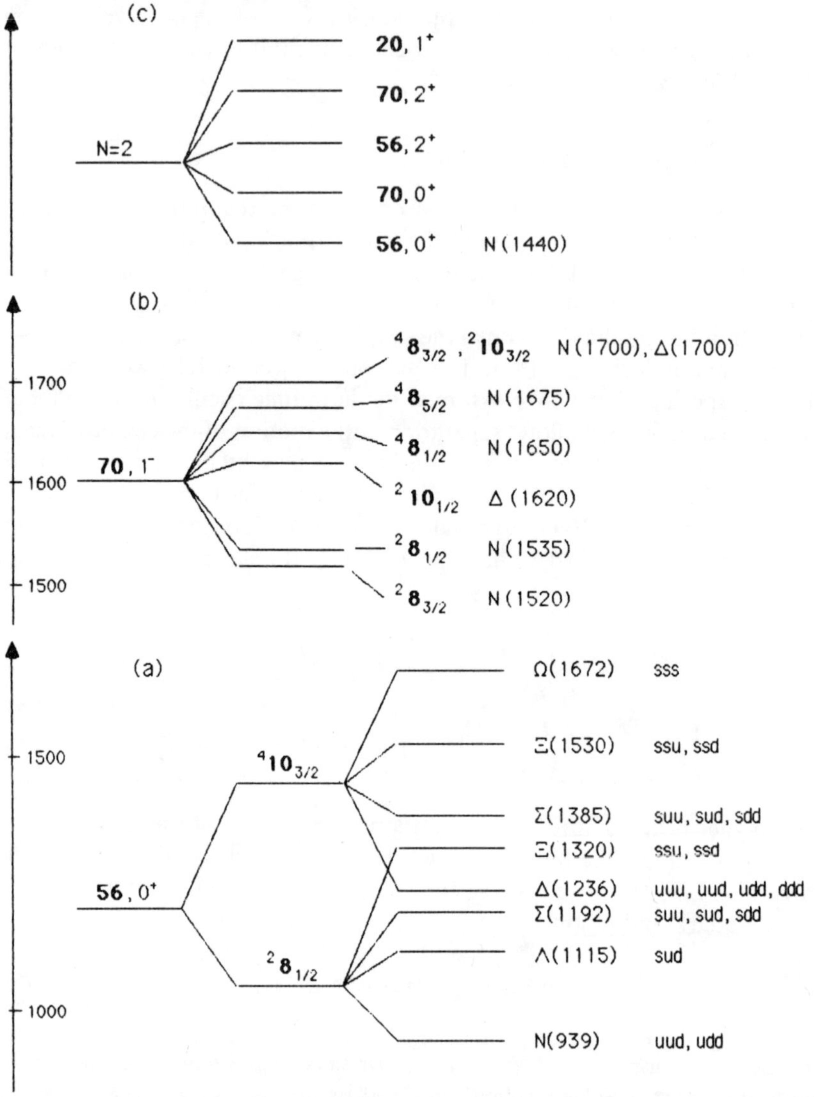

Figure 1.4 The spectrum of baryons made up of light quarks u, d and s, according to the SU(6)×O(3) classification. Each SU(6)×O(3) multiplet splits into ${}^{2S+1}\boldsymbol{\alpha}_J$, where S is the total 3q spin ($S=\frac{1}{2}$ or $\frac{3}{2}$), $\boldsymbol{\alpha}$ is the SU(3) representation and J is the spin of the baryon. In (a) the quark content of all ground-state baryons is made explicit. In (b) only the splitting of the **70**, $L=1^-$ multiplet into ${}^{2S+1}\boldsymbol{\alpha}_J$ is given, following the mass pattern of the nonstrange states. There is mixing between states of the same J^P. In (c) the SU(6)×O(3) multiplets with two harmonic-oscillator excitations are shown (the scale of breaking is not realistic, and many states are mixed). The usual classification of the Roper resonance N(1440) is shown.

$$V(r) = \left\{ \begin{array}{ll} -\frac{4}{3}\frac{\alpha_s}{r} & \text{for } q\bar{q} \text{ singlet,} \\ \\ -\frac{2}{3}\frac{\alpha_s}{r} & \text{for } qq \text{ in } \bar{\mathbf{3}}. \end{array} \right\} \tag{1.98}$$

However, QCD tells us much more about the nature of the spin-independent part of the potential, both at *short distances* and at *large distances*. As we have already mentioned in Section 1.1, QCD has the crucial property of asymptotic freedom. Owing to the vacuum-polarization effects associated with the self-couplings of gluons, we have here a situation that is opposite to that in QED. In the latter we have screening of the charges owing to the fermion vacuum polarization, and the effective coupling decreases at large distances. In contrast, in QCD the vacuum polarization due to the colored gluons overcomes the fermion vacuum polarization, and the effective coupling $\alpha_s(Q^2)$ decreases at small distances, or at large Q^2. One thus obtains an effective coupling at large Q^2,

$$\alpha_s(Q^2) = \frac{12\pi}{(33 - 2n_f)\ln\left(\dfrac{Q^2}{\Lambda^2}\right)}, \tag{1.99}$$

where n_f is the number of flavors (giving the contribution of the fermions to the vacuum polarization) and $\Lambda \sim 100$–200 MeV is the fundamental scale in QCD that fixes the coupling. If n_f is not too large—as appears to be the case—then $\alpha_s(Q^2) \to 0$ as $Q^2 \to \infty$, or equivalently as $r \to 0$. An important consequence of (1.99) is that, in the true short-distance potential, the Coulomb singularity as $r \to 0$ in (1.98) will be smoothed by the logarithmic decrease of the effective coupling at small r.

Concerning the behavior of the potential at large distances, we have, within the framework of QCD, two main approaches that give some insight into the phenomenon of confinement. These investigations were motivated by the empirical evidence that quarks do not exist as free particles—at least at presently obtainable energies—and by the success of the naive quark model in describing the hadron spectrum with confining potentials.

It must first be emphasized that the medium- and long-distance ($r \gtrsim 0.1$ fm) behavior of the potential is necessarily a nonperturbative phenomenon in QCD. The theory is only simple at short distances, where pertubation theory can be applied, or equivalently for large-Q^2 phenomena, such as deep-inelastic lepton–nucleon scattering.

As we have already mentioned in Section 1.1.3, one non-perturbative approach relies on lattice QCD. QCD is formulated on a discrete lattice, the lattice spacing a being the inverse ultraviolet cut-off necessary to make the theory finite. This permits numerical calculations starting from first principles, i.e. the exact QCD action (Wilson, 1974; Kogut and Susskind, 1975; Susskind, 1977). The result of interest here is that, for heavy quarks, one finds a string-like structure, with the color sources (the quarks) at its ends. The interquark potential is linear for large r.

$$V(r)=\kappa r, \tag{1.100}$$

where κ is the so-called *string tension*, computed by numerical methods from L_{QCD}. Although these results are still not very accurate, they give a nice illustration of the confinement phenomenon in QCD.

The other main theoretical framework for understanding confinement in QCD is the analogy with the theory of superconductivity (Nielsen and Olesen, 1973). In the latter the superconducting phase of a metal is taken as being due to the condensation of electron pairs due to attraction mediated by phonon exchange. One main effect is that a *magnetic* field is *confined* outside the volume of the superconductor (Meissner effect) or can penetrate only by a flux amount that is quantized (in type II superconductors). In the same way, one assumes ('t Hooft, 1979, 1980; Mandelstam, 1979) that in QCD one has a phenomenon that is dual to the Meissner effect. The vacuum is understood as a perfect color-*dielectric* medium (Friedberg and Lee, 1978; Lee, 1979, 1981) analogous to the diamagnetic property of a superconductor, which repels color-electric flux lines. A $q\bar{q}$ state is in a hole (the bag) in which color-electric flux lines are confined. This color-dielectric property is related to a *magnetic gluon condensate*. This leads also to string-like structures (Baker *et al.*, 1985) of the type found in lattice-QCD calculations.

Indeed, one finds in the analysis of the hadron spectrum by QCD sum rules (Shifman *et al.*, 1979) that, besides the $q\bar{q}$ condensate responsible for dynamical chiral-symmetry breaking, one needs a gluon condensate

$$\frac{\alpha_s}{\pi}\langle F^2\rangle=\frac{\alpha_s}{\pi}\langle 0|\sum_a F^a_{\mu\nu}F^{\mu\nu a}|0\rangle\neq 0, \tag{1.101}$$

with a value $(\alpha_s/\pi)\langle F^2\rangle\approx(330\text{ MeV})^4$. This corresponds to a *magnetic* color configuration of the vacuum, which is responsible for the confinement of the electric flux (Baker *et al.*, 1985; Leutwyler, 1979). In terms of the magnetic and electric color fields (1.101) becomes

$$\frac{\alpha_s}{\pi}\langle F^2\rangle=2\frac{\alpha_s}{\pi}\langle 0|\sum_a[(\boldsymbol{B}^a)^2-(\boldsymbol{E}^a)^2]|0\rangle>0. \tag{1.102}$$

In this scheme the string tension κ (1.100) is related to the gluon condensate, and, in principle, is computable in terms of it (Hansson, Johnson and Peterson, 1982). This relation is explicity found in the strong-coupling expansion of lattice QCD (Wilson, 1974).

From these considerations we have for example a $q\bar{q}$ potential of the form

$$V(r) = -\frac{4}{3}\frac{\alpha_s(r)}{r} + \kappa r, \tag{1.103}$$

where $\alpha_s(r)$ decreases logarithmically as $r \to 0$ according to asymptotic freedom:

$$\alpha_s(r) = \frac{12\pi}{(33-2n_f)\ln\left(\frac{1}{r^2\Lambda^2}\right)}. \tag{1.104}$$

A number of phenomenological potentials have been proposed to describe the hadron spectrum (mainly the ψ, $c\bar{c}$ and the Υ, $b\bar{b}$ families) that follow more or less the behavior suggested by (1.103). We plot in Figure 1.5 a few of these

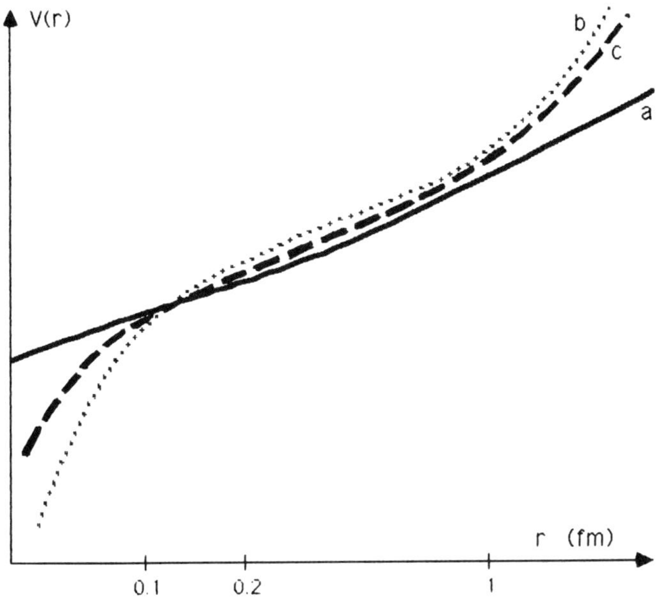

Figure 1.5 Quark–quark potentials, close in the intermediate-distance region, but differing at short distances (from Martin, 1984). (a) Phenomenological potential $A+Br^{0.1}$ (full line). (b) Linear potential with a short-distance Coulomb piece, $-a/r+br$ (dotted line). (c) Linear potential with a short-distance piece satisfying asymptotic freedom (broken line).

proposals: the Kraseman–Ono (1979) and the Richardson (1979) potentials, which incorporate asymptotic freedom at small distances, the purely phenomenological Martin (1980) potential and the potential of the Cornell group (Eichten *et al.*, 75), which behaves like $-1/r$ at small distances. Their common feature is that they are very similar at intermediate distances, for $0.1\,\text{fm} < r < 1\,\text{fm}$. This is the range that is crucial for the description of the ψ and Υ spectra. The quarks in these cases are massive enough for a nonrelativistic description, but not massive enough to test the very-small-distance behavior of (1.103). On the other hand, we shall show in Chapter 4 that the details of the potential at short distance may be important for e^+e^- annihilation of the ψ and Υ. Especially interesting, as it follows closely the QCD behaviour at small distances, is the Richardson (1979) potential.

$$\tilde{V}(\boldsymbol{q}^2) = -\frac{4}{3}\frac{12\pi}{33-2n_f}\frac{1}{\boldsymbol{q}^2}\frac{1}{\ln\left(1+\dfrac{\boldsymbol{q}^2}{\Lambda^2}\right)}. \tag{1.105}$$

If the t quark is found as a mass $m_t \approx 40\,\text{GeV}$, only the ground state of the $t\bar{t}$ system will be sensitive to the shape of the potential at $r < 0.1\,\text{fm}$ (Martin, 1984).

1.3.5 Spin-Dependent Effects and Relativistic Potentials

As we have seen in the context of the Schrödinger equation, the spectrum predicted by a rotationally invariant and spin-independent potential $V(|\boldsymbol{x}-\boldsymbol{y}|)$ is rather different from the experimental one. In fact, the spin and orbital angular-momentum symmetry is badly broken. While the breaking of flavour symmetry is easily explained by the mass difference between quarks, the breaking of these spin and orbital symmetries is a much more delicate question. There are two possible interpretations of the breaking. In the first one remains in the framework of the Schrödinger equation and introduces spin and orbital angular-momentum dependence into the Schrödinger potential phenomenologically. Alternatively, one considers relativistic effects or a relativistic framework and then derives the breaking from the potentials described in the previous sections. Although our subject is not a detailed understanding of the spectrum, it is necessary to explain the main ideas behind these developments in order to discuss logically the relativistic effects in transitions.

The first approach, which maintains the naive quark-model framework, has been the object of considerable phenomenological work. Special attention has been given to the complicated case of baryons (Greenberg and Resnikoff, 1967; Horgan and Dalitz, 1973; Dalitz, 1975). We shall here just list the types of forces that must be introduced. It is important to know them because they are less dependent on theoretical assumptions than the subsequent interpretation by QCD. We shall assume in the following that, rather than Dirac spinors, as considered in Section 1.3.1, we are now dealing with Pauli spinors, which are nothing but the upper or lower components of the Dirac spinors for quarks or antiquarks respectively.

Spin–spin forces of the form

$$V_{ss}(r)\boldsymbol{S}_1\cdot\boldsymbol{S}_2, \tag{1.106}$$

where $\boldsymbol{S}=\frac{1}{2}\boldsymbol{\sigma}$ is the quark spin, are necessary to describe a very important category of splittings, namely, $\frac{3}{2}^+-\frac{1}{2}^+$ ground-state baryon mass differences (e.g.Δ–N) or the 1^--0^- ground-state meson mass differences (e.g. ρ–π, K*–K).

One also needs spin–orbit forces. A potential of the form

$$V_{SO}(r)\,\boldsymbol{L}\cdot(\boldsymbol{S}_q+\boldsymbol{S}_{\bar{q}}) \tag{1.107}$$

partially describes the L–S splittings of $L=1$ meson multiplets $2^{++}, 1^{++}, 1^{+-}, 0^{++}$. From $\boldsymbol{J}=\boldsymbol{L}+\boldsymbol{S}$, $\boldsymbol{S}=\boldsymbol{S}_q+\boldsymbol{S}_{\bar{q}}$, we obtain

$$2\boldsymbol{L}\cdot\boldsymbol{S}=\boldsymbol{J}^2-\boldsymbol{L}^2-\boldsymbol{S}^2=J(J+1)-L(L+1)-S(S+1), \tag{1.108}$$

which gives $2\boldsymbol{L}\cdot\boldsymbol{S}=1, 0, -1, -2$ respectively for the states ${}^3P_2$, ${}^3P_1$, ${}^1P_1$ and ${}^3P_0$. The ordering predicted with $V_{SO}>0$ is well verified for the χ $c\bar{c}$ states, but the experimental splitting is not uniform. For light mesons, the mass of the 1^{++} state A_1 is controversial, but the ordering with $V_{SO}>0$ still seems to be verified. The nonuniformity of the splitting between $L=1$, $S=1$ states can be interpreted by the presence of a new spin-dependent force, the so-called tensor force:

$$V_T(r)\left[\frac{(\boldsymbol{r}\cdot\boldsymbol{S}_1)(\boldsymbol{r}\cdot\boldsymbol{S}_2)}{r^2}-\frac{\boldsymbol{S}_1\cdot\boldsymbol{S}_2}{3}\right]. \tag{1.109}$$

For baryons there are several types of spin–orbit terms that are possible in principle, but the main forces needed seem to be the spin–spin ones. The Σ–Λ splitting requires a correlation of spin and flavor dependence.

This area of the fine details of the hadron spectrum deserves a whole book in itself, and we shall only give a few indications about the QCD interpretation of the spin-dependent potential. De Rújula, Georgi and Glashow (1975) made a breakthrough in this problem with the hypothesis of attributing the spin-dependent part of the potential to its short-distance piece, interpreted by one-gluon exchange. This is a bold hypothesis, because we are not at small distances in the hadron spectrum, and one-gluon exchange, justified only at small coupling, cannot be expected to be a good approximation. However, the results are in surprising good agreement with the observation. One-gluon exchange in a nonrelativistic expansion leads (like one-photon exchange in positronium), in addition to the Coulomb potential examined above, to the Breit–Fermi interaction (see e.g. Landau and Lifshitz, 1972)

$$H_{\mathrm{BF}} = k\alpha_s \sum_{i>j} S_{ij}, \tag{1.110}$$

where $k = -\frac{4}{3}$ for a $q\bar{q}$ singlet and $-\frac{2}{3}$ for a qq in $\bar{\mathbf{3}}$, and

$$\begin{aligned} S_{ij} = {} & \frac{1}{|\boldsymbol{r}|} - \frac{1}{2m_i m_j}\left(\frac{\boldsymbol{p}_i\cdot\boldsymbol{p}_j}{|\boldsymbol{r}|} + \frac{\boldsymbol{r}\cdot(\boldsymbol{r}\cdot\boldsymbol{p}_i)\boldsymbol{p}_j}{|\boldsymbol{r}|^3}\right) \\ & - \frac{\pi}{2}\delta(\boldsymbol{r})\left(\frac{1}{m_i^2} + \frac{1}{m_j^2} + \frac{16\boldsymbol{S}_i\cdot\boldsymbol{S}_j}{3m_i m_j}\right) \\ & - \frac{1}{2|\boldsymbol{r}|^3}\left\{\frac{1}{m_i^2}(\boldsymbol{r}\times\boldsymbol{p}_i)\cdot\boldsymbol{S}_i - \frac{1}{m_j^2}(\boldsymbol{r}\times\boldsymbol{p}_j)\cdot\boldsymbol{S}_j \right. \\ & \left. + \frac{1}{m_i m_j}\left[2(\boldsymbol{r}\times\boldsymbol{p}_i)\cdot\boldsymbol{S}_j - 2(\boldsymbol{r}\times\boldsymbol{p}_j)\cdot\boldsymbol{S}_i - 2\boldsymbol{S}_i\cdot\boldsymbol{S}_j + 6\frac{(\boldsymbol{S}_i\cdot\boldsymbol{r})(\boldsymbol{S}_j\cdot\boldsymbol{r})}{|\boldsymbol{r}|^2}\right]\right\} + \ldots \end{aligned} \tag{1.111}$$

Equation (1.110) is treated as a perturbation to the Schrödinger energy. There are a number of nice features in this expression. The spin–spin force has the right sign to explain Δ–N, ρ−π and J/ψ–η_c. Also, as it is inversely proportional to the quark masses, the hyperfine splitting will be smaller for heavier quarks, as is observed. The spin–spin part of the Breit–Fermi potential (1.111) also describes correctly the gross features of the splittings within the (**70**, 1^-) SU(6) × O(3) multiplet, as well as within the positive-parity $L=2$ multiplets. Extensive phenomenological studies of the baryon spectrum have been carried out by the Canadian School (Isgur and Karl, 1978; Isgur, Karl and Koniuk, 1978) using a spin-independent confining potential plus a short-distance QCD-inspired spin–spin part. The Σ–Λ splitting is naturally explained as a combined effect of the spin–spin force and the s–d mass difference (De Rújula, 1975; Le Yaouanc *et al.*, 1978a,b).

The spin–spin force $\boldsymbol{S}_1\cdot\boldsymbol{S}_2$ does not contribute to P-wave *mesons* because of the δ-function. The spin–orbit term, together with the tensor term, have the right sign to explain the χ-states ordering. However, there is a quantitative failure; in particular, the predicted splittings are too large. This has led to the hypothesis of a contribution from the confining potential, which should be relativistically a Lorentz scalar to explain the splittings.

But we are here entering upon the subject of *relativistic* potentials. In fact, such a notion was already implied in the Breit–Fermi interaction. Indeed (1.110), (1.111) is a correction of order v^2/c^2 to the Coulomb interaction. It could be derived from a relativistic equation in the following way. The Hamiltonian (1.43) of the Schrödinger equation is essentially nonrelativistic.

One can introduce some relativistic features through a Dirac-type kinetic energy and by completing the product of color-charge-density operators in the potential term by their space counterparts, to form a Lorentz-invariant product of γ-matrices:

$$\begin{aligned} H = &\sum_i \int \mathrm{d}\boldsymbol{x}\, q_i^\dagger(\boldsymbol{x})(-\mathrm{i}\boldsymbol{\alpha}\cdot\boldsymbol{\nabla}+\beta m_i)q_i(\boldsymbol{x}) \\ &+\frac{1}{2}\sum_{i,j,a,\mu}\int \mathrm{d}\boldsymbol{x}\int \mathrm{d}\boldsymbol{y}\, V_0(\boldsymbol{x}-\boldsymbol{y})\,\bar{q}_i(\boldsymbol{x})\gamma^\mu\frac{\lambda^a}{2}q_i(\boldsymbol{x})\bar{q}_j(\boldsymbol{y})\gamma_\mu\frac{\lambda^a}{2}q_j(\boldsymbol{y}). \end{aligned} \tag{1.112}$$

What is still noncovariant in (1.112) is the presence of the instantaneous interaction. Note that when $V_0=\alpha_s/r$ one can deduce from (1.112) the Breit–Fermi interaction by making a nonrelativistic expansion. This corresponds to one-gluon exchange up to $O(mv^4/c^4)$, the retardation effects being of higher order. The deduction goes as follows. V_0 must be considered as being of the same order as the kinetic energy, i.e. $O(mv^2/c^2)$, as is obvious from the Schrödinger equation. Since the first two terms in (1.112), which constitute the free-quark Hamiltonian, are larger, $O(mv/c)$ and $O(m)$ respectively, one can see that, up to and including $O(v^2/c^2)$, the Hamiltonian will be put in the non relativistic form (1.43) by expressing it in terms of modified quark-field operators $q'(\boldsymbol{x})$, related to the $q(\boldsymbol{x})$ by a transformation that is canonical up to $O(v^2/c^2)$:

$$\left.\begin{aligned} q'(\boldsymbol{x}) &\approx \left(1-\mathrm{i}\frac{\beta\boldsymbol{\alpha}\cdot\boldsymbol{\nabla}}{2m}+\frac{\boldsymbol{\nabla}^2}{8m^2}\right)q(\boldsymbol{x}), \\ q(\boldsymbol{x}) &\approx \left(1+\mathrm{i}\frac{\beta\boldsymbol{\alpha}\cdot\boldsymbol{\nabla}}{2m}+\frac{\boldsymbol{\nabla}^2}{8m^2}\right)q'(\boldsymbol{x}), \end{aligned}\right\} \tag{1.113}$$

This is because

$$\begin{aligned} &\left(1-\mathrm{i}\frac{\beta\boldsymbol{\alpha}\cdot\boldsymbol{\nabla}}{2m}+\frac{\boldsymbol{\nabla}^2}{8m^2}\right)(-\mathrm{i}\boldsymbol{\alpha}\cdot\boldsymbol{\nabla}+\beta m)\left(1+\mathrm{i}\frac{\beta\boldsymbol{\alpha}\cdot\boldsymbol{\nabla}}{2m}+\frac{\boldsymbol{\nabla}^2}{8m^2}\right) \\ &=\beta m-\beta\frac{\boldsymbol{\nabla}^2}{2m}+O\left(\frac{mv^3}{c^3}\right). \end{aligned} \tag{1.114}$$

In fact, applying (1.113) to the nonrelativistic quark-field expansion (1.44), one gets for $q(\boldsymbol{x})$ the expansion into free-quark Dirac spinors, which is considered at the start of the next chapter, with the additional approximation of keeping only the second-order terms. To this order, the Hamiltonian written using (1.113) is *identical* with the Schrödinger Hamiltonian (1.43), specified by $V_0(\boldsymbol{x}-\boldsymbol{y})=\alpha_s/|\boldsymbol{x}-\boldsymbol{y}|$. However, there are additional terms of higher order. To evaluate the kinetic energy up to $O(mv^4/c^4)$, one has to push the expansion (1.113) up to $O(v^4/c^4)$. With regard to the terms involving the potential (which is $O(mv^2/c^2)$), (1.113) is enough to reach $O(mv^4/c^4)$: there are $O(mv^3/c^3)$ terms, which correspond to pair creation and are considered in the Cornell model of strong decays (Section 2.2.3(d)); there are also terms $O(mv^4/c^4)$, which preserve the number of quarks and induce simply a perturbation of the energy that can be computed in first-order pertubation theory.

One finds, as concerns energy levels,

(i) the expected relativistic correction to the kinetic energy:

$$H'_{\text{kin}} = -\frac{p^4}{8m_i^3}; \tag{1.115}$$

(ii) the Breit–Fermi interaction (1.110), which is indeed $O(mv^4/c^4)$, since it is a correction $O(v^2/c^2)$ with respect to the Coulomb potential, $O(mv^2/c^2)$.

The terms $O(mv^3/c^3)$ that change the number of quarks do not contribute in first-order perturbation theory to the energy, since their average over states with a fixed number of quarks and antiquarks is zero. The effects of such terms will be further discussed in Section 3.3.

We have thus seen how to deal with the relativistic corrections to the Coulomb potential. The prescription is identical with one-gluon exchange, except for retardation effects. The same framework will allow a discussion of the *confining* part. But we now encounter the embarrassing question of the Lorentz structure of the confining potential. In contrast with gluon exchange, we have no immediate hint, and, as we shall see below, the theoretical indications are not compelling. In fact, two simple answers are considered. Either it is a vector, as in (1.112) or a scalar, as in (1.54), which corresponds to dropping the γ^μ in (1.112). These two choices share the property that *spin-dependent effects vanish in the nonrelativistic limit*, a trend confirmed by heavy quarks. Their differences are as follows.

(i) As we have already indicated in Section 1.3.1, they relate meson and baryon potentials differently. The scalar potential should be repulsive within baryons if it is attractive within mesons.

(ii) The vector potential will generate a long-distance spin–spin force.

(iii) The spin–orbit terms in mesons will have opposite signs. The vector confining potential will reinforce the effect of one-gluon exchange, the scalar will diminish it.

The consequences (ii) and (iii) are obtained by a procedure entirely similar to that described for obtaining the Breit–Fermi interaction from (1.112) using (1.113).

Unhappily, the three points above, when confronted with experiment, do not give the same answer. (i), if taken seriously, would exclude the scalar "solution" from the beginning. (ii) would exclude the vector solution, since it is well verified that the spin–spin force is a short-distance effect. (iii) would exclude the vector solution because they χ splittings disagree with a large spin–orbit term. One usually discards (ii) by the rather odd proposal that the vector could have only a timelike component, and one usually forgets (i). Then the "solution" is the scalar, but (iii) may seem to be weakly compelling in view of the uncertainties of phenomenology. Anyway, none of the "solutions" is really a solution.

What is learned from theory? On the one hand, if confinement is attributed to a modification of the gluon propagator at small q^2 (which would correspond to the transformation of α_s into the running $\alpha_s(q^2)$ in the large-q^2 region) then the long-distance interaction should remain a vector. On the other hand, one has attempted to extract directly the spin dependence of the $q\bar{q}$ energy by a generalization of the Wilson-loop procedure, used for static quarks. Spin dependence should be obtained by making an expansion in $1/M$ as seen through (1.111) (Eichten and Feinberg, 1981). It is claimed (Gromes, 1984a,b), from rather model-independent arguments, that the sign of the L–S terms corresponding to the long-range force is as predicted by a Lorentz scalar. Finally, lattice studies of the same quantities by Monte-Carlo numerical methods give contradictory results about the presence of spin-dependent long-range forces (De Forcrand and Stack, 1985; Michael 1986).

Chiral symmetry for light quarks would give an indication in favor of the vector potential if one follows recent proposals (Finger *et al.*, 1979; Amer *et al.*, 1983; Adler and Davis, 1984) to implement chiral symmetry and dynamical breaking of chiral symmetry in the relativistic-potential approach. Chiral symmetry would be preserved in (1.112) if $m_i = 0$. It can be shown that the interaction induces a dynamical breakdown of this symmetry; constituent-quark masses are generated by this mechanism and the pion is obtained as a Goldstone boson. A remarkable fact is that a large ρ–π splitting is automatically generated even if one drops the space components of the vector interaction, which generate the spin–spin Fermi interaction in (1.111) (Le Yaouanc *et al.*, 1985a,b). This suggests that the phenomenological analysis should be redone in this new context. It is clear that if one considers instead of (1.112) a scalar interaction (1.54) then chiral symmetry would be submitted to a large *explicit* breaking and the whole scheme would be destroyed. It is, however, fair to say that this scheme does not offer a solution to the spin–orbit problem described above.

Whatever the final answer to this delicate question, it is now time to draw the essential conclusions of this Chapter for the treatment of transitions. The main feature that is emerging is that spin-dependent effects are *relativistic corrections to the Schrödinger energy*. As such, consistency requires that they should be included only if the other corrections of the same order in v/c are taken into account (although the latter may be practically smaller). In the

Schrödinger equation, which is the lowest order of the nonrelativistic expansion, the potential must be considered as spin-independent, and SU(6) is exact. The corresponding wave functions, which serve everywhere in the next chapter as a starting point are, as described in Sections 1.3.2 and 1.3.3, in accordance with SU(6). If, on the other hand, we consider, as in Chapter 3, Dirac-type wave equations and Hamiltonians, the conclusion is still the same: V_V, V_S (cf. Chapter 3) are to be identified, at least in their sum, with the nonrelativistic, spin-independent potential of Section 1.3.5. More precisely, V_V must include α_s/r, while it will remain undetermined whether κr is a part of V_V or V_S. The spin-dependent effects do not have to be added separately, but are now generated by the relativistic effects.

CHAPTER 2

Nonrelativistic quark model of transitions

Übrigens brauchen wir uns keineswegs darüber zu erschrecken, dass die Erkenntnistufe, auf der wir heute stehn, ebensowenig endgültig ist als alle vorhergegangenen.†

F. ENGELS, *Anti-Dühring*

La meilleure manière de prendre un autobus en marche, c'est d'attendre qu'il s'arrête.‡

P. DAC, *L'os à moëlle*

Sans facile pas de difficile...

We enter now into the main subject of this book, namely hadron transitions in the quark model. This Chapter is restricted to the deduction within the nonrelativistic model of the main formulae concerning a number of electroweak and strong transitions. The relativistic corrections will be studied in Chapter 3. Chapter 4 will be devoted to phenomenological applications, including the discussion of the parameters of the nonrelativistic quark model introduced in the present Chapter.

2.1 ELECTROWEAK TRANSITIONS

Many processes can be described by the emission or absorption of an elementary quantum by a quark. This quantum may be truly elementary like the photon or the W with which we are concerned here, or it may in fact be composite, like the pion, in which case this is only a simplified representation (see Section 2.2.1). Moreover, the quantum may remain virtual, as in electroproduction or in semileptonic weak interactions, where the vector

† Besides, we should not be alarmed at the fact that the present level of knowledge is no more definitive than all the previous ones.

‡ The best way to get in a moving bus, is to wait until it stops.

boson is very heavy and one observes only a leptonic pair. In any of these cases, the interaction involves a trilinear coupling of the quarks with the field of the elementary quantum, like γqq or Wqq. The quarks, on the other hand, are confined within hadrons by the strong interaction. The problem is to calculate the matrix elements of the trilinear interaction, considered at first order, between composite hadron states. In the nonrelativistic approximation the calculation does not explicitly involve the interaction responsible for the binding of quarks, but only the assumed wave functions. The results are very simple (Sections 2.1.1–2.1.3).

In general, calculations involving the electroweak interaction at second or higher order are much more complicated. In the case of *nonleptonic weak decays*, which are such second-order processes, there is, however, a very important simplifying feature: the second-order weak interaction, which proceeds through W exchange between quarks, can be very well described by an effective four-quark local Hamiltonian, by taking advantage of the fact that the W is very heavy; this is the so-called current–current interaction. In addition, it is shown (see Section 4.6 for details) that the physical process can be related to matrix elements of the local effective Hamiltonian between one-hadron states. These matrix elements are still tractable through the simple methods of this chapter (Section 2.1.4).

2.1.1 Radiative Decays

2.1.1(a) The transition Hamiltonian

We shall present in detail the case of emission and absorption of a photon (Figure 2.1).

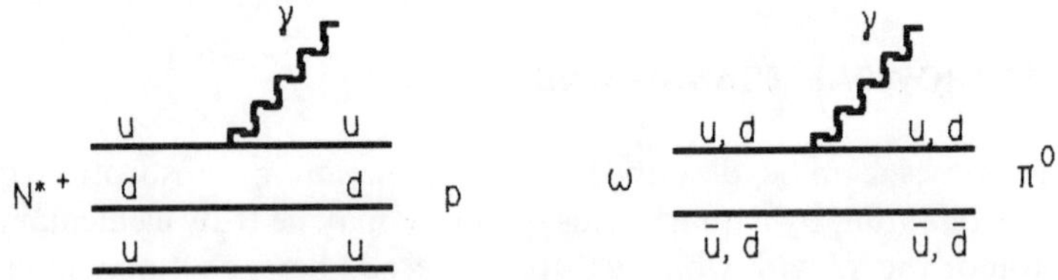

Figure 2.1 Examples of radiative decays of baryons ($N^{*+} \to p\gamma$) and mesons ($\omega \to \pi^0\gamma$).

The interaction Hamiltonian is

$$H_I = e\int dx\, j^{\mu}_{em}(x) A_{\mu}(x), \tag{2.1}$$

where the quark current j^{μ} is

$$j^{\mu}_{em}(x) = \bar{q}(x) Q \gamma^{\mu} q(x) \tag{2.2}$$

and $A_\mu(\boldsymbol{x})$ is the electromagnetic field, $q(\boldsymbol{x})$ is the quark field, and Q is the quark charge matrix in flavor space, in e units. The decomposition of the electromagnetic and quark fields into creation an annihilation operators of plane waves is defined by

$$A_\mu(\boldsymbol{x}) = \int \frac{\mathrm{d}\boldsymbol{k}}{(2\pi)^{3/2}} \frac{1}{(2k^0)^{1/2}} [a_\mu(\boldsymbol{k})\mathrm{e}^{\mathrm{i}k\cdot x} + a_\mu^\dagger(\boldsymbol{k})\mathrm{e}^{-\mathrm{i}k\cdot x}], \tag{2.3a}$$

$$q(\boldsymbol{x}) = \int \frac{\mathrm{d}\boldsymbol{p}}{(2\pi)^{3/2}} \left(\frac{m}{p^0}\right)^{1/2} \sum_s [u_s(\boldsymbol{p})b_s(\boldsymbol{p})\mathrm{e}^{\mathrm{i}p\cdot x} + v_s(\boldsymbol{p})d_s^\dagger(\boldsymbol{p})\mathrm{e}^{-\mathrm{i}p\cdot x}], \tag{2.3b}$$

with the nonvanishing commutation and anticommutation relations given respectively by

$$[a_\mu(\boldsymbol{k}), a_\nu^\dagger(\boldsymbol{k}')] = -g_{\mu\nu}\delta(\boldsymbol{k}-\boldsymbol{k}'), \tag{2.4a}$$

$$\{b_s(\boldsymbol{p}), b_{s'}^\dagger(\boldsymbol{p}')\} = \{d_s(\boldsymbol{p}), d_{s'}^\dagger(\boldsymbol{p}')\} = \delta_{ss'}\delta(\boldsymbol{p}-\boldsymbol{p}'), \tag{2.4b}$$

and the noncovariant normalization of states

$$\langle \boldsymbol{k}, \mu \,|\, \boldsymbol{k}', \nu\rangle = -g_{\mu\nu}\delta(\boldsymbol{k}-\boldsymbol{k}'), \tag{2.5a}$$

$$\langle \boldsymbol{p}, s \,|\, \boldsymbol{p}', s'\rangle = \delta_{ss'}\delta(\boldsymbol{p}-\boldsymbol{p}'). \tag{2.5b}$$

The spinors have in contrast a covariant normalization to fit with the usual conventions,

$$u_s^\dagger(\boldsymbol{p})u_{s'}(\boldsymbol{p}) = v_s^\dagger(\boldsymbol{p})v_{s'}(\boldsymbol{p}) = \frac{p^0}{m}\delta_{ss'}. \tag{2.6}$$

The preceding conventions for the interaction Hamiltonian and the electromagnetic field imply that the electric charge e is expressed in Heaviside–Lorentz units: one has $e^2/4\pi = \alpha = \frac{1}{137}$ and the Coulomb potential is $\alpha/r = e^2/4\pi r$.

2.1.1(b) General presentation of the nonrelativistic approximation

Up to now, we have been completely general. Let us now define the nonrelativistic approximation. With the risk of being a bit pedantic, we shall try to make more precise than is usually done the various steps that are involved in this approximation. The problem is not of course to define nonrelativistic quantum mechanics *in itself*, a rather well-known subject, but to explain the passage from an *a priori* fully relativistic theory to an approximate nonrelativistic one. Of course, strictly speaking, this could be done only by discussing the strong-interaction QCD Hamiltonian. Instead, we shall have to assume first that it is replaced by a model Hamiltonian like

(1.112), where gluons have already been integrated out, but which still bears the essential feature of having a *Dirac-like free part*. This free part is completely diagonalized by the expansion of the fields (2.3b). Secondly, we want to benefit from the simplicity of the nonrelativistic Hamiltonian defined by (1.43); its oustanding feature is that it has eigenstates with a definite number of quarks and antiquarks, the so-called *valence*-quark approximation. For instance, a baryon eigenstate is defined in this framework by

$$|\mathrm{B}\rangle = \frac{1}{\sqrt{6}}\int \prod \mathrm{d}\boldsymbol{p}_i\, \Psi^{\mathrm{tot}}(\boldsymbol{p}_1,\boldsymbol{p}_2,\boldsymbol{p}_3)\, b^\dagger_{\boldsymbol{p}_1} b^\dagger_{\boldsymbol{p}_2} b^\dagger_{\boldsymbol{p}_3}|0\rangle, \tag{2.7}$$

where Ψ^{tot} is here taken to be the nonrelativistic *wave function*, a solution of the Schrödinger equation. Not only is the description of states then very simple, but the calculation of matrix elements of an interaction Hamiltonian, like the electromagnetic one, responsible for transitions would be very simple. For instance, in the case of a radiative transition, we have simply to select the terms that conserve the number of quarks and antiquarks.

Happily enough, we have found in Section 1.3.5 that it is possible to transform the strong-interaction Hamiltonian with a Dirac-like free part into the Schrödinger Hamiltonian provided one stops at order mv^2/c^2, the strong-interaction potential being counted as $O(mv^2/c^2)$. This is accomplished through the canonical transformation (1.113), which amounts to expanding the Dirac spinors u and v in (2.3b) in powers of p/m. Had we stopped at a lower order we would only have obtained the mass term without any dynamics. Therefore the Schrödinger *wave functions* are taken as the starting point for the v/c expansion of the wave functions. As to the calculation of *transition-matrix elements*, we should just have to borrow the eigenstates of (1.43), such as the baryon state (2.7), and to perform on the fields in H_{I} the indicated substitution (1.113), or equivalently to perform the indicated v/c expansion of the free-quark Dirac spinors in (2.3b).

A consistent v/c expansion of the transition-matrix element needs a simultaneous expansion of the operator and the wave functions, starting from the lowest nontrivial order. In (1.113) we are writing terms of order $(v/c)^0$, v/c and v^2/c^2. Some current-matrix elements, for example the charge density $q^\dagger q$ or the axial vector $\bar{q}\gamma\gamma_5 q$ give nonvanishing matrix elements at order $(v/c)^0$. We therefore start the expansion of the fields in H_{I} from zeroth order. This consists in taking the expression (1.44) for the fields in the transition operator and taking the matrix element between the eigenstates of (1.43). We cannot keep only this order—this would be too drastic since it gives no radiative transitions, i.e. no $\bar{q}\gamma q$ transition. Happily we can use the same lowest-order *eigenstates* (e.g. (2.7)), while consistently retaining also the first-order term of the field expansion (1.113) in the *transition operator* (leading to a nonvanishing radiative transition); this is because there is no v/c correction to the eigenstates of (1.43). On the contrary, there are v^2/c^2 corrections to the latter eigenstates, more precisely the three-quark Schrödinger wave function in (2.7) is corrected by v^2/c^2 terms; therefore, if we want to keep the Schrödinger wave functions, for consistency we must not retain the v^2/c^2 terms of (1.113) in the transition operator. In conclusion, we write

$$q(\boldsymbol{x}) = \left(1 + \mathrm{i}\frac{\beta\boldsymbol{\alpha}\cdot\boldsymbol{\nabla}}{2m}\right) q'(\boldsymbol{x}), \tag{2.8}$$

which is easily recognized as the *Pauli approximation* in the second-quantized form. Equations (2.7) and (2.8) constitute the *basis of our nonrelativistic approximation.* In addition, we shall not retain the second-order operator terms coming through the Pauli approximation in the transition operators.

The higher order corrections to the eigenstates come as follows (for details see Sections 3.2.5 and 3.3). We have shown in Section 1.3.5 that the nonrelativistic Hamiltonian (1.43) is corrected at order $mO(v^3/c^3)$ by pair-creation terms, and at order $mO(v^4/c^4)$ by terms that do not change the number of quarks or antiquarks. To estimate the corrections induced on the eigenstates of (1.43), we use perturbation theory for eigenstates, and we thus have to divide the perturbation by some energy-level difference (Messiah, 1959). Since the energy difference between states differing by one pair is approximately $2m$, the correction induced by the $mO(v^3/c^3)$ term is $O(v^3/c^3)$. On the other hand, the energy differences between levels with the same number of quarks and antiquarks are $mO(v^2/c^2)$; therefore the correction induced by the $mO(v^4/c^4)$ term is $O(v^2/c^2)$. We find the announced conclusion that the corrections to (2.7) begin with $O(v^2/c^2)$ and that they consist in a modification of the three-quark wave function. Additional pairs come in only at the next order.

2.1.1(c) Practical calculations

To make clearer the connection with the standard nonrelativistic formalism, we now separate out the quark part in (2.8), dropping the antiquark part, and use Pauli spinors. If we denote the large and small components by q_1 and q_2, we have

$$q_2(\boldsymbol{x}) = -\mathrm{i}\frac{\boldsymbol{\sigma}\cdot\boldsymbol{\nabla}}{2m} q_1(\boldsymbol{x}). \tag{2.9}$$

Substituting (2.9) into the general expression for the current density (2.2), we obtain

$$j^0_{\mathrm{em}}(\boldsymbol{x}) = q_1^\dagger Q q_1, \tag{2.10a}$$

$$\boldsymbol{j}_{\rm em}(\boldsymbol{x}) = -\frac{\mathrm{i}}{2m}[q_1^\dagger Q(\boldsymbol{\nabla} q_1) - (\boldsymbol{\nabla} q_1^\dagger)Q q_1 + \mathrm{i}\boldsymbol{\nabla}\times(q_1^\dagger Q\boldsymbol{\sigma} q_1)]. \tag{2.10b}$$

Had we substituted the whole of (2.8) in (2.2), we would have obtained also the antiquark current (see below) and a term $q'^\dagger \boldsymbol{\alpha} q'$ in the space current, which corresponds to pair creation or annihilation, and which does not contribute in this approximation to matrix elements between two baryons or two mesons. This term is responsible for the vector-meson annihilation studied in Section 2.1.3.

If we want to calculate matrix elements with ordinary wave functions, i.e. nonrelativistic spin–flavor–space wave functions, we have to find the corresponding first-quantization operators. We do this by imposing the condition that their mean value for one-quark states of the ordinary wave function $\psi(\boldsymbol{r})$ is the same as calculated in the second-quantization formalism; this mean value is given precisely by the same expressions as in (2.10a) and (2.10b) after replacing the field q_1 with the ordinary wave function ψ. For antiquarks it is necessary to exchange annihilation and creation operators, which introduces a minus sign (see Appendix). The desired operators are, for quarks,

$$j^0_{\rm em}(\boldsymbol{x}) = Q\delta(\boldsymbol{x}-\boldsymbol{r}), \tag{2.11a}$$

$$\boldsymbol{j}_{\rm em}(\boldsymbol{x}) = \frac{1}{2m}Q[\delta(\boldsymbol{x}-\boldsymbol{r})\boldsymbol{p} + \boldsymbol{p}\delta(\boldsymbol{x}-\boldsymbol{r})] - \frac{1}{2m}Q\boldsymbol{\sigma}\times\boldsymbol{\nabla}_x\delta(\boldsymbol{x}-\boldsymbol{r}), \tag{2.11b}$$

where Q is the quark charge matrix, for example for u, d and s

$$Q = \begin{pmatrix} \frac{2}{3} & 0 & 0 \\ 0 & -\frac{1}{3} & 0 \\ 0 & 0 & -\frac{1}{3} \end{pmatrix}, \tag{2.12}$$

and $\boldsymbol{r}$ is the position operator while $\boldsymbol{x}$ labels a spatial point.

One finally obtains the nonrelativistic interaction Hamiltonian in first-quantized form for quarks:

$$H_{\rm I} = eQ\left\{A^0(\boldsymbol{r}) - \frac{1}{2m}[\boldsymbol{A}(\boldsymbol{r})\cdot\boldsymbol{p} + \boldsymbol{p}\cdot\boldsymbol{A}(\boldsymbol{r})] - \frac{1}{2m}\boldsymbol{\sigma}\cdot[\boldsymbol{\nabla}\times\boldsymbol{A}(\boldsymbol{r})]\right\} \tag{2.13}$$

Conventionally, one calls the first term the *charge interaction*, the second the *convective-current interaction* and the third the *magnetic interaction.* We must distinguish this terminology from that of the multipole expansion, where one speaks of electric and magnetic radiation, as well as of scalar and longitudinal multipoles for virtual radiation (electroproduction).

For antiquarks, the calculation shows that one has just to change the signs in the matrix Q, so that the form (2.13) is preserved. This is true with the convention for antiquark spinors $\chi^{\rm C} = -\mathrm{i}\sigma^2\chi^*$ (see (1.66)).

Let us consider hereafter the *absorption* of a *real* photon of momentum $\boldsymbol{k}$ and polarization ε_μ; $A_\mu(\boldsymbol{r})$ must be replaced by its matrix element to yield the operator between quarks only. From

$$\langle 0|\, A_\mu(\boldsymbol{r})\, |\boldsymbol{k},\varepsilon\rangle = \frac{1}{(2\pi)^{3/2}} \frac{1}{(2k^0)^{1/2}} \varepsilon_\mu e^{i\boldsymbol{k}\cdot\boldsymbol{r}}, \tag{2.14}$$

taking the partial matrix element on the photon degrees of freedom, one gets

$$\langle 0|\, H_\mathrm{I}\, |\boldsymbol{k},\varepsilon\rangle$$

$$= \frac{1}{(2\pi)^{3/2}} \frac{1}{(2k^0)^{1/2}} eQ \left\{ \varepsilon^0 e^{i\boldsymbol{k}\cdot\boldsymbol{r}} - \frac{1}{2m}[e^{i\boldsymbol{k}\cdot\boldsymbol{r}}\boldsymbol{\varepsilon}\cdot\boldsymbol{p} + \boldsymbol{\varepsilon}\cdot\boldsymbol{p}e^{i\boldsymbol{k}\cdot\boldsymbol{r}}] - \frac{i}{2m}\boldsymbol{\sigma}\cdot(\boldsymbol{k}\times\boldsymbol{\varepsilon})e^{i\boldsymbol{k}\cdot\boldsymbol{r}} \right\}. \tag{2.15}$$

In fact $\varepsilon_0 = 0$ and $\boldsymbol{\varepsilon}\cdot\boldsymbol{k} = 0$ for a real photon, but it will be useful to calculate the expression (2.15) for an arbitrary ε_μ because one encounters all the components of the current in electroproduction. We rewrite (2.15) as

$$\langle 0|\, H_\mathrm{I}\, |\boldsymbol{k},\varepsilon\rangle = \frac{1}{(2\pi)^{3/2}} \frac{1}{(2k^0)^{1/2}} eQ\varepsilon_\mu \tilde{j}^\mu_{\mathrm{em}}(\boldsymbol{k}), \tag{2.16}$$

with the help of the Fourier-transformed current density

$$\tilde{j}^\mu_{\mathrm{em}}(\boldsymbol{k}) = \int d\boldsymbol{x} e^{i\boldsymbol{k}\cdot\boldsymbol{x}} j^\mu_{\mathrm{em}}(\boldsymbol{x}), \tag{2.17}$$

$$\tilde{j}^0_{\mathrm{em}}(\boldsymbol{k}) = Qe^{i\boldsymbol{k}\cdot\boldsymbol{r}}, \tag{2.18a}$$

$$\tilde{\boldsymbol{j}}_{\mathrm{em}}(\boldsymbol{k}) = Q\left[\frac{e^{i\boldsymbol{k}\cdot\boldsymbol{r}}\boldsymbol{p} + \boldsymbol{p}e^{i\boldsymbol{k}\cdot\boldsymbol{r}}}{2m} + i\frac{\boldsymbol{\sigma}\times\boldsymbol{k}}{2m}e^{i\boldsymbol{k}\cdot\boldsymbol{r}}\right]. \tag{2.18b}$$

when $\boldsymbol{k}$ is small, the first term in (2.18b) contributes to electric-dipole transitions between $L=1$ orbital excitations and the ground state, like $N^* \to N\gamma$ or $\chi \to J/\psi\,\gamma$, and the second term contributes to magnetic moments or magnetic-dipole transitions like $\omega \to \pi^0\gamma$, $J/\psi \to \eta_c\gamma$ or $\Delta \to N\gamma$. In practice $\boldsymbol{k}$ is not small and the second term also contributes to $L=1$ decays as we shall see in Section 4.3.1.

In order now to calculate S-matrix elements, we have to calculate the matrix elements of $\tilde{j}^\mu(\boldsymbol{k})$ between hadron states. This requires some definitions and conventions. First, we decide to normalize the hadron states according to $\langle \boldsymbol{P}'|\boldsymbol{P}\rangle = \delta(\boldsymbol{P}' - \boldsymbol{P})$ as we did for quarks. In terms of ordinary wave functions, this is, for baryons,

$$\int d\boldsymbol{r}_1\, d\boldsymbol{r}_2\, d\boldsymbol{r}_3\, \Psi^{\mathrm{tot}}_{\boldsymbol{P}'}(\boldsymbol{r}_1,\boldsymbol{r}_2,\boldsymbol{r}_3)^\dagger \Psi^{\mathrm{tot}}_{\boldsymbol{P}}(\boldsymbol{r}_1,\boldsymbol{r}_2,\boldsymbol{r}_3) = \delta(\boldsymbol{P}' - \boldsymbol{P}). \tag{2.19}$$

We pass to the momentum representation by expanding in normalized quark plane waves:

$$\Psi^{\text{tot}}(\boldsymbol{r}_1,\boldsymbol{r}_2,\boldsymbol{r}_3) = \int \prod_i \frac{\mathrm{d}\boldsymbol{p}_i}{(2\pi)^{3/2}} \Psi^{\text{tot}}(\boldsymbol{p}_1,\boldsymbol{p}_2,\boldsymbol{p}_3) \exp\left(\mathrm{i}\sum_i \boldsymbol{p}_i \cdot \boldsymbol{r}_i\right). \tag{2.20}$$

Then we have also

$$\int \prod_i \mathrm{d}\boldsymbol{p}_i\, \Psi^{\text{tot}}_{\boldsymbol{P}'}(\boldsymbol{p}_i)^\dagger \Psi^{\text{tot}}_{\boldsymbol{P}}(\boldsymbol{p}_i) = \delta(\boldsymbol{P}' - \boldsymbol{P}). \tag{2.21}$$

The important step now is to factorize the total wave function Ψ^{tot} into an internal wave function ψ and the plane wave $\frac{1}{(2\pi)^{3/2}} e^{\mathrm{i}\boldsymbol{P}\cdot\boldsymbol{R}}$ describing the center-of-mass motion. This is more easily done in the momentum representation

$$\Psi^{\text{tot}}_{\boldsymbol{P}}(\boldsymbol{p}_i) = \delta\left(\sum_i \boldsymbol{p}_i - \boldsymbol{P}\right)\psi(\boldsymbol{p}_i - \tfrac{1}{3}\boldsymbol{P}) \tag{2.22}$$

where $\boldsymbol{p}_i - \frac{1}{3}\boldsymbol{P}$ represents the impulsion of the quark in the center of mass, and the factor $\frac{1}{3}$ comes from the assumption of equal masses.

The normalization of $\psi(\boldsymbol{p}_i)$ is easily found:

$$\int \prod_i \mathrm{d}\boldsymbol{p}_i\, \delta\left(\sum_i \boldsymbol{p}_i\right) |\psi(\boldsymbol{p}_i)|^2 = 1. \tag{2.23}$$

When one calculates a matrix element of an operator like $e^{\mathrm{i}\boldsymbol{k}\cdot\boldsymbol{r}_3}$, one finds, in terms of the internal wave function ψ,

$$\langle f | e^{\mathrm{i}\boldsymbol{k}\cdot\boldsymbol{r}_3} | i \rangle$$

$$= \delta(\boldsymbol{P}' - \boldsymbol{k} - \boldsymbol{P}) \int \prod_i \mathrm{d}\boldsymbol{p}_i\, \delta\left(\sum_i \boldsymbol{p}_i\right) \psi_f(\boldsymbol{p}_i)^\dagger\, \psi_i(\boldsymbol{p}_1 + \tfrac{1}{3}\boldsymbol{k}, \boldsymbol{p}_2 + \tfrac{1}{3}\boldsymbol{k}, \boldsymbol{p}_3 - \tfrac{2}{3}\boldsymbol{k}). \tag{2.24}$$

The δ-function in front of the integral ensures momentum conservation. In general, one finds

$$\langle \mathrm{f} | H_{\mathrm{I}} | \mathrm{i} \rangle = \delta(\boldsymbol{P}' - \boldsymbol{k} - \boldsymbol{P}) M, \tag{2.25}$$

where M is free of the δ-function ensuring total momentum conservation. The S-matrix element, which is given in terms of the Hamiltonian by

$$S_{\mathrm{fi}} = \delta_{\mathrm{fi}} - 2\pi\mathrm{i}\delta(E_{\mathrm{f}} - E_{\mathrm{i}}) \langle \mathrm{f} | H_{\mathrm{I}} | \mathrm{i} \rangle, \tag{2.26}$$

is then expressed in terms of the finite M by

$$S_{\mathrm{fi}} = \delta_{\mathrm{fi}} - 2\pi\mathrm{i}\delta^{(4)}(\mathrm{P}_{\mathrm{f}} - \mathrm{P}_{\mathrm{i}}) M. \tag{2.27}$$

In the case of emission, one has to replace $\boldsymbol{k}$ by $-\boldsymbol{k}$ in the previous expressions, $\boldsymbol{k}$ being the momentum of the photon, which is now in the final state.

Starting from the expression for the transition rate

$$W = 2\pi\delta(E_f - E_i)\,|\langle f|\,H_I\,|i\rangle|^2, \tag{2.28}$$

one finds the transition rate w for one particle by dividing W by the total number of particles corresponding to the plane wave $(2\pi)^{-3/2}e^{i\boldsymbol{P}\cdot\boldsymbol{R}}$, which is $V/(2\pi)^3$, V being the volume. On the other hand, on squaring the factor $\delta(\boldsymbol{P}_f - \boldsymbol{P}_i)$ in (2.25), one gets $\delta^3(0)\delta(\boldsymbol{P}_f - \boldsymbol{P}_i)$, where $\delta^3(0)$ must be interpreted as $V/(2\pi)^3$, and one therefore gets the Fermi formula

$$w = 2\pi\delta(E_f - E_i)\,\delta(\boldsymbol{P}_f - \boldsymbol{P}_i)|M|^2 \tag{2.29}$$

The decay width is then obtained by integrating over the phase space of the final state:

$$\Gamma = 2\pi\rho_f|M|^2, \tag{2.30}$$

where, for a process A→B+C, one has, in the rest frame of A,

$$\rho_f = \int d\boldsymbol{p}_B\, d\boldsymbol{p}_C\, \delta(m_A - E_B - E_C)\,\delta(\boldsymbol{p}_B + \boldsymbol{p}_C) = 4\pi\frac{E_B E_C}{m_A}k. \tag{2.31a}$$

In the present particular case of a photon transition, $m_C = 0$, $E_C = k^0 = k$, and this formula simplifies to

$$\rho_f = 4\pi\frac{E_B}{m_A}k^2. \tag{2.31b}$$

Note that one factor k will always be cancelled by the square of $(k^0)^{-1/2}$ in $|M|^2$.

Let us now return to the practical calculation of M. The wave functions ψ are products of a color wave function and of a part depending on the other degrees of freedom. Since the current operator, in first-quantization formalism, is just the identity matrix on the color indices, and since the color wave function is always the same (for baryons or mesons), the color can be simply omitted. The treatment of color is not so simple when there is quark annihilation ($\rho^0 \to e^+e^-$) or a change in the quark number (proton decay) or when the color wave function is not factorizable (as in the decay of diquonia $qq\bar{q}\bar{q}$). Let us keep for the moment to the case of γ emission or absorption.

For baryons, the singlet color wave function is antisymmetric, and therefore the remaining part must be symmetric. This property is useful to simplify the calculation of baryon current matrix elements:

$$\langle\Psi|\sum_{i=1}^{3} j_{em}^{(i)}|\Psi\rangle = 3\langle\Psi|j_{em}^{(3)}|\Psi\rangle, \tag{2.32}$$

where $j_{em}^{(i)}$ is the electromagnetic current operator in first quantization acting on the ith quark.

For the structure of the wave functions ψ we refer to Chapter 1, except for comments directly relevant to the calculation of matrix elements. A general step after performing the transformation (2.24) consists in choosing a system of relative coordinates, so as to eliminate $\delta(\sum_i \boldsymbol{p}_i)$. The usual choice is

$$\left.\begin{aligned} \boldsymbol{p}_\rho &= \frac{1}{\sqrt{2}}(\boldsymbol{p}_1 - \boldsymbol{p}_2), \\ \boldsymbol{p}_\lambda &= \frac{1}{\sqrt{6}}(\boldsymbol{p}_1 + \boldsymbol{p}_2 - 2\boldsymbol{p}_3). \end{aligned}\right\} \tag{2.33}$$

Concerning the third combination, one chooses

$$\boldsymbol{P} = \sum_i \boldsymbol{p}_i, \tag{2.34}$$

which is conjugate to the center-of-mass coordinate

$$\boldsymbol{R} = \sum_i \tfrac{1}{3}\boldsymbol{r}_i. \tag{2.35}$$

With (2.34) and (2.35), we finally have the result

$$\prod_i \mathrm{d}\boldsymbol{p}_i = \frac{1}{3\sqrt{3}} d\boldsymbol{p}_\rho \, \mathrm{d}\boldsymbol{p}_\lambda \mathrm{d}\boldsymbol{P}. \tag{2.36}$$

In configuration space, we find similarly

$$\prod_i \mathrm{d}\boldsymbol{r}_i = 3\sqrt{3}\, \mathrm{d}\boldsymbol{R}\, \mathrm{d}\boldsymbol{\rho}\, \mathrm{d}\boldsymbol{\lambda}. \tag{2.37}$$

It is possible to factorize out a $\delta(\frac{1}{3}\sum_i \boldsymbol{r}_i)$ of spatial coordinates and at the same time the $\delta(\boldsymbol{P}' - \boldsymbol{k} - \mathrm{P})$ representing momentum conservation. One starts with the momentum expression (2.24) and applies the formulae that relate the internal wave functions:

$$\psi(\boldsymbol{p}_i) = \frac{1}{(2\pi)^3} \int \prod_i \mathrm{d}\boldsymbol{r}_i \exp\left(-\mathrm{i}\sum_i \boldsymbol{p}_i \cdot \boldsymbol{r}_i\right) \delta\left(\sum_i \tfrac{1}{3}\boldsymbol{r}_i\right) \psi(\boldsymbol{r}_i), \tag{2.38a}$$

$$\delta\left(\sum_i \boldsymbol{p}_i\right) \psi(\boldsymbol{p}_i) = \frac{1}{(2\pi)^6} \int \prod_i \mathrm{d}\boldsymbol{r}_i \exp\left(-\mathrm{i}\sum_i \boldsymbol{p}_i \cdot \boldsymbol{r}_i\right) \psi(\boldsymbol{r}_i). \tag{2.38b}$$

One then gets the normalization condition

$$\int \prod_i \mathrm{d}\boldsymbol{r}_i\,\delta\left(\tfrac{1}{3}\sum_i \boldsymbol{r}_i\right)|\psi(\boldsymbol{r}_i)|^2 = 1 \tag{2.39}$$

and the matrix element of the operator $\mathrm{e}^{\mathrm{i}\boldsymbol{k}\cdot\boldsymbol{r}_3}$,

$$\begin{aligned}&\langle \mathrm{f}|\,\mathrm{e}^{\mathrm{i}\boldsymbol{k}\cdot\boldsymbol{r}_3}|i\rangle\\ &= \delta(\boldsymbol{P}'-\boldsymbol{k}-\boldsymbol{P})\int \prod_i \mathrm{d}\boldsymbol{r}_i\,\mathrm{e}^{\mathrm{i}\boldsymbol{k}\cdot\boldsymbol{r}_3}\,\delta\left(\sum_i \tfrac{1}{3}\boldsymbol{r}_i\right)\psi_{\mathrm{f}}(\boldsymbol{r}_i)^{\dagger}\,\psi_{\mathrm{i}}(\boldsymbol{r}_i).\end{aligned} \tag{2.40}$$

We note here a technical point. It is usual in the literature to normalize the internal wave function relatively to the measure $\mathrm{d}\boldsymbol{\rho}\mathrm{d}\boldsymbol{\lambda}$ instead of the correct measure $\prod_i \mathrm{d}\boldsymbol{r}_i\,\delta(\frac{1}{3}\sum_i \boldsymbol{r}_i)$. Both results differ by a factor $3\sqrt{3}$, compensated by its inverse in the calculation of the matrix elements. Although harmless in this case, the use of the measure $\mathrm{d}\boldsymbol{\rho}\mathrm{d}\boldsymbol{\lambda}$ can lead to incorrect results when there is no conservation of the number of quarks, as in proton decay, e.g. $\mathrm{p}\to\mathrm{e}^{+}\pi^{0}$ (for a correct treatment, see Gavela *et al.*, 1981a).

The treatment of the operator $\{\boldsymbol{p}_3, \mathrm{e}^{\mathrm{i}\boldsymbol{k}\cdot\boldsymbol{r}_3}\}$ in (2.15) can be done along similar lines in both representations. In momentum representation, we propose a formula that is explicitly symmetric with respect to initial and final states, exhibiting the global convective current $(\boldsymbol{P}+\boldsymbol{P}')/6m$ (where $6m$ is the baryon mass):

$$\begin{aligned}&\langle \mathrm{f}|\boldsymbol{p}_3\mathrm{e}^{\mathrm{i}\boldsymbol{k}\cdot\boldsymbol{r}_3}+\mathrm{e}^{\mathrm{i}\boldsymbol{k}\cdot\boldsymbol{r}_3}\boldsymbol{p}_3\,|\mathrm{i}\rangle\\ &=\delta(\boldsymbol{P}'-\boldsymbol{k}-\boldsymbol{P})\int \prod_i \mathrm{d}\boldsymbol{p}_i\,\delta\left(\sum_i \boldsymbol{p}_i\right)\psi_{\mathrm{f}}(\boldsymbol{p}_1-\tfrac{1}{6}\boldsymbol{k},\boldsymbol{p}_2-\tfrac{1}{6}\boldsymbol{k},\boldsymbol{p}_3+\tfrac{1}{3}\boldsymbol{k})^{\dagger}\\ &\quad\times[2\boldsymbol{p}_3+\tfrac{1}{3}(\boldsymbol{P}+\boldsymbol{P}')]\,\psi_{\mathrm{i}}(\boldsymbol{p}_1+\tfrac{1}{6}\boldsymbol{k},\boldsymbol{p}_2+\tfrac{1}{6}\boldsymbol{k},\boldsymbol{p}_3-\tfrac{1}{3}\boldsymbol{k}).\end{aligned} \tag{2.41}$$

Such a symmetric form is obtained from the previous asymmetric formula by performing a translation of the integration variables. It also has the advantage of giving simpler calculations for harmonic-oscillator wave functions. In configuration space, we have forms similar to (2.40):

$$\begin{aligned}&\langle \mathrm{f}|\boldsymbol{p}_3\mathrm{e}^{\mathrm{i}\boldsymbol{k}\cdot\boldsymbol{r}_3}+\mathrm{e}^{\mathrm{i}\boldsymbol{k}\cdot\boldsymbol{r}_3}\boldsymbol{p}_3\,|\mathrm{i}\rangle\\ &=\delta(\boldsymbol{P}'-\boldsymbol{k}-\boldsymbol{P})\int \prod_i \mathrm{d}\boldsymbol{r}_i\,\delta\left(\sum_i \tfrac{1}{3}\boldsymbol{r}_i\right)\\ &\quad\left\{\left[2\mathrm{i}\boldsymbol{\nabla}_3-\frac{2}{3}\boldsymbol{k}+\frac{1}{3}(\boldsymbol{P}+\boldsymbol{P}')\right]\psi_{\mathrm{f}}(\boldsymbol{r}_i)^{\dagger}\right\}\mathrm{e}^{\mathrm{i}\boldsymbol{k}\cdot\boldsymbol{r}_3}\,\psi_{\mathrm{i}}(\boldsymbol{r})\end{aligned} \tag{2.42}$$

$$= \delta(\boldsymbol{P}' - \boldsymbol{k} - \boldsymbol{P}) \int \prod_i \mathrm{d}\boldsymbol{r}_i \, \delta\left(\tfrac{1}{3}\sum_i \boldsymbol{r}_i\right) \psi_{\mathrm{f}}(\boldsymbol{r}_i)^\dagger \, \mathrm{e}^{\mathrm{i}\boldsymbol{k}\cdot\boldsymbol{r}_3} \left\{\left[-2\mathrm{i}\boldsymbol{\nabla}_3 + \frac{2}{3}\boldsymbol{k} + \frac{1}{3}(\boldsymbol{P}+\boldsymbol{P}')\right] \psi_{\mathrm{i}}(\boldsymbol{r}_i)\right\} \tag{2.43}$$

One of the two forms is the more useful, according to which state (i or f) is the ground state. Particular forms used in the literature are found from (2.41) and (2.42), (2.43) by specifying the reference frame (e.g. $\boldsymbol{P}=0$ or $\boldsymbol{P}'=0$). It is important to realize that the expression for the space current depends explicitly on the reference frame after this last step of fixing the frame.

A final remark: up to now, we have calculated matrix elements of the Fourier transform of the current density according to (2.15) and (2.16). In fact they are simply related to the matrix elements of the density itself through

$$\langle \mathrm{f} | \tilde{j}^{\mu}_{\mathrm{em}}(\boldsymbol{k}) | \mathrm{i} \rangle = (2\pi)^3 \delta(\boldsymbol{P}' - \boldsymbol{k} - \boldsymbol{P}) \langle \mathrm{f} | j^{\mu}_{\mathrm{em}}(\boldsymbol{0}) | \mathrm{i} \rangle \tag{2.44}$$

One therefore sees that the δ-function would be automatically factorized in the expression of $\langle \mathrm{f}|H_{\mathrm{I}}|\mathrm{i}\rangle$ by using the current density, according to (2.44). One finds

$$M = \frac{1}{(2\pi)^{3/2}} \frac{1}{(2k^0)^{1/2}} \, e\varepsilon_\mu (2\pi)^3 \langle \mathrm{f} | j^{\mu}_{\mathrm{em}}(\boldsymbol{0}) \; | \mathrm{i} \rangle, \tag{2.45}$$

and therefore sees that the interesting quantity for the calculation of widths and cross-sections is the matrix element of the current density. The factor $(2\pi)^3$ in (2.45) is always cancelled by a similar factor, since the density of particles is $1/(2\pi)^3$. The expression for $(2\pi)^3 \langle \mathrm{f}|j^{\mu}_{\mathrm{em}}(\boldsymbol{0})|\mathrm{i}\rangle$ is given by dropping the δ-factor in front of the previous expressions (2.24) and (2.40)–(2.43).

2.1.2 Semileptonic Weak Decays

In the neutrino production of resonances and in semileptonic decays, the basic interaction, analogous to (2.1), is

$$H_{\mathrm{I}} = \frac{G_{\mathrm{F}}}{\sqrt{2}} \int \mathrm{d}\boldsymbol{x} \, J^{(+)\mu}_{\mathrm{had}}(\boldsymbol{x}) \, J^{(-)}_{\mathrm{lept}\mu}(\boldsymbol{x}) + \mathrm{h.c.}, \tag{2.46}$$

where J_μ has a V–A structure:

$$J_\mu = j_\mu - j_{5\mu}, \tag{2.47}$$

with j_μ and $j_{5\mu}$ the vector and axial-vector currents respectively. More specifically,

$$J_{\text{lept}}^{(-)\mu} = \bar{\mu}\gamma^{\mu}(1-\gamma_5)\nu_{\mu} + \bar{e}\gamma^{\mu}(1-\gamma_5)\nu_e + \text{heavier leptons,} \tag{2.48}$$

$$J_{\text{had}}^{(+)\mu} = \cos\theta_C\,\bar{u}\gamma^{\mu}(1-\gamma_5)d + \sin\theta_C\,\bar{u}\gamma^{\mu}(1-\gamma_5)s + \text{heavier quarks.} \tag{2.49}$$

We do not consider neutral currents; this is just to simplify the discussion: we are interested here in quark-model methods. The extension to the neutral-current sector of the electroweak Glashow–Salam–Weinberg theory is straightforward.

The problem of calculating S-matrix elements can be reduced, by steps similar to those performed in the previous Section, to the calculation of matrix elements of the current J_{had}^{μ}, the pair of leptons playing the role of the external electromagnetic field A_{μ} (Figure 2.2).

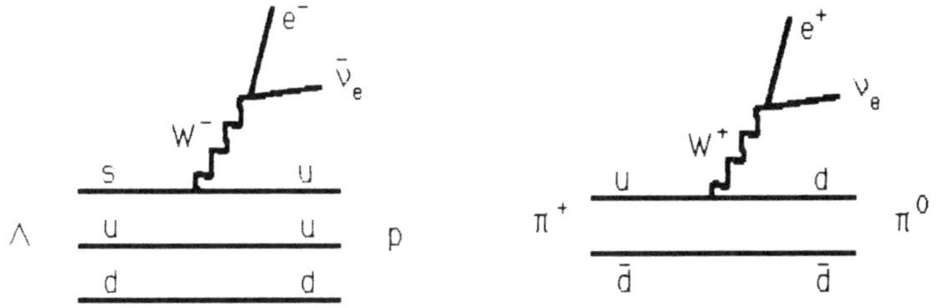

Figure 2.2 Examples of semileptonic decays of baryons ($\Lambda \to pe^-\bar{\nu}_e$) and mesons ($\pi^+ \to \pi^0 e^+ \nu_e$).

Let us write the first-quantization operators corresponding to (2.49) in the nonrelativistic approximation. As to the vector part j^{μ}, we have simply to replace Q in (2.11a,b) by the corresponding flavor operator

$$\cos\theta_C\,\tau^{(+)} + \sin\theta_C\,v^{(+)} \tag{2.50}$$

The $SU(3)_F$ matrices τ and v in (2.50) act on quark states as follows: $\tau^{(+)}d = u$, $v^{(+)}s = u$.

The axial part is given by the reduction of $\bar{q}\gamma_{\mu}\gamma_5 q$:

$$j_5^0(\boldsymbol{x}) = (\cos\theta_C\tau^{(+)} + \sin\theta_C v^{(+)})\frac{1}{2m}[\delta(\boldsymbol{x}-\boldsymbol{r})\boldsymbol{\sigma}\cdot\boldsymbol{p} + \boldsymbol{\sigma}\cdot\boldsymbol{p}\,\delta(\boldsymbol{x}-\boldsymbol{r})], \tag{2.51a}$$

$$\boldsymbol{j}_5(\boldsymbol{x}) = (\cos\theta_C\tau^{(+)} + \sin\theta_C v^{(+)})\,\boldsymbol{\sigma}\,\delta(\boldsymbol{x}-\boldsymbol{r}). \tag{2.51b}$$

It can be seen that the matrix elements of the parts of these operators acting on space have already been calculated when considering j_{em}^{μ}. The spatial part in j_5^0 is like the convective piece in $\boldsymbol{j}$ (2.11b); in $\boldsymbol{j}_5$ it is the same as in j^0 (2.11a). As we shall see in Section 4.2.1, the operator $\boldsymbol{j}_5(\boldsymbol{x})$ will give, at the bound-state level, the axial-vector coupling $G_A = -g_1(0)$, involved for instance in neutron β-decay, in the same way as the four-component vector part in (2.50) will give the vector coupling $G_V = f_1(0)$.

Now, we want for instance to calculate widths of semileptonic transitions. We start again from the Fermi formula (2.29):

$$w = 2\pi\delta(E_f - E_i)\delta(\boldsymbol{P}_f - \boldsymbol{P}_i)|M|^2.$$

In the present case, the major difference is that the process

$$A(\boldsymbol{P}) \to B(\boldsymbol{P}') + \ell(\boldsymbol{p}_\ell) + \bar{\nu}(\boldsymbol{p}_\nu) \tag{2.52}$$

has a three-body phase space, and, instead of (2.30) and (2.31), we have

$$\Gamma = 2\pi \int |M|^2 \delta(E_f - E_i)\delta(\boldsymbol{P}_f - \boldsymbol{P}_i)\, d\boldsymbol{P}'\, d\boldsymbol{p}_\ell\, d\boldsymbol{p}_\nu \tag{2.53}$$

with a nontrivial integration of $|M|^2$ on the momenta, M being defined as in (2.25) by

$$\langle f|H_I|i\rangle = M\,\delta(\boldsymbol{P}_f - \boldsymbol{P}_i). \tag{2.54}$$

We make explicit the leptonic part of H_I:

$$\begin{aligned}\langle f|\,H_I\,|i\rangle = &\frac{1}{(2\pi)^3}\frac{1}{(2p_\ell^0)^{1/2}}\frac{1}{(2p_\nu^0)^{1/2}}\frac{G_F}{\sqrt{2}}\bar{u}_\ell\gamma^\mu(1-\gamma_5)v_\nu \\ &\times \int d\boldsymbol{x}\, e^{-i(\boldsymbol{p}_l+\boldsymbol{p}_\nu)\cdot\boldsymbol{x}}\langle B|\,J^{(+)}_{\text{had}\,\mu}(\boldsymbol{x})|A\rangle\end{aligned} \tag{2.55}$$

The Fourier transform of the current again appears:

$$\tilde{J}^\mu_{\text{had}}(\boldsymbol{q}), \qquad \text{with} \qquad \boldsymbol{q} = -(\boldsymbol{p}_\ell + \boldsymbol{p}_\nu) \tag{2.56}$$

whose matrix elements can be written

$$\langle B|\,\tilde{J}^\mu_{\text{had}}(\boldsymbol{q})\,|A\rangle = (2\pi)^3\delta(\boldsymbol{P}_f - \boldsymbol{P}_i)\langle B|\,J^\mu_{\text{had}}(\boldsymbol{0})\,|A\rangle \tag{2.57}$$

Therefore,

$$M = \frac{1}{(2\pi)^3}\frac{1}{(2p_\ell^0)^{1/2}}\frac{1}{(2p_\nu^0)^{1/2}}\frac{G_F}{\sqrt{2}}\bar{u}_\ell\gamma_\mu(1-\gamma_5)v_\nu(2\pi)^3\langle B|\,J^\mu_{\text{had}}(\boldsymbol{0})\,|A\rangle \tag{2.58}$$

The leptonic spinors have the particular normalization adapted to zero-mass fermions:

$$v^\dagger v = u^\dagger u = 2p^0, \tag{2.59}$$

which explains the factors $1/(2p^0)^{1/2}$ in (2.55) instead of $(m/p^0)^{1/2}$. Let us now consider the unpolarized rate relevant for the width:

$$\sum_{\substack{\text{spins i, f}\\ \text{average for i}}} |M|^2. \tag{2.60}$$

It can be written in terms of the usual leptonic and hadronic tensors

$$T^{\mu\nu}\ell_{\mu\nu}, \tag{2.61}$$

with the definitions

$$\ell^{\mu\nu} = \sum_{\text{spin}} [\bar{u}_\ell(\boldsymbol{p}_\ell)\gamma^\mu(1-\gamma_5)v_\nu(\boldsymbol{p}_\nu)]\,[\bar{u}_\ell(\boldsymbol{p}_\ell)\gamma^\nu(1-\gamma_5)v_\nu(\boldsymbol{p}_\nu)]^*, \tag{2.62}$$

$$T^{\mu\nu} = \sum_{\substack{\text{spins i, f}\\ \text{average for i}}} (2\pi)^6 \langle B|J^\mu_{\text{had}}(\mathbf{0})|A\rangle \langle B|J^\nu_{\text{had}}(\mathbf{0})|A\rangle^* \tag{2.63}$$

A simple calculation gives

$$\ell^{\mu\nu} = 8[p^\mu_\ell p^\nu_\nu + p^\nu_\ell p^\mu_\nu - g^{\mu\nu}(p_\ell \cdot p_\nu) + \mathrm{i}\varepsilon^{\mu\nu\sigma\tau} p_{\nu\sigma} p_{\ell\tau}]. \tag{2.64}$$

If instead of $\ell^-\bar{\nu}_\ell$ emission we have $\ell^+\nu_\ell$ emission, i will be replaced by $-$i in this formula. Practical calculations are simplified by using a particular space–time basis of polarization vectors to define the components of the tensors. This basis is defined relatively to the time-like momentum transfer four-vector ($q^2 > 0$) by

$$\left.\begin{aligned}
\varepsilon^{(\mathrm{T})\mu} &= \frac{q^\mu}{(q^2)^{1/2}} = \frac{1}{(q^2)^{1/2}}(q^0, 0, 0, |\boldsymbol{q}|),\\
\varepsilon^{(\mathrm{S})\mu} &= \frac{1}{(q^2)^{1/2}}(|\boldsymbol{q}|, 0, 0, q^0),\\
\varepsilon^{(\mathrm{R})\mu} &= \frac{1}{\sqrt{2}}(0, -1, -\mathrm{i}, 0),\\
\varepsilon^{(\mathrm{L})\mu} &= \frac{1}{\sqrt{2}}(0, 1, -\mathrm{i}, 0).
\end{aligned}\right\} \tag{2.65}$$

The three space like components S, R, L are orthogonal to q^μ or $\varepsilon^{(T)\mu}$ and more generally

$$\left.\begin{aligned}
&\varepsilon^{(\alpha)}\varepsilon^{(\beta)*} = g^{\alpha\beta} = \delta_{\alpha\beta}\, g^{\alpha\alpha},\\
&g^{\mathrm{TT}} = 1, \qquad g^{\mathrm{SS}} = g^{\mathrm{RR}} = g^{\mathrm{LL}} = -1.
\end{aligned}\right\} \tag{2.66}$$

We accordingly define the new components of the tensors (2.62) and (2.63):

$$\ell_{\alpha\beta} = \varepsilon^\mu_{(\alpha)}\ell_{\mu\nu}\varepsilon^{\nu*}_{(\beta)} \tag{2.67}$$

with

$$\varepsilon_{(\alpha)} = g_{\alpha\beta}\varepsilon^{(\beta)*}, \tag{2.68}$$

$$T^{\alpha\beta} = \varepsilon^{(\alpha)\mu} T_{\mu\nu} \varepsilon^{(\beta)\nu*}. \tag{2.69}$$

We then have

$$T^{\mu\nu}\ell_{\mu\nu} = T^{\alpha\beta}\ell_{\alpha\beta}. \tag{2.70}$$

After integration over phase space according to (2.53), there remain only the diagonal components of $T^{\alpha\beta}$. Moreover, the quasiconservation of the leptonic current (the lepton masses being small), implies the disappearance of the component $\alpha = T$. The space-like components corresponding to $T^{\alpha\alpha}$ with $\alpha = S$, R, L are denoted by $|f^S|^2$, $|f^R|^2$ and $|f^L|^2$, and one finds

$$\Gamma = \frac{1}{(2\pi)^4} G_F^2 \frac{m_B}{3} \int \frac{d\boldsymbol{q}}{E_B} q^2(|f^S|^2 + |f^R|^2 + |f^L|^2). \tag{2.71}$$

This is the formula for a decay width $A \to B\ell^-\bar{\nu}_\ell$, where $\ell^- = e^-, \mu^-, \ldots$, and A and B are hadrons, as in $n \to pe^-\bar{\nu}_e$, $K^- \to \pi^0\mu^-\bar{\nu}_\mu$, ... If instead of having lepton–antineutrino emission we have antilepton–neutrino emission as in $\Sigma^+ \to \Lambda e^+\nu_e$, or $K^+ \to \pi^0\mu^+\nu_\mu$ then we get the same formula (2.71), but the leptonic tensor (2.64) has a minus sign in front of $\varepsilon^{\mu\nu\sigma\tau}$.

2.1.3 Quark–Antiquark Annihilation into Lepton Pairs

We are now considering a category of processes that are based on the same current interaction but with it involved in different combinations. Instead of having one quark emitting or absorbing a photon or a lepton pair (Figures 2.1 and 2.2) with the other quarks of the same hadron remaining spectators, we consider now that the q and $\bar{q}$ of a meson interact with each other and annihilate, producing a lepton pair (Figures 2.3 and 2.4). They will interact through exchange of photons or by the weak four-fermion interaction (2.46).

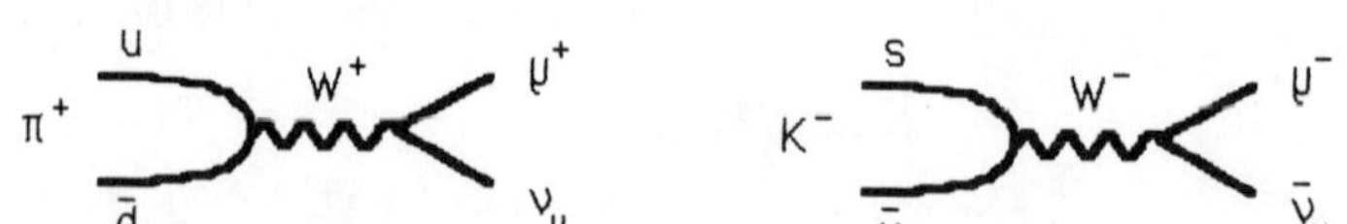

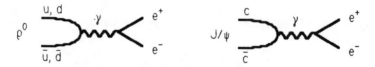

Figure 2.3 Examples of electromagnetic annihilation of mesons into lepton pairs ($\rho^0 \to e^+e^-$, $J/\psi \to e^+e^-$).

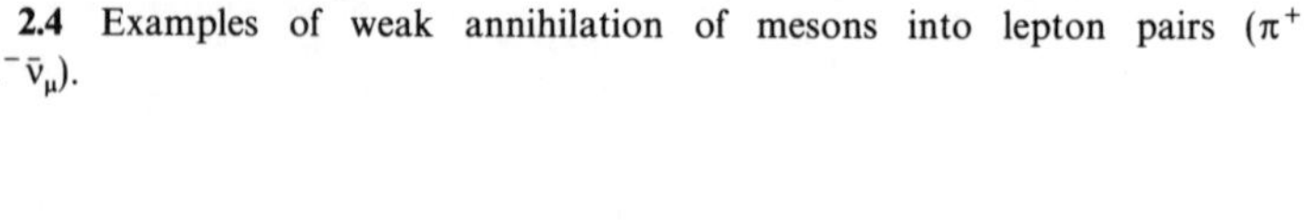

Figure 2.4 Examples of weak annihilation of mesons into lepton pairs ($\pi^+ \to \mu^+\nu_\mu$, $K^- \to \mu^-\bar{\nu}_\mu$).

A distinctive feature is to be found in this case as well as in the process of Section 2.1.4. The amplitude of emission or absorption processes depends essentially on the overlap of initial and final hadrons. It can be argued that such an overlap depends only on the intermediate-distance region of the spatial wave functions, which could be called *peripheral*, extending from around 0.5 fm. This intermediate region can be described by potentials that are similar in this region but that could have a very different behavior in the center and outer regions. This is why the harmonic-oscillator model, although certainly unrealistic at both ends, can be expected to reproduce emission and absorption processes rather well. In contrast, in the processes with which we are now concerned, the amplitude depends essentially on the values of the wave function for short interquark distances. The interaction can only occur when the two quarks are at the same point, the amplitude being proportional to the wave function at the center, $\psi(0)$. These values depend on the *central region* of the potential. There is the hope that this region can be easily described by QCD owing to asymptotic freedom, and in fact Coulomb exchange corrected by asymptotic-freedom effects improves the results for $|\psi(0)|^2$ of heavy quarkonia. The wave function at short interquark distances can also be determined indirectly from spectroscopic splittings induced by the electromagnetic Fermi contact force proportional to $\boldsymbol{\sigma}(1)\cdot\boldsymbol{\sigma}(2)\delta(\boldsymbol{r}_1-\boldsymbol{r}_2)$.

In processes such as $\rho^0\to e^+e^-$, $\pi\to\mu\nu$, one encounters then a new type of hadronic matrix element. As in the previous processes, these can be reduced to matrix elements of densities multiplied by momentum-conservation δ-functions. Supposing that the necessary steps have been performed, one is faced with expressions such as

$$(2\pi)^{3/2}\langle 0|j^\mu(\mathbf{0})|\mathrm{C}\rangle, \tag{2.72}$$

where $j^\mu(\mathbf{0})$ is any vector or axial current, C is the decaying meson and $(2\pi)^{3/2}$ has been introduced for convenience as it cancels a factor coming from the calculation of the matrix element.

For definiteness, let $j^\mu(\mathbf{0})$ be the electromagnetic current (for example in $\rho^0\to e^+e^-$):

$$j^\mu_{\mathrm{em}}=\bar{q}Q\gamma^\mu q.$$

One has to consider here the annihilating contribution of $j^\mu(\mathbf{0})$, and the recourse to the first-quantized operators is now not so direct as in the preceding sections. We have to make things very explicit for this new situation. Let us work in momentum space. We write the state C according to

$$\begin{aligned}|C(\boldsymbol{p}_{\mathrm{C}})\rangle = \sum\frac{\delta_{\alpha_1\alpha_2}}{\sqrt{3}}\varphi_{i_1i_2}\chi_{s_1s_2}\int \mathrm{d}\boldsymbol{p}_1\mathrm{d}\boldsymbol{p}_2\,\psi(\boldsymbol{p}_1-\tfrac{1}{2}\boldsymbol{p}_{\mathrm{C}},\boldsymbol{p}_2-\tfrac{1}{2}\boldsymbol{p}_{\mathrm{C}})\\ \times\,\delta(\boldsymbol{p}_1+\boldsymbol{p}_2-\boldsymbol{p}_{\mathrm{C}})b^\dagger_{\alpha_1i_1s_1\boldsymbol{p}_1}d^\dagger_{\alpha_2i_2s_2\boldsymbol{p}_2}|0\rangle,\end{aligned} \tag{2.73}$$

where we denote quark and antiquark by 1 and 2 respectively, and the sum extends over the repeated indices as usual; α_1, α_2 are color indices, and φ, χ and ψ are the flavor, spin and spatial wave functions.

On the other hand, Q being the matrix (2.12), the relevant part of the current is

$$j^{\mu}_{\mathrm{em}}(\mathbf{0}) = \frac{1}{(2\pi)^3} \sum \delta_{\alpha_1\alpha_2} Q_{i_2 i_1} \int \mathrm{d}\boldsymbol{p}_1 \left(\frac{m}{E_1}\right)^{1/2} \mathrm{d}\boldsymbol{p}_2 \left(\frac{m}{E_2}\right)^{1/2} \times \bar{v}_{s_2}(\boldsymbol{p}_2)\gamma^{\mu} u_{s_1}(\boldsymbol{p}_1) d_{\alpha_2 i_2 s_2 \boldsymbol{p}_2} b_{\alpha_1 i_1 s_1 \boldsymbol{p}_1} + \dots \tag{2.74}$$

In this case there is only one possible way of contracting operators, and we find,

$$(2\pi)^3 \langle 0 | j^{\mu}_{\mathrm{em}}(\mathbf{0}) | C \rangle = \sqrt{3} \sum_{s_1 s_2 i_1 i_2} \int \mathrm{d}\boldsymbol{p}_1 \left(\frac{m}{E_1}\right)^{1/2} \mathrm{d}\boldsymbol{p}_2 \left(\frac{m}{E_2}\right)^{1/2} Q_{i_2 i_1} \bar{v}_{s_2}(\boldsymbol{p}_2)\gamma^{\mu} u_{s_1}(\boldsymbol{p}_1) \times \chi_{i_1 i_2} \varphi_{s_1 s_2} \psi(\boldsymbol{p}_1 - \tfrac{1}{2}\boldsymbol{p}_{\mathrm{C}}, \boldsymbol{p}_2 - \tfrac{1}{2}\boldsymbol{p}_{\mathrm{C}})\, \delta(\boldsymbol{p}_1 + \boldsymbol{p}_2 - \boldsymbol{p}_{\mathrm{C}}). \tag{2.75}$$

Note the important factor $\sqrt{3}$ coming from color. In the $q\bar{q}$ annihilation we no longer have a scalar product of two identical color wave functions, as in the previous processes.

It now remains to make the usual nonrelativistic expansion of the Dirac spinors. But to proceed, we have to make some warnings concerning the choice of antiquark spinors. One has to use charge-conjugate spinors for antiquarks, as defined for example in Bjorken and Drell (1964):

$$v_s = \mathrm{i}\gamma^2 u^*{}_s, \qquad v_s^{\dagger} = \mathrm{i}\tilde{u}_s \gamma^2. \tag{2.76}$$

This corresponds to the conventions in (1.66). However, there is another natural way of treating antiparticles, taking

$$v_s = \gamma_5 u_s. \tag{2.77}$$

In this latter choice, the antiparticle spin states behave according to the complex conjugate of the representation of the particle under the symmetry group SO(3), because the v spinors are the coefficients of the antiparticle creation operators $d^{\dagger}$, while the u spinors are the coefficients of the particle annihilation operators b, and because γ_5 commutes with the rotation generators. For SO(3), complex-conjugate representations are equivalent, and the convention (2.76) is the choice of a standard basis in the complex-conjugate representation. However, this possibility does not exist for other symmetry groups like SU(3), and therefore (2.77) gives a more homogeneous convention.

This latter convention gives easier calculations when one has to deal with Dirac bilinears and we want to reduce them to a nonrelativistic form. Through (2.77), the vector bilinear between $q\bar{q}$ is reduced to the axial-vector one between quark states, whose nonrelativistic reduction has already been considered. Moreover, the spin wave functions are much simpler.

If we adopt this convention, the spin scalar wave function is written

$$\chi_{s_1 s_2} = \frac{\delta_{s_1 s_2}}{\sqrt{2}} \tag{2.78}$$

while the spin-vector wave function, with polarization vector $\boldsymbol{\varepsilon}$, is

$$\chi^m_{s_1 s_2} = \frac{(\sigma^m)_{s_1 s_2}}{\sqrt{2}} = \frac{(\boldsymbol{\sigma}\cdot\boldsymbol{\varepsilon}^m)_{s_1 s_2}}{\sqrt{2}} \tag{2.79}$$

Applying (2.77) to the vector current

$$\bar{v}_{s_2}\gamma_\mu u_{s_1} = \bar{u}_{s_2}\gamma_\mu\gamma_5 u_{s_1}, \tag{2.80}$$

we obtain

$$\bar{v}_{s_2}\gamma_0 u_{s_1} \approx \chi^\dagger_{s_2}\frac{\boldsymbol{\sigma}\cdot(\boldsymbol{p}_1+\boldsymbol{p}_2)}{2m}\chi_{s_1}, \tag{2.81}$$

$$\bar{v}_{s_2}\boldsymbol{\gamma} u_{s_1} \approx \chi^\dagger_{s_2}\boldsymbol{\sigma}\chi_{s_1} \tag{2.82}$$

and, on the other hand, $\left(\frac{m}{E}\right)^{1/2} = 1$ in the nonrelativistic approximation.

Finally, one can write (2.75), when C is a vector meson of polarization $\boldsymbol{\varepsilon}$, as

$$\begin{aligned}(2\pi)^{3/2}\langle 0|j^0_{\rm em}(\mathbf{0})|{\rm C}\rangle &= \sqrt{6}\,{\rm Tr}\,(Q\varphi)\frac{\boldsymbol{p}_{\rm C}\cdot\boldsymbol{\varepsilon}}{2m} \\ &\times \int {\rm d}\boldsymbol{p}_1\,{\rm d}\boldsymbol{p}_2\,\psi(\boldsymbol{p}_1-\tfrac{1}{2}\boldsymbol{p}_{\rm C},\boldsymbol{p}_2-\tfrac{1}{2}\boldsymbol{p}_{\rm C})\,\delta(\boldsymbol{p}_1+\boldsymbol{p}_2-\boldsymbol{p}_{\rm C}),\end{aligned} \tag{2.83}$$

$$\begin{aligned}(2\pi)^{3/2}\langle 0|\boldsymbol{j}_{\rm em}(\mathbf{0})|{\rm C}\rangle &= \sqrt{6}\,{\rm Tr}\,(Q\varphi)\boldsymbol{\varepsilon} \\ &\times \int {\rm d}\boldsymbol{p}_1\,{\rm d}\boldsymbol{p}_2\,\psi(\boldsymbol{p}_1-\tfrac{1}{2}\boldsymbol{p}_{\rm C},\boldsymbol{p}_2-\tfrac{1}{2}\boldsymbol{p}_{\rm C})\,\delta(\boldsymbol{p}_1+\boldsymbol{p}_2-\boldsymbol{p}_{\rm C}).\end{aligned} \tag{2.84}$$

The integral represents the wave function at the center:

$$\int {\rm d}\boldsymbol{p}_1\,{\rm d}\boldsymbol{p}_2\,\psi(\boldsymbol{p}_1-\tfrac{1}{2}\boldsymbol{p}_{\rm C},\boldsymbol{p}_2-\tfrac{1}{2}\boldsymbol{p}_{\rm C})\,\delta(\boldsymbol{p}_1+\boldsymbol{p}_2-\boldsymbol{p}_{\rm C}) = (2\pi)^{3/2}\,\psi(\boldsymbol{r}_1=0,\boldsymbol{r}_2=0). \tag{2.85}$$

This relation comes from the definition of the Fourier transform between *total* wave functions,

$$\Psi^{\rm tot}(\boldsymbol{r}_1,\boldsymbol{r}_2) = \frac{1}{(2\pi)^3}\int {\rm d}\boldsymbol{p}_1\,{\rm d}\boldsymbol{p}_2 \exp({\rm i}\sum_i \boldsymbol{p}_i\boldsymbol{r}_i)\,\Psi^{\rm tot}(\boldsymbol{p}_1,\boldsymbol{p}_2), \tag{2.86}$$

and from the definitions of the internal wave functions in terms of total wave functions,

$$\Psi_P^{\rm tot}(\boldsymbol{r}_1,\boldsymbol{r}_2)=\frac{1}{(2\pi)^{3/2}}\exp\left(\tfrac{1}{2}\mathrm{i}\boldsymbol{P}\cdot\sum_i \boldsymbol{r}_i\right)\psi(\boldsymbol{r}_1,\boldsymbol{r}_2), \tag{2.87}$$

$$\Psi_P^{\rm tot}(\boldsymbol{p}_1,\boldsymbol{p}_2)=\delta(\boldsymbol{p}_1+\boldsymbol{p}_2-\boldsymbol{P})\,\psi(\boldsymbol{p}_1,\boldsymbol{p}_2), \tag{2.88}$$

taken at $\boldsymbol{r}_1=\boldsymbol{r}_2=0$ and $\boldsymbol{P}=0$.

The results (2.83) and (2.84) then translate into

$$\begin{aligned}(2\pi)^{3/2}\langle 0|j^0_{\rm em}(\boldsymbol{0})\,|\mathrm{C}\rangle&=\sqrt{6}\,\mathrm{Tr}\,(Q\varphi)\frac{\boldsymbol{p}_{\rm C}\cdot\boldsymbol{\varepsilon}}{2m}\,\psi(\boldsymbol{r}_1=0,\boldsymbol{r}_2=0),\\ (2\pi)^{3/2}\langle 0|\boldsymbol{j}_{\rm em}(\boldsymbol{0})\,|\mathrm{C}\rangle&=\sqrt{6}\,\mathrm{Tr}\,(Q\varphi)\,\boldsymbol{\varepsilon}\,\psi(\boldsymbol{r}_1=0,\boldsymbol{r}_2=0).\end{aligned} \tag{2.89}$$

The matrix element of pion decay $\pi^+\to\mu^+\nu_\mu$ is quite similar, though now the axial-vector current is involved:

$$\bar{v}_{s_2}\gamma_\mu\gamma_5 u_{s_1}=\bar{u}_{s_2}\gamma_\mu u_{s_1}, \tag{2.91}$$

$$\bar{u}_{s_2}\gamma_0 u_{s_1}\approx\chi^\dagger_{s_2}\chi_{s_1}, \tag{2.92}$$

$$\bar{u}_{s_2}\boldsymbol{\gamma} u_{s_1}\approx\chi^\dagger_{s_2}\frac{\boldsymbol{p}_1+\boldsymbol{p}_2}{2m}\chi_{s_1} \tag{2.93}$$

and we obtain

$$(2\pi)^{3/2}\langle 0|j^0_5(\boldsymbol{0})\,|\mathrm{C}\rangle=\sqrt{6}\,\mathrm{Tr}\,(F\varphi)\,\psi(\boldsymbol{r}_1=0,\boldsymbol{r}_2=0), \tag{2.94}$$

$$(2\pi)^{3/2}\langle 0|\boldsymbol{j}_5(\boldsymbol{0})\,|\mathrm{C}\rangle=\sqrt{6}\,\mathrm{Tr}\,(F\varphi)\frac{\boldsymbol{p}_{\rm C}}{2m}\psi(\boldsymbol{r}_1=0,\boldsymbol{r}_2=0), \tag{2.95}$$

where F denotes the flavor matrix (2.50).

A word must be said about the normalization of the spatial wave function and about the relation with other definitions in order to understand the formulae encountered in the literature.

Our spatial wave functions are normalized just as in (2.39), which reads for mesons,

$$\int \mathrm{d}\boldsymbol{r}_1\,\mathrm{d}\boldsymbol{r}_2\,\delta\left(\tfrac{1}{2}\sum_i \boldsymbol{r}_i\right)\psi(\boldsymbol{r}_1,\boldsymbol{r}_2)^*\,\psi(\boldsymbol{r}_1,\boldsymbol{r}_2)=1. \tag{2.96}$$

If we introduce the relative coordinate

$$\boldsymbol{r}=\boldsymbol{r}_1-\boldsymbol{r}_2 \tag{2.97}$$

along with

$$\boldsymbol{R}=\tfrac{1}{2}(\boldsymbol{r}_1+\boldsymbol{r}_2), \tag{2.98}$$

then we find a Jacobian equal to unity,

$$\mathrm{d}\boldsymbol{r}_1\,\mathrm{d}\boldsymbol{r}_2=\mathrm{d}\boldsymbol{R}\,\mathrm{d}\boldsymbol{r}, \tag{2.99}$$

whence

$$\int d\boldsymbol{r}\, |\psi(\boldsymbol{r}_1 = \tfrac{1}{2}\boldsymbol{r}, \boldsymbol{r}_2 = -\tfrac{1}{2}\boldsymbol{r})|^2 = 1. \tag{2.100}$$

Therefore, if we call

$$\psi_{\mathrm{Rel}}(\boldsymbol{r}) = \psi(\boldsymbol{r}_1 = \tfrac{1}{2}\boldsymbol{r}, \boldsymbol{r}_2 = -\tfrac{1}{2}\boldsymbol{r}), \tag{2.101}$$

this wave function ψ_{Rel} is normalized to unity, and we can write

$$\psi(\boldsymbol{r}_1 = 0, \boldsymbol{r}_2 = 0) = \psi_{\mathrm{Rel}}(0) = \psi(0), \tag{2.102}$$

where $\psi(0)$ is the central value of the meson wave function, found usually in the literature, normalized to unity,

$$\int |\psi(\boldsymbol{r})|^2\, d\boldsymbol{r} = 1. \tag{2.103}$$

It is now easy to calculate a decay width. Take for instance the weak leptonic decay $\pi^- \to \mu^- \bar{\nu}_\mu$. The relevant Hamiltonian is given by (2.46), and the matrix element M is, by a straightforward analogy with (2.58), in the π rest frame,

$$M = \frac{1}{(2\pi)^3} \frac{1}{(2p^0_\mu)^{1/2}} \frac{1}{(2p^0_\nu)^{1/2}} \frac{G_F}{\sqrt{2}} \bar{u}_\mu \gamma^\lambda (1-\gamma_5) v_\nu (2\pi)^{3/2} \langle 0 | J_\lambda(\mathbf{0}) | C \rangle \tag{2.104}$$

$$= \frac{1}{(2\pi)^{3/2}} \frac{1}{(2p^0_\mu)^{1/2}} \frac{1}{(2p^0_\nu)^{1/2}} \frac{G_F}{\sqrt{2}} u^\dagger_\mu (1-\gamma_5) v_\nu \sqrt{6}\, \mathrm{Tr}\,(F\varphi)\, \psi(0). \tag{2.105}$$

The width here corresponds to a two-body phase space: the corresponding treatment has been given in (2.28)–(2.31b). Since the neutrino has zero mass, $p^0_\nu = k$,

$$\begin{aligned} \Gamma &= 2\pi \times 4\pi \frac{p^0_\mu}{m_\pi} k^2 |M|^2 \\ &= \frac{1}{\pi} \frac{k}{m_\pi} \frac{1}{4} \left(\frac{G_F}{\sqrt{2}} \right)^2 [\sqrt{6}\, \mathrm{Tr}\,(F\varphi)\, \psi(0)]^2 |u^\dagger_\mu (1-\gamma_5) v_\nu|^2. \end{aligned} \tag{2.106}$$

It only remains to evaluate $|u^\dagger_\mu(1-\gamma_5)v_\nu|^2$. This can be done directly with explicit spinors or by following the usual procedure: putting the modulus squared into the form of a trace and summing over polarizations (although there is actually only one polarization state for the μ and the ν owing to the projector $\frac{1}{2}(1-\gamma_5)$). One finds

$$|u^\dagger_\mu (1-\gamma_5) v_\nu|^2 = 8k \frac{m^2_\mu}{m_\pi}. \tag{2.107}$$

Remember that the leptonic spinors have been normalized to

$$u^\dagger u = v^\dagger v = 2p^0 \tag{2.108}$$

Finally, using $k=(m_\pi^2-m_\mu^2)/2m_\pi$ in the rest frame of the π, and $\mathrm{Tr}(F\varphi_\pi)=\cos\theta_C$,

$$\Gamma=\frac{1}{\pi}\left(\frac{G_F}{\sqrt{2}}\right)^2 m_\mu^2\frac{(m_\pi^2-m_\mu^2)^2}{m_\pi^4}3\cos^2\theta_C|\psi(0)|^2. \tag{2.109}$$

Similar formulae can be found for *Cabbibo-suppressed* (proportional to $\sin\theta_C$, see (2.49)) processes like $K^+\to\mu^+\nu_\mu$.

We have shown here how to calculate decays of the type $\pi\to\mu\nu$, $\rho\to e^+e^-$. The basic process underlying these decays is an annihilation, $q\bar{q}\to\mu\nu$ or $q\bar{q}\to e^+e^-$, directly induced by the electroweak Hamiltonian at first order. One encounters other decays also corresponding to a $q\bar{q}$ annihilation, but involving more complicated interactions. For instance, the electromagnetic interaction may be involved at higher orders, as in $q\bar{q}\to\gamma\gamma$ or $q\bar{q}\to\gamma\gamma\gamma$. This is the case of $\pi^0\to 2\gamma$, the analogue of parapositronium decay (Figure 2.5). This

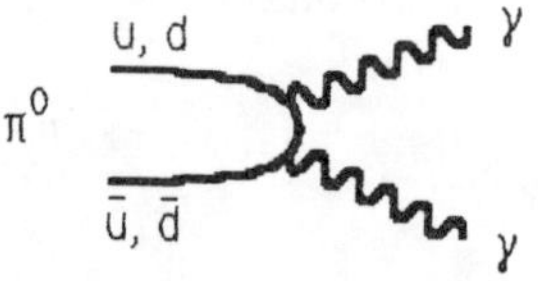

Figure 2.5 The decay $\pi^0\to 2\gamma$, the quark analogue of the parapositronium decay ${}^1S_0\to 2\gamma$.

particular decay, and also $\eta\to 2\gamma$, are related to an important phenomenon of quantum field theory, the *chiral anomaly*, and will be treated in detail in a fully relativistic way in Section 3.4.4.

2.1.4 Nonleptonic Weak Matrix Elements

The analysis of weak nonleptonic decays like $\Lambda\to N\pi$ and $K\to 2\pi$ through Current Algebra and pole diagrams—which will be described later, in Section 4.6—allows us to relate the whole amplitude to the matrix element of the current–current Hamiltonian between two hadron states, for instance hyperon and nucleon.

We do not want to be involved at this stage in a phenomenological discussion. We are aware that there is no consensus at present and no general satisfactory model describing hyperon and kaon decays together. We only address at present the question of calculating quantities of the type (Figure 2.6)

$$\langle B|H_W|A\rangle, \tag{2.110}$$

Figure 2.6 The matrix element $\langle n | H_W^{pc} | \Lambda \rangle$ in the limit of large W-boson mass.

where A and B are baryons and where H_W is the purely hadronic counterpart of the semileptonic hamiltonian (2.46):

$$H_W = \frac{G_F}{\sqrt{2}} \int dx \frac{1}{2} \left\{ J^{(-)}_{\text{had}\mu}(x), J^{(+)\mu}_{\text{had}}(x) \right\}, \tag{2.111}$$

where now the charge-lowering and charge-raising hadronic currents are coupled together.

A very important feature of this interaction is that it is of *contact* character: two quarks interact when they have the same position. This feature is similar to what we have commented on for meson annihilation in Section 2.1.3. Once more, we shall have to know the wave function at short interquark distance, and this requires a better knowledge of the wave function than in the peripheral processes of Sections 2.1.1 and 2.1.2. Once more, the wave function at such distances can be related to the spectroscopic splittings induced by the electromagnetic Fermi contact force, this time for baryons (Le Yaouanc *et al.*, 1977b).

What is new here is that we are faced with a product of four quark operators. We want to reduce the matrix element to an expression involving only wave functions and first-quantized operators. We therefore have to consider the expressions for states $|A\rangle$ and $|B\rangle$ and for H_W in terms of creation and annihilation operators. For baryons,

$$|A\rangle = \frac{1}{\sqrt{6}} \sum \Psi_{Ai_1i_2i_3} b^\dagger_{i_1} b^\dagger_{i_2} b^\dagger_{i_3} |0\rangle,$$
$$|B\rangle = \frac{1}{\sqrt{6}} \sum \Psi_{Bi_1i_2i_3} b^\dagger_{i_1} b^\dagger_{i_2} b^\dagger_{i_3} |0\rangle, \tag{2.112}$$

where the $\Psi_{A,B}$ are normalized as before. We recall that the indices represent every possible label: color, spin, flavor and momentum.

As to H_W, we can write, according to parity behavior,

$$H_W = H_W^{pc} + H_W^{pv}, \tag{2.113}$$

where

$$H_{\mathrm{W}}^{\mathrm{pc}} = \frac{G_{\mathrm{F}}}{\sqrt{2}} \int \mathrm{d}\boldsymbol{x}\, [\mathrm{j}_{\mu}^{(-)}(\boldsymbol{x})\mathrm{j}^{(+)\mu}(\boldsymbol{x}) + \mathrm{j}_{5\mu}^{(-)}(\boldsymbol{x})\mathrm{j}_{5}^{(+)\mu}(\boldsymbol{x})], \tag{2.114}$$

$$H_{\mathrm{W}}^{\mathrm{pv}} = \frac{G_{\mathrm{F}}}{\sqrt{2}} \int \mathrm{d}\boldsymbol{x}\, [\mathrm{j}_{\mu}^{(-)}(\boldsymbol{x})\mathrm{j}_{5}^{(+)\mu}(\boldsymbol{x}) + \mathrm{j}_{5\mu}^{(-)}(\boldsymbol{x})\mathrm{j}^{(+)\mu}(\boldsymbol{x})], \tag{2.115}$$

where $H_{\mathrm{W}}^{\mathrm{pc}}$ ($\mathrm{H}_{\mathrm{W}}^{\mathrm{pv}}$) correspond to the parity-conserving (parity-violating) parts.

Let us deal for the moment only with $H_{\mathrm{W}}^{\mathrm{pc}}$—this is the only operative part for matrix elements between two states of the same parity like the ground-state baryons $\frac{1}{2}^+$, as in $\langle \mathrm{p}|H_{\mathrm{W}}^{\mathrm{pc}}|\Sigma^+\rangle$. We can use the nonrelativistic expressions obtained for the currents in the preceding sections. If we retain the lowest v/c order, we have to neglect $\boldsymbol{j}$(space) and j_5^0. It would not be consistent to retain $\boldsymbol{j}\cdot\boldsymbol{j}$ or $j_5^0 j_5^0$ terms, which are of order $(v/c)^2$, without taking into account at the same time the second-order corrections to the main components $\boldsymbol{j}_5$ and j^0. We therefore have

$$H_{\mathrm{W}}^{\mathrm{pc}} \approx \frac{G_{\mathrm{F}}}{\sqrt{2}} \int \mathrm{d}\boldsymbol{x}\, [\mathrm{j}^{(-)0}(\boldsymbol{x})\mathrm{j}^{(+)0}(\boldsymbol{x}) - \boldsymbol{j}_5^{(-)}(\boldsymbol{x}) \cdot \boldsymbol{j}_5^{(+)}(\boldsymbol{x})], \tag{2.116}$$

with

$$j^0 = \cos\theta_{\mathrm{C}}\, u_1^\dagger d_1 + \sin\theta_{\mathrm{C}}\, u_1^\dagger s_1, \tag{2.117}$$

$$\boldsymbol{j}_5 = \cos\theta_{\mathrm{C}}\, u_1^\dagger \boldsymbol{\sigma} d_1 + \sin\theta_{\mathrm{C}}\, u_1^\dagger \boldsymbol{\sigma} s_1; \tag{2.118}$$

the index 1 denotes, as previously, the large components of the quark-field operators.

The operators (2.116) contain four quark fields and therefore we have to carefully consider the action of these fields on the hadronic states. In principle, the calculation of a matrix element like (2.110) always amounts to writing the states $|\mathrm{A}\rangle$ and $|\mathrm{B}\rangle$ in the form (2.112). At the same time, the fields in the interaction Hamiltonian are expanded into creation and annihilation operators according to (2.3b). The calculation then consists in making all the possible contractions between creation and annihilation operators. The momentum-space expression of the matrix element is then obtained in terms of ordinary wave functions. It is not desirable at this stage to specialize the expression so much. However the counting of contractions allows determination of the number of similar terms appearing in the first-quantized formalism. In the case of the matrix elements of one current the answer is intuitively obvious: one has (for baryons) three identical terms corresponding to the action of the current on each of the three quarks. This is equivalent to the notion of *additivity*, which states that one has to sum the first-quantized quark current operators of the three quarks. Here the counting of contractions is a more secure method than intuition. One finds 36 possible ways of making the contractions that lead to a similar term. These 36 terms correspond to *two-quark interactions.* We discard other possible ways of contracting, which are not physically significant in a transition. These correspond either to contributions to the *vacuum energy* or correspond to a divergent *one-quark self-energy.* On the whole, the retained terms imply that the Hamiltonian (2.111) has already been put into the *normal-ordered* form. On the other hand, the expression (2.112) for the states $|A\rangle$ and $|B\rangle$ in terms of the usual wave functions introduces a factor $(1/\sqrt{6})^2=\frac{1}{6}$. One finds finally, by reduction to a typical term representing one of the possible ways of making the contractions,

$$\langle B|H_W^{pc}|A\rangle = 6\langle \Psi_B^{tot}(1,2,3)|\, H_W^{pc}(1,2)\,|\Psi_A^{tot}(1,2,3)\rangle, \tag{2.119}$$

where

$$H_W^{pc}(1,2)\approx\frac{G_F}{\sqrt{2}}\int dx\,[j^{(-)0}(x)(1)j^{(+)0}(x)(2)-\boldsymbol{j}_5^{(-)}(x)(1)\cdot\boldsymbol{j}_5^{(+)}(x)(2)], \tag{2.120}$$

the currents now being the *first-quantized operators*, and the second label (1) or (2) indicating the quark on which they act. More explicitly, we have in configuration space

$$j^{(\pm)0}(x)=[\cos\theta_C\,\tau^{(\pm)}+\sin\theta_C v^{(\pm)}]\,\delta(\boldsymbol{x}-\boldsymbol{r}), \tag{2.121}$$

$$\boldsymbol{j}_5^{(\pm)}(x)=[\cos\theta_C\,\tau^{(\pm)}+\sin\theta_C v^{(\pm)}]\,\boldsymbol{\sigma}\,\delta(\boldsymbol{x}-\boldsymbol{r}), \tag{2.122}$$

whence, retaining only the Cabibbo-suppressed part for illustration,

$$H_W^{pc}(1,2)=\frac{G_F}{\sqrt{2}}\cos\theta_C\sin\theta_C\,[\tau^{(-)}(1)v^{(+)}(2)+v^{(-)}(1)\tau^{(+)}(2)]\,(1-\boldsymbol{\sigma}(1)\cdot\boldsymbol{\sigma}(2))$$
$$\times\int d\boldsymbol{x}\,\delta(\boldsymbol{x}-\boldsymbol{r}_1)\,\delta(\boldsymbol{x}-\boldsymbol{r}_2), \tag{2.123}$$

$$H_W^{pc}(1,2)=\frac{G_F}{\sqrt{2}}\cos\theta_C\sin\theta_C\,[\tau^{(-)}(1)v^{(+)}(2)+v^{(-)}(1)\tau^{(+)}(2)](1-\boldsymbol{\sigma}(1)\cdot\boldsymbol{\sigma}(2))$$
$$\times\,\delta(\boldsymbol{r}_1-\boldsymbol{r}_2). \tag{2.124}$$

In the matrix element of $\delta(\boldsymbol{r}_1-\boldsymbol{r}_2)$ between total spatial wave functions, the center-of-mass motion can be treated straightforwardly:

$$\langle \Psi_B^{tot}\,|\,\delta(\boldsymbol{r}_1-\boldsymbol{r}_2)\,|\,\Psi_A^{tot}\rangle$$
$$=\frac{1}{(2\pi)^3}\int\prod_i d\boldsymbol{r}_i \exp\left[\,i(\boldsymbol{P}_A-\boldsymbol{P}_B)\cdot\left(\tfrac{1}{3}\sum_i \boldsymbol{r}_i\right)\right]\psi_B^*\,\delta(\boldsymbol{r}_1-\boldsymbol{r}_2)\psi_A. \tag{2.125}$$

We can now consider $\boldsymbol{R}$ as an independent variable by introducing a $\delta(\boldsymbol{R}-\frac{1}{3}\Sigma\boldsymbol{r}_i)$:

$$\int\prod_i \mathrm{d}\boldsymbol{r}_i \exp\left[\mathrm{i}(\boldsymbol{P}_{\mathrm{A}}-\boldsymbol{P}_{\mathrm{B}})\cdot\left(\frac{1}{3}\sum_i \boldsymbol{r}_i\right)\right]\psi_{\mathrm{B}}^* \delta(\boldsymbol{r}_1-\boldsymbol{r}_2)\psi_{\mathrm{A}}$$
$$=\int \mathrm{d}\boldsymbol{R}\,\mathrm{e}^{\mathrm{i}(\boldsymbol{P}_{\mathrm{A}}-\boldsymbol{P}_{\mathrm{B}})\cdot\boldsymbol{R}}\int\prod_i \mathrm{d}\boldsymbol{r}_i\,\delta\left(\boldsymbol{R}-\frac{1}{3}\sum_i \boldsymbol{r}_i\right)\psi_{\mathrm{B}}^*\delta(\boldsymbol{r}_1-\boldsymbol{r}_2)\psi_{\mathrm{A}}. \qquad (2.126)$$

But the translational invariance of the internal wave functions ψ_{A} and ψ_{B} can now be used to eliminate the $\boldsymbol{R}$ dependence of the integral, making the change of variables $\boldsymbol{r}_i \rightarrow \boldsymbol{r}_i+\frac{1}{3}\boldsymbol{R}$:

$$\int\prod_i \mathrm{d}\boldsymbol{r}_i\,\delta\left(\boldsymbol{R}-\frac{1}{3}\sum_i \boldsymbol{r}_i\right)\psi_{\mathrm{B}}^*\delta(\boldsymbol{r}_1-\boldsymbol{r}_2)\psi_{\mathrm{A}}=\int\prod_i \mathrm{d}\boldsymbol{r}_i\,\delta\left(\frac{1}{3}\sum_i \boldsymbol{r}_i\right)\psi_{\mathrm{B}}^*\delta(\boldsymbol{r}_1-\boldsymbol{r}_2)\psi_{\mathrm{A}}, \qquad (2.127)$$

while the integral over $\boldsymbol{R}$ yields the desired momentum-conservation factor $(2\pi)^3\delta(\boldsymbol{P}_{\mathrm{A}}-\boldsymbol{P}_{\mathrm{B}})$:

$$\langle \Psi_{\mathrm{B}}^{\mathrm{tot}}|\,\delta(\boldsymbol{r}_1-\boldsymbol{r}_2)|\Psi_{\mathrm{A}}^{\mathrm{tot}}\rangle=\delta(\boldsymbol{P}_{\mathrm{B}}-\boldsymbol{P}_{\mathrm{A}})\int\prod_i \mathrm{d}\boldsymbol{r}_i\,\delta\left(\frac{1}{3}\sum_i \boldsymbol{r}_i\right)\psi_{\mathrm{B}}^*\delta(\boldsymbol{r}_1-\boldsymbol{r}_2)\psi_{\mathrm{A}} \qquad (2.128)$$

As to the SU(3) operators in (2.125), only one of the two terms will be operative, according to whether $\Delta S=+1$ or -1.

Finally, Y being for example Σ or Λ, we obtain (Le Yaouanc *et al.*, 1977b; Schmid, 1977)

$$\langle \mathrm{N}|H_{\mathrm{W}}^{\mathrm{pc}}|\mathrm{Y}\rangle=\delta(\boldsymbol{P}_{\mathrm{f}}-\boldsymbol{P}_{\mathrm{i}})\frac{G_{\mathrm{F}}}{\sqrt{2}}\cos\theta_{\mathrm{C}}\sin\theta_{\mathrm{C}}$$
$$\times\, 6\langle\psi_{\mathrm{N}}|\,\tau^{(-)}(1)v^{(+)}(2)(1-\boldsymbol{\sigma}(1)\cdot\boldsymbol{\sigma}(2))\,\delta(\boldsymbol{r}_1-\boldsymbol{r}_2)\,|\psi_{\mathrm{Y}}\rangle. \qquad (2.129)$$

The first nonvanishing terms in the nonrelativistic expansion of $H_{\mathrm{W}}^{\mathrm{pv}}$ are of order v/c, thus involving momentum operators. The calculation of the matrix elements of $H_{\mathrm{W}}^{\mathrm{pv}}$ is then more easily done in the momentum representation (Le Yaouanc *et al.*, 1979). The extraction of the center-of-mass motion can be made along similar lines to that used for elementary emission models. We display the treatment of one typical operator matrix element that appears in the product $j^0(1)j_5^0(2)$:

$$\int \mathrm{d}\boldsymbol{x}\,\langle \mathrm{B}|\,\delta(\boldsymbol{x}-\boldsymbol{r}_1)\delta(\boldsymbol{x}-\boldsymbol{r}_2)\boldsymbol{p}_2+\delta(\boldsymbol{x}-\boldsymbol{r}_1)\boldsymbol{p}_2\,\delta(\boldsymbol{x}-\boldsymbol{r}_2)\,|\mathrm{A}\rangle$$
$$=\langle \mathrm{B}|\,\delta(\boldsymbol{r}_1-\boldsymbol{r}_2)\boldsymbol{p}_2+\boldsymbol{p}_2\delta(\boldsymbol{r}_1-\boldsymbol{r}_2)\,|\mathrm{A}\rangle$$
$$=\frac{1}{(2\pi)^3}\int\prod_{i=1}^{2}\mathrm{d}\boldsymbol{p}_i\,\mathrm{d}\boldsymbol{p}_i'\,\mathrm{d}\boldsymbol{p}_3\,\delta(\boldsymbol{p}_1'+\boldsymbol{p}_2'-\boldsymbol{p}_1-\boldsymbol{p}_2)(\boldsymbol{p}_2'+\boldsymbol{p}_2)$$
$$\Psi_{\mathrm{B}}^{\mathrm{tot}}(\boldsymbol{p}_1',\boldsymbol{p}_2',\boldsymbol{p}_3)^*\,\Psi_{\mathrm{A}}^{\mathrm{tot}}(\boldsymbol{p}_1,\boldsymbol{p}_2,\boldsymbol{p}_3), \qquad (2.130)$$

where we have performed a Fourier transformation.

We then have to give the plane wave center-of-mass motion wave functions explicitly, and by using translational invariance make the change of variables $\boldsymbol{p} \to \boldsymbol{p} + \frac{1}{3}\boldsymbol{P}(\boldsymbol{P} = \boldsymbol{P}_{\mathrm{A}} = \boldsymbol{P}_{\mathrm{B}})$:

$$\langle \mathrm{B}|\, \delta(\boldsymbol{r}_1 - \boldsymbol{r}_2)\boldsymbol{p}_2 + \boldsymbol{p}_2\delta(\boldsymbol{r}_1 - \boldsymbol{r}_2)\,|\mathrm{A}\rangle$$

$$= \delta(\boldsymbol{P}_{\mathrm{B}} - \boldsymbol{P}_{\mathrm{A}})\frac{1}{(2\pi)^3}\int \prod_{i=1}^{2} \mathrm{d}\boldsymbol{p}_i\, \mathrm{d}\boldsymbol{p}_i'\, \mathrm{d}\boldsymbol{p}_3\, \delta(\boldsymbol{p}_1' + \boldsymbol{p}_2' - \boldsymbol{p}_1 - \boldsymbol{p}_2)\,\delta\left(\sum_i \boldsymbol{p}_i\right) \tag{2.131}$$

$$\times \left(\boldsymbol{p}_2' + \boldsymbol{p}_2 + \frac{2\boldsymbol{P}}{3}\right)\psi_{\mathrm{B}}(\boldsymbol{p}_1', \boldsymbol{p}_2', \boldsymbol{p}_3)^*\psi_{\mathrm{A}}(\boldsymbol{p}_1, \boldsymbol{p}_2, \boldsymbol{p}_3).$$

We can now write the result of performing these transformations on the full $\mathrm{H}_{\mathrm{W}}^{\mathrm{pv}}$ (1,2) (with $\Delta S = 1$):

$$\langle \mathrm{B}|H_{\mathrm{W}}^{\mathrm{pv}}(1,2)\,|\mathrm{A}\rangle = \frac{G_{\mathrm{F}}}{\sqrt{2}}\cos\theta_{\mathrm{C}}\sin\theta_{\mathrm{C}}\,\delta(\boldsymbol{P}_{\mathrm{B}} - \boldsymbol{P}_{\mathrm{A}})$$

$$\times \int \prod_i \mathrm{d}\boldsymbol{p}_i\, \mathrm{d}\boldsymbol{p}_i'\, \delta\left(\sum_i \boldsymbol{p}_i\right)\psi_{\mathrm{B}}(\boldsymbol{p}_1', \boldsymbol{p}_2', \boldsymbol{p}_3')^\dagger\, \mathcal{O}^{\mathrm{pv}}(\boldsymbol{p}_1', \boldsymbol{p}_2'; \boldsymbol{p}_1, \boldsymbol{p}_2)\,\psi_{\mathrm{A}}(\boldsymbol{p}_1, \boldsymbol{p}_2, \boldsymbol{p}_3), \tag{2.132}$$

where the scalar product over spin and flavor labels is understood, and the operator $\mathcal{O}^{\mathrm{pv}}$ is given by

$$\mathcal{O}^{\mathrm{pv}}(\boldsymbol{p}_1', \boldsymbol{p}_2'; \boldsymbol{p}_1, \boldsymbol{p}_2) = \frac{1}{(2\pi)^3}\delta(\boldsymbol{p}_1' + \boldsymbol{p}_2' - \boldsymbol{p}_1 - \boldsymbol{p}_2)\,\delta(\boldsymbol{p}_3' - \boldsymbol{p}_3)$$

$$\times \frac{1}{2m}\{-(\boldsymbol{\sigma}_1 - \boldsymbol{\sigma}_2)\cdot[(\boldsymbol{p}_1 - \boldsymbol{p}_2) + (\boldsymbol{p}_1' - \boldsymbol{p}_2')] + \mathrm{i}(\boldsymbol{\sigma}_1 \times \boldsymbol{\sigma}_2)\cdot[(\boldsymbol{p}_1 - \boldsymbol{p}_2) \tag{2.133}$$

$$-(\boldsymbol{p}_1' - \boldsymbol{p}_2')]\}\,\tau_1^{(-)}v_2^{(+)}.$$

For comparison, we write the corresponding expression for the parity-conserving case:

$$\mathcal{O}^{\mathrm{pc}}(\boldsymbol{p}_1', \boldsymbol{p}_2'; \boldsymbol{p}_1, \boldsymbol{p}_2) = \frac{1}{(2\pi)^3}\delta(\boldsymbol{p}_1' + \boldsymbol{p}_2' - \boldsymbol{p}_1 - \boldsymbol{p}_2)\,\delta(\boldsymbol{p}_3' - \boldsymbol{p}_3) \\ \times (1 - \boldsymbol{\sigma}_1 \cdot \boldsymbol{\sigma}_2)\,\tau_1^{(-)}v_2^{(+)}. \tag{2.134}$$

The integral in (2.132) can be done in a symmetric way by setting

$$\left.\begin{aligned} \boldsymbol{p}_1' - \boldsymbol{p}_1 &= \boldsymbol{q}, \\ \boldsymbol{p}_2' - \boldsymbol{p}_2 &= -\boldsymbol{q} \end{aligned}\right\} \tag{2.135}$$

Then the integral is written

$$\frac{1}{(2\pi)^3}\int d\boldsymbol{q}\int\prod_i d\boldsymbol{p}_i\,\delta\Big(\sum_i \boldsymbol{p}_i\Big)\,\psi_B(\boldsymbol{p}_1+\boldsymbol{q},\boldsymbol{p}_2-\boldsymbol{q},\boldsymbol{p}_3)^\dagger\,\mathcal{O}^{pv}(1,2)\,\psi_A(\boldsymbol{p}_1,\boldsymbol{p}_2,\boldsymbol{p}_3), \quad (2.136)$$

where we now have

$$\mathcal{O}^{pv}(1,2) = -\frac{1}{m}\{(\boldsymbol{\sigma}_1-\boldsymbol{\sigma}_2)\cdot(\boldsymbol{p}_1-\boldsymbol{p}_2+\boldsymbol{q})+i(\boldsymbol{\sigma}_1\times\boldsymbol{\sigma}_2)\cdot\boldsymbol{q}\}\tau_1^{(-)}v_2^{(+)}. \quad (2.137)$$

It is now easy to return to configuration space with the help of (2.38a,b):

$$\langle B|\,H_W^{pv}(1,2)\,|A\rangle = \frac{G_F}{\sqrt{2}}\cos\theta_C\sin\theta_C\,\delta(\boldsymbol{P}_B-\boldsymbol{P}_A)\langle B|\mathcal{O}^{pv}(1,2)|A\rangle, \quad (2.138)$$

with

$$\begin{aligned}\mathcal{O}^{pv}(1,2) = &-\frac{1}{m}[\delta(\boldsymbol{r}_1-\boldsymbol{r}_2)(\boldsymbol{\sigma}_1-\boldsymbol{\sigma}_2)\cdot(\boldsymbol{p}_1-\boldsymbol{p}_2)]\,\tau_1^{(-)}v_2^{(+)}\\ &-\frac{1}{m}\{[\boldsymbol{\sigma}_1-\boldsymbol{\sigma}_2+i(\boldsymbol{\sigma}_1\times\boldsymbol{\sigma}_2)]\cdot[-i\nabla_{r_1}\delta(\boldsymbol{r}_1-\boldsymbol{r}_2)]\}\,\tau_1^{(-)}v_2^{(+)}. \quad (2.139)\end{aligned}$$

In all the preceding calcuations since (2.119), we have singled out a particular combination of two quarks within the set of three quarks — the first two. Just as for elementary emission, it is possible to write a more symmetric formula by dropping the factor 6 in the expression (2.119) and writing instead

$$\langle B|\,H_W\,|A\rangle = \langle\Psi_B|\sum_{i\neq j}H_W(i,j)\,|\Psi_A\rangle. \quad (2.140)$$

As to the color quantum numbers, we have simply omitted them, since the operator $H_W(1,2)$ is a product of first-quantized current operators that act as the identity on color labels; we are therefore left with the scalar product of identical color wave functions.

It is important here to emphasize that the contact character of the interaction we are considering is reflected in the presence of a $\delta(\boldsymbol{r}_1-\boldsymbol{r}_2)$ in (2.129) and (2.139). Nonleptonic matrix elements like $\langle N|H_W^{pc}|\Sigma\rangle$ will therefore depend on the square of the baryon wave function when two quarks are at the same point (Figure 2.6), $|\psi(\boldsymbol{r}_1=\boldsymbol{r}_2,\,\boldsymbol{r}_3)|^2$.

Finally, one must be aware that the effective four-fermion weak interaction, deriving from the unified theory of weak and electromagnetic interactions through the short-distance expansion, is not purely of the (V−A)(V−A) current–current form. There may also be, for instance, the exchange of the weak neutral Z^0 boson, which couples to fermions through a different combination of vector and axial currents. There is no difficulty in including these other types of interaction, which enter for example the parity-violating part of the nucleon–nucleon interaction. One has simply to consider the corresponding nonrelativistic expansions of the weak neutral current $J_\mu^3 - \sin^2\theta_W J_\mu^{em}$. Also QCD corrections induce (V − A)(V + A) interactions in the charged nonleptonic part of the interaction, as we shall see in Chapter 4. To compute these types of four-fermion interactions, all the necessary steps are given above.

2.2 STRONG DECAYS

As we pointed out in Chapter 1, the description of strong decays has a very different status from electroweak processes, in which the effective interaction responsible for the transition is known from the start. Moreover, for the simplest electroweak matrix elements or processes considered in Section 2.1, we have to calculate the matrix elements of known operators (determined by the electroweak theory) between known wave functions, through the well-defined recipes of the nonrelativistic quark model. The only source of uncertainty, once we have admitted these quark-model methods, lies in the wave functions, which can be determined independently from a study of hadron spectroscopy. It is only through these wave functions, and therefore through the underlying potential model, that the rather poorly known strong interactions are playing a role.

For strong decays we have a more difficult situation. We do not know the transition operator itself from theoretical principles, although we know the basic Lagrangian of strong interactions, which is that of QCD. This is of course because we cannot always treat QCD perturbatively. In fact, the only case where we know the transition operator is when we can calculate it by perturbative QCD; this is, as we already explained, when we deal with hard processes such as OZI-forbidden decays of heavy quarkonia (see Section 2.2.2 below). Most often, we have to recourse to *phenomenological models*, which are certainly based on theoretical ideas but with a speculative or tentative character, and which must therefore receive their main support from empirical successes. Indeed, this is what has already been done for spectroscopy, and it is at the heart of the methodology of quark models; but here, in the domain of strong decays, we have to formulate new hypotheses in addition to those that have been required in order to understand spectroscopy. The situation is also complicated in two respects: there is no unified model for all processes, and, even for the same category of processes (e.g. OZI-allowed quasi-two-body decays), there are various competing quark models, not to mention other approaches such as Current Algebra or phenomenological Lagrangians. This is partly connected with the fact that, although the amount of experimental data is considerable, it is still too crude and scarce to discriminate models on a purely empirical basis; moreover, models are an attempt to give only a partial description of the facts.

The elaboration of phenomenological quark models relies first on a sort of classification of the wide domain of strong decays into several categories, rather like what happened in the past in botany and zoology. A large-scale classification is one that separates OZI-forbidden and allowed process (for the definition see Section 2.2.2). Another general category is that of two-body decays, i.e. decays into two hadrons only, as opposed to decays into many hadrons; this category of two-body decays is considerably enlarged by the fact that many hadron final states can frequently be analysed as two unstable hadrons with subsequent decay, for example $K\pi\pi$ can be often viewed as $K^*\pi$ and $K\rho$, and $K\pi\pi\pi$ as $K^*\rho$. The quark model provides a clear insight into the mechanism of these two-body decays and a tool for their analysis. The decay of an ordinary hadron (meson or baryon) into two ordinary hadrons corresponds necessarily to the creation of an additional $q\bar{q}$ pair. Multiquark hadrons like a diquonium $qq\bar{q}\bar{q}$ may decay in contrast to two mesons by a simple rearrangement of the quarks; finally a hybrid containing a constituent gluon may decay to two ordinary hadrons by the process gluon$\rightarrow q\bar{q}$. In fact, it is easier to formulate unambiguous models for such decays of exotic states: the rearrangement can be logically derived from the interquark potential (see Section 2.2.4); the gluon$\rightarrow q\bar{q}$ process can be naturally described by a natural extension of the QCD coupling to constituent gluons.

The main domain of phenomenological applications consists in the decay of ordinary hadrons into two ordinary hadrons, and we shall concentrate mainly on this topic (Sections 2.2.1 and 2.2.3). The elementary emission model (Section 2.2.1) has a rather general scope, while the pair-creation models (Section 2.2.3) are concerned specifically with the so-called OZI-allowed decays.

2.2.1 Elementary-Meson Emission

In spite of the fact that mesons are composed of quarks, their emission and absorption are often described in terms of an elementary quantum. This type of description has been mainly developed for pion emission, although it can be developed for other pseudoscalars and for any meson. It describes the two-body decays of ordinary hadrons in analogy with the radiative or semileptonic decays of the previous chapter: one hadron, composed of quarks, emits a meson through one of them; the meson is not resolved into quarks and the number of quarks is conserved; the pair creation that is strictly speaking necessary for such a decay is concealed by the formalism. The main advantage of this model is *simplicity*: one remains within a framework where quarks and antiquarks are conserved. Moreover, a relativistic kinematics can be maintained for this elementary quantum, while its resolution into quarks would naturally restrict us to the nonrelativistic approximation. As to its theoretical basis, it is first of all phenomenological, but it may seem a natural extension of the analogous baryon–meson trilinear couplings (the prototype of which is the Yukawa coupling), baryons being substituted by quarks; in the case of pseudoscalar and vector mesons, the analogy with radiative or semileptonic decays was given a theoretical sense in the sixties with the ideas of current-field identities used to express pion dominance of the divergence of the axial current (PCAC) (see e.g. Adler and Dashen, 1968) or vector-meson dominance (VMD) of the electromagnetic current (Sakurai, 1969). The modern understanding of PCAC and VMD implies a relation between meson couplings and matrix elements of electroweak currents; however, this relation does not imply the rather simple-minded emission models that we now present. The implications of PCAC and VMD are a subtle question, which we shall leave aside.

We shall adopt the point of view of considering the elementary-meson Hamiltonian as an effective interaction simply fitted to describe the amplitudes in the lowest order in the coupling constant, and we shall not consider higher orders in this scheme.

For pion emission, a natural idea would be to assume, in analogy with the nucleon pseudoscalar coupling, that

$$H_{\mathrm{I}} = -\mathrm{i}g_{\pi qq} \int \mathrm{d}\boldsymbol{x}\, \bar{q}(\boldsymbol{x})\gamma_5 \tau^a q(\boldsymbol{x})\varphi^a(\boldsymbol{x}). \tag{2.141}$$

Let us introduce

$$j_5^a(\boldsymbol{x}) = -\mathrm{i}\bar{q}(\boldsymbol{x})\gamma_5 \tau^a q(\boldsymbol{x}). \tag{2.142}$$

Performing the nonrelativistic approximation for quarks (the extension to the meson emission from an antiquark is done as in Section 2.1.1(c)),

$$q_2 = -\mathrm{i}\frac{\boldsymbol{\sigma}\cdot\boldsymbol{\nabla}}{2m}q_1, \tag{2.143}$$

we get,

$$j_5^a(\boldsymbol{x}) = -\boldsymbol{\nabla}\cdot\left[\bar{q}_1(\boldsymbol{x})\frac{\boldsymbol{\sigma}}{2m}\tau^a q_1(\boldsymbol{x})\right], \tag{2.144}$$

or, in first-quantized form,

$$j_5^a(\boldsymbol{x}) = -\tau^a \frac{\boldsymbol{\sigma}}{2m}\cdot\boldsymbol{\nabla}_x \delta(\boldsymbol{x}-\boldsymbol{r}). \tag{2.145}$$

This density will appear in $\langle \mathrm{f}|H_{\mathrm{I}}|\mathrm{i}\rangle$ through its Fourier transform:

$$\tilde{j}_5^a(\boldsymbol{k}) = \mathrm{i}\tau^a \frac{\boldsymbol{\sigma}\cdot\boldsymbol{k}}{2m}\mathrm{e}^{\mathrm{i}\boldsymbol{k}\cdot\boldsymbol{r}} \tag{2.146}$$

The pion momentum is $\boldsymbol{k}$ in the case of absorption, $-\boldsymbol{k}$ in the case of emission.

An alternative possibility would be to introduce the axial-vector coupling, with the Lagrangian density

$$L_{\rm I}(x) = -\frac{g_{\pi qq}}{2m}\bar{q}(x)\gamma^\mu\gamma_5\tau^a q(x)\partial_\mu\varphi^a(x). \tag{2.147}$$

We have now a derivative coupling, and the derivation of the Hamiltonian of interaction is less trivial than in the nonderivative case. Since $L_{\rm I}(x)$ contains $\partial_0\varphi$, which enters in the definition of the canonical conjugate momentum π^0, one cannot set $H_{\rm I}(x) = -L_{\rm I}(x)$. One must return to the total Lagrangian containing the free parts, and recalculate π^0, which is found to be (omitting the isospin for the sake of simplicity)

$$\pi^0 = \partial^0\varphi - \frac{g_{\pi qq}}{2m}\bar{q}\gamma^0\gamma_5 q. \tag{2.148}$$

With this point in mind, the Hamiltonian density becomes

$$H(\boldsymbol{x}) = H_{\rm D}(\boldsymbol{x}) + \frac{1}{2}(\pi^0)^2 + \frac{1}{2}(\boldsymbol{\nabla}\varphi)^2 + \frac{1}{2}m^2\varphi^2$$
$$+\frac{g_{\pi qq}}{2m}(\bar{q}\gamma^0\gamma_5 q\pi^0 - \bar{q}\boldsymbol{\gamma}\gamma_5 q\cdot\boldsymbol{\nabla}\varphi) + \frac{1}{2}\left(\frac{g_{\pi qq}}{2m}\bar{q}\gamma^0\gamma_5 q\right)^2, \tag{2.149}$$

$H_{\rm D}(\boldsymbol{x})$ being the usual free Dirac Hamiltonian density. Suppose now that we are working in the interaction representation. The free-meson Hamiltonian density is the quadratic part in π^0 and φ. Calling it $H_0(\boldsymbol{x})$, we have: $\partial\varphi(x)/\partial t = \delta H_0/\delta\pi^0 = \pi^0(x)$. Then the Hamiltonian interaction density will be

$$H_{\rm I}(x) = \frac{g_{\pi qq}}{2m}(\bar{q}\gamma^0\gamma_5 q\partial^0\varphi - \bar{q}\boldsymbol{\gamma}\gamma_5 q\cdot\boldsymbol{\nabla}\varphi) + \frac{1}{2}\left(\frac{g_{\pi qq}}{2m}\bar{q}\gamma^0\gamma_5 q\right)^2. \tag{2.150}$$

We finally have a result that differs from $H_{\rm I}(\boldsymbol{x}) = -L_{\rm I}(\boldsymbol{x})$ only by a four-fermion term. But this four-fermion term is irrelevant here because we want to consider only pion emission and we are working in first order in the coupling constant. To sum up, we can finally set,

$$H_{\rm I}^{\rm eff} = \frac{g_{\pi qq}}{2m}\int d\boldsymbol{x}\, \bar{q}(\boldsymbol{x})\gamma^\mu\gamma_5\tau^a q(\boldsymbol{x})\partial_\mu\varphi^a(\boldsymbol{x}) \tag{2.151}$$

Here the relevant density is the axial current

$$j_5^{a\mu}(\boldsymbol{x}) = \bar{q}(\boldsymbol{x})\gamma^\mu\gamma_5\tau^a q(\boldsymbol{x}), \tag{2.152}$$

which is completely analogous to the weak axial-vector current except for flavor matrices. Because of the derivative affecting the pseudoscalar field, we have to replace $\mathrm{e}^{\mathrm{i}\boldsymbol{k}\cdot\boldsymbol{r}}$, appearing in the nonderivative case, with either $-\mathrm{i}\omega_\pi \mathrm{e}^{\mathrm{i}\boldsymbol{k}\cdot\boldsymbol{r}}$ or $\mathrm{i}k_i \mathrm{e}^{\mathrm{i}\boldsymbol{k}\cdot\boldsymbol{r}}$ for $\partial_0\varphi(x)$ or $\partial_i\varphi(x)$.

We then get for the operator between quarks in the pseudoscalar coupling case (2.141) (for pion absorption)

$$\langle 0|\, H_{\mathrm{I}}\, |\boldsymbol{k},a\rangle = \frac{i}{(2\pi)^{3/2}} \frac{1}{(2\omega_\pi)^{1/2}} g_{\pi qq} \tau^a \frac{\boldsymbol{\sigma}\cdot\boldsymbol{k}}{2m} \mathrm{e}^{\mathrm{i}\boldsymbol{k}\cdot\boldsymbol{r}}. \tag{2.153}$$

Note the presence of a factor i necessary to ensure Hermiticity. In the axial-vector coupling case we obtain instead

$$\langle 0|\, H_{\mathrm{I}}\, |\boldsymbol{k},a\rangle = \frac{i}{(2\pi)^{3/2}} \frac{1}{(2\omega_\pi)^{1/2}} g_{\pi qq} \tau^a \left[\frac{\boldsymbol{\sigma}\cdot\boldsymbol{k}}{2m} \mathrm{e}^{\mathrm{i}\boldsymbol{k}\cdot\boldsymbol{r}} - \frac{\omega_\pi}{2m} \frac{\boldsymbol{\sigma}\cdot(\boldsymbol{p}\mathrm{e}^{\mathrm{i}\boldsymbol{k}\cdot\boldsymbol{r}} + \mathrm{e}^{\mathrm{i}\boldsymbol{k}\cdot\boldsymbol{r}}\boldsymbol{p})}{2m} \right]. \tag{2.154}$$

We note that the results are not identical. We first see that the additional term, which is a recoil term (Mitra and Ross, 1967) in opposition to the direct term (2.153), corresponds to a higher order in the expansion of the matrix element in the quark momenta. This is so because ω_π is itself of second order, being the difference of the two energy levels between the initial and final hadrons, for instance in the decay $\mathrm{N}^* \to \mathrm{N}\pi$. To recover such a term, in the pseudoscalar coupling (2.141), we should have to go beyond the lowest approximation (2.143). Indeed, the systematic expansion in v/c allows us to show that the next order gives exactly the result found in (2.154). On the other hand, beyond this order, a systematic difference appears between the two types of coupling, owing to the Lorentz-scalar part of the binding potential.

One could analyse in the same way the emission of vector mesons by introducing a vector current parallel to the electromagnetic current. Anyway, there are definite drawbacks in this treatment of strong-interaction processes: it gives a rather asymmetric treatment of the decay products, while all the hadrons present in the process are composites of quarks. This unwanted asymmetry becomes particularly blatant in the case of mesons decaying into two pions. Moreover, since we have to introduce specific couplings for each type of emitted mesons, the number of *ad hoc* couplings would become intolerably large. Finally, it is not possible in general to disregard the spatial extension of the wave function of the emitted meson in OZI-allowed decays. We shall therefore in this latter case turn to the more satisfactory quark-pair-creation model (Section 2.2.3).

2.2.2 OZI Rule, OZI-Forbidden Decays

2.2.2(a) General discussion

The collisions and decays of hadrons, represented in terms of quark-line diagrams, can be classified in two main categories according to the *quark-line topology* (Harari, 1969; Rosner, 1969). A process is said to be forbidden by the Okubo–Zweig–Iizuka (OZI) rule (Okubo, 1963; Zweig, 1964; Iizuka, 1966) when its quark-line diagram can be divided into parts containing only complete hadrons (color singlets) without cutting a quark line; otherwise it is said to be OZI-allowed. The classical examples are the two modes in which the ϕ meson, composed of a $s\bar{s}$ pair, decays. There are two modes, $K\bar{K}$ and $\pi\pi\pi$. Since the ϕ meson is composed of s quarks, and K mesons (pions) contain (do not contain) strange quarks, the two decays look very different in terms of quark lines. This is shown in Figure 2.7: $\phi \to K\bar{K}$ proceeds by propagation of

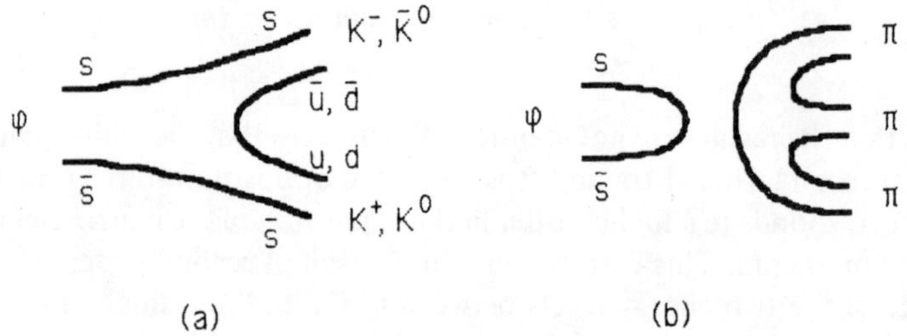

Figure 2.7 The strong decays of the ϕ meson: (a) $\phi \to K\bar{K}$, allowed by the OZI rule; (b) $\phi \to 3\pi$, forbidden by the OZI rule.

the s and $\bar{s}$ quarks, which combine with a created $q\bar{q}$ pair to form the final state $K\bar{K}$. In contrast, in the case of $\phi \to 3\pi$, the initial $s\bar{s}$ pair annihilates first and does not propagate into the final state. This final state 3π, composed of only nonstrange quarks, occurs by creation of three $q\bar{q}$ pairs. Empirically, it occurs that this latter class of diagrams are strongly suppressed relatively to the former. This rule has been spectacularly confirmed by the discovery of the J/ψ, $c\bar{c}$ meson, which decays by the strong interactions only through annihilation of the initial $c\bar{c}$ pair (Figure 2.8). Its decay width is found to be very small in comparison with hadrons decaying strongly through OZI-allowed processes.

There is no general explanation of the OZI rule; one can give a *partial* explanation of it in the case of annihilation of heavy quarks (see below), but it does not explain why the rule has a very general validity. A general *interpretation* of the rule is found in terms of the $1/N_c$ expansion (N_c is the number of colors) (see especially Witten, 1979); the quark-line diagrams that

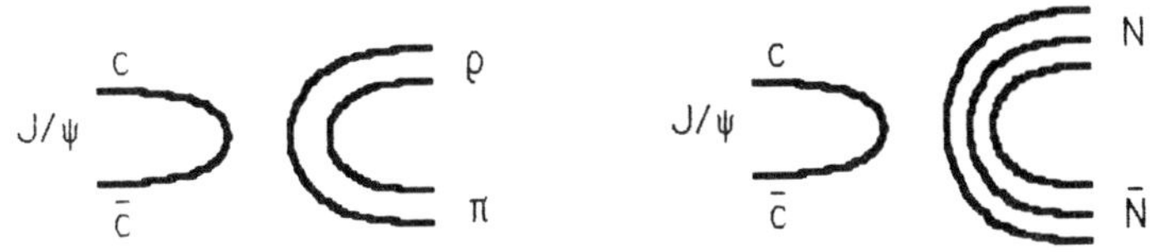

Figure 2.8 Examples of decays of the $c\bar{c}$ system J/ψ: J/ψ→ρπ, J/ψ→N$\bar{N}$, . . ., all suppressed by the OZI rule.

dominate according to the OZI rule are shown in QCD to correspond to the leading contributions in the $1/N_c$ expansion; i.e. they are shown to dominate for N_c very large. However, it is not known why this result can be extrapolated to such a low number as $N_c = 3$; therefore this idea must be considered as a very interesting theoretical formulation of the OZI rule, but not as an explanation.

The *annihilation* decay processes of mesons, where the $q\bar{q}$ pair constituting the meson annihilates itself, are certainly one of the most important subcategories of OZI-forbidden processes; the two examples given above, $\phi \rightarrow 3\pi$ and ψ decays, are of this type. There are other possible types of OZI-forbidden processes. For example the ψ′ excitation of the ψ can decay through $\psi' \rightarrow \psi\pi\pi$; in this case the $c\bar{c}$ pair is conserved; however, it is still an OZI-forbidden process according to the above definition. From now on, we concentrate on annihilation processes, because their mechanism is the best elucidated, especially in the case of *heavy* quarkonia (corresponding to the c and heavier quarks). These processes occur by a mechanism that is close in structure to the annihilation of mesons into lepton pairs, analysed in Section 2.1.3, or into two photons. In QCD the *total* strong decay rates, i.e. the sum over all possible final channels, may be interpreted as $q\bar{q} \rightarrow$gluons; at a later stage, the gluons produce a final state of ordinary hadrons by $q\bar{q}$ creation (Figure 2.9). Since we are considering total rates, we do not have to bother about this second stage. The only thing that counts is the probability of annihilation into gluons. And here comes the very important contribution of Appelquist and Politzer (1975). They have shown that, because of the asymptotic freedom of QCD, the emission of gluons in the annihilation of heavy quarks can be described by perturbation in an effective coupling constant α_s,

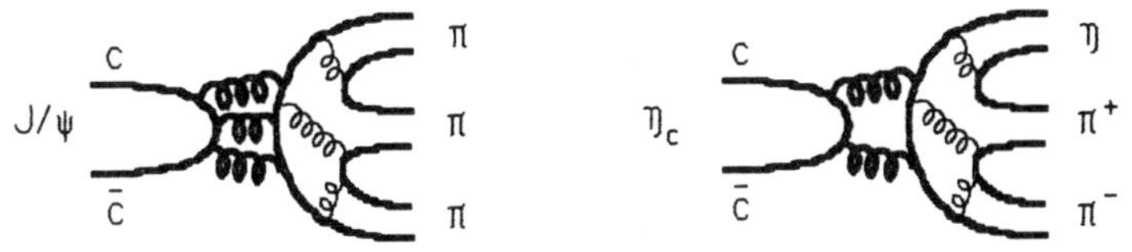

Figure 2.9 Examples of OZI-suppressed decays in QCD, J/ψ→3π, $\eta_c \rightarrow \eta\pi^+\pi^-$. The curly lines are gluons.

which is small. This coupling constant α_s determines the emission of gluons by the quarks through an interaction very similar to the QED emission of photons by electrons, the fine-structure constant α being essentially replaced by α_s (to see this quantitatively, one has to return to the QCD Lagrangian (1.1)). This is true at least for the lowest-order graphs, because at higher order in α_s, new interactions appear between the gluons, which have no parallel in QED. Precisely, because of the smallness of α_s, the main contribution to the probability comes from the emission of the *minimum possible number* of gluons by the quark line representing the annihilating pair, without any radiative correction. In QED the minimum possible number of emitted photons is one or two according to the C parity of the state (one virtual photon for ρ decay, two photons for the π^0). In QCD, because the meson is color-neutral, the decay to one gluon, which is color octet, is forbidden. The minimum possible is two or three gluons. The annihilation into two or three gluons is closely related to the one into photons, with the replacement of α by α_s, and with due account being taken of the additional color quantum number.

2.2.2(b) Relation between cross-sections and annihilation widths

It will be useful to have a general formula giving the decay amplitude and rate in terms of the amplitude or cross-section for the annihilation process $q\bar{q} \to X$(X being any final state) and the hadronic wave function describing the $q\bar{q}$ decaying state. Such a formula is proposed in Landau and Lifshitz (1972). We give here the demonstration with a few more details. Let us first reinterpret the formula given in Section 2.1.3 for the decay $\pi \to \mu\nu$. We observe that in the rest frame of the π only one component of the current ($\mu = 0$) contributes. We shall deal only with this term, assuming that we are in the rest frame of the decaying meson C. Tracing back the origin of (2.105), we see that we get it through the following steps.

(i) We take the expression given by the Feynman rules for the amplitude $M(q\bar{q} \to \mu\nu)$, the usual invariant amplitude, multiply it by the suitable normalization factors for the outgoing particles ($(2k^0)^{-1/2}$ for bosons) and for the ingoing quarks, $(m/E)^{1/2}$. Also, for the ingoing quarks, we have a factor $(2\pi)^{-3/2}$ for each quark owing to the normalization of the wave function, and an overall factor $(2\pi)^3$ coming from the integration over space of the plane waves $e^{i\boldsymbol{p}\cdot\boldsymbol{r}}$, which gives $(2\pi)^3\delta(\boldsymbol{P}_f - \boldsymbol{P}_i)$. This $(2\pi)^3$ cancels with the quark factors $[(2\pi)^{-3/2}]^2$.

(ii) We take the nonrelativistic limit $v_1, v_2 \to 0$ for the two ingoing quarks. This selects the nonzero components of the amplitude in the nonrelativistic limit, while dropping the others. Here, "nonrelativistic" is taken in the sense of *static*, zero quark velocities, which implies only S-wave scattering.

(iii) We multiply by the hadronic wave functions. The multiplication by the spatial wave function $\psi(0)$ is straightforward; while the spin, isospin and color indices have to be contracted with the corresponding labels of $q\bar{q}$ in the amplitude $M(q\bar{q}\to X)$. An additional factor $(2\pi)^{3/2}$ multiplies $\psi(0)$.

The first step gives, for the annihilation into lepton pairs,

$$M(q_{\alpha_1 i_1 s_1}\,\bar{q}_{\alpha_2 i_2 s_2}\to \mu\nu)$$
$$=\frac{1}{(2\pi)^3}\frac{1}{(2p_\mu^0)^{1/2}}\frac{1}{(2p_\nu^0)^{1/2}}\frac{G_F}{\sqrt{2}}\bar{u}_\mu\gamma^\lambda(1-\gamma_5)v_\nu\delta_{\alpha_2\alpha_1}Q_{i_2 i_1}\left(\frac{m}{E_2}\right)^{1/2}\left(\frac{m}{E_1}\right)^{1/2}\bar{v}_{s_2}\gamma_\lambda u_{s_1}, \tag{2.155}$$

with $(2\pi)^3$ coming from the two factors $(2\pi)^{3/2}$ for the two leptons (states normalized to $\delta(\boldsymbol{p}-\boldsymbol{p}')$).

Step (ii) reduces (2.155) to

$$M(q_{\alpha_1 i_1 s_1}\,\bar{q}_{\alpha_2 i_2 s_2}\to \mu\nu)$$
$$=\frac{1}{(2\pi)^3}\frac{1}{(2p_\mu^0)^{1/2}}\frac{1}{(2p_\nu^0)^{1/2}}\frac{G_F}{\sqrt{2}}\bar{u}_\mu\gamma^0(1-\gamma_5)v_\nu\delta_{\alpha_2\alpha_1}Q_{i_2 i_1}\,\chi^\dagger_{s_2}\chi_{s_1} \tag{2.156}$$

The last step gives a factor $\sqrt{6}$, coming from a factor 6 from the traces on color and spin, and a factor $1/\sqrt{6}$ representing the product of color and spin normalizations.

Quite generally, we obtain, for S-wave bound states,

$$M(C\to X)$$
$$=(2\pi)^{3/2}\psi(0)\sum_{\substack{\alpha_1 i_1 s_1\\ \alpha_2 i_2 s_2}}\omega_C(\alpha_1,\alpha_2)\varphi_C(i_1,i_2)\chi_C(s_1,s_2)\,M(q_{\alpha_1 i_1 s_1}\bar{q}_{\alpha_2 i_2 s_2}\to X)_{\text{static}}. \tag{2.157}$$

From (2.157) we deduce the formula for the decay rate into a final state X:

$$\Gamma=2\pi\int\delta(E_f-E_i)\delta(\boldsymbol{P}_f-\boldsymbol{P}_i)|M(C\to X)|^2\prod_f d\boldsymbol{p}. \tag{2.158}$$

It is easy to recognize in (2.158) a cross-section for the scattering $q\bar{q}\to X$. Let us calculate the cross-section for the scattering $(q\bar{q})_C\to X$, where $(q\bar{q})_C$ denotes the $q\bar{q}$ state having quantum numbers C:

$$M[(q\bar{q})_C\to X]$$
$$=\sum_{\substack{\alpha_1 i_1 s_1\\ \alpha_2 i_2 s_2}}\omega_C(\alpha_1,\alpha_2)\varphi_C(i_1,i_2)\chi_C(s_1,s_2)\,M(q_{\alpha_1 i_1 s_1}\bar{q}_{\alpha_2 i_2 s_2}\to X)_{\text{static}}. \tag{2.159}$$

The nonrelativistic limit automatically selects S-wave scattering.

The cross section is defined as

$$\sigma = \frac{\begin{array}{c}\text{transition rate by}\\ \text{unit of volume and time}\end{array}}{\begin{array}{c}\text{density of flux of the}\\ \text{projectile beam}\end{array} \times \begin{array}{c}\text{number of targets}\\ \text{per unit volume}\end{array}}. \tag{2.160}$$

The transition rate per unit time is

$$\mathrm{d}W = 2\pi\delta(E_\mathrm{f} - E_\mathrm{i})\delta(\boldsymbol{P}_\mathrm{f} - \boldsymbol{P}_\mathrm{i})\,|M[(\mathrm{q}\bar{\mathrm{q}})_\mathrm{C} \to \mathrm{X}]|^2 \frac{V}{(2\pi)^3} \prod_\mathrm{f} \mathrm{d}\boldsymbol{p}, \tag{2.161}$$

whence, for unit volume,

$$\mathrm{d}w = 2\pi\delta(E_\mathrm{f} - E_\mathrm{i})\delta(\boldsymbol{P}_\mathrm{f} - \boldsymbol{P}_\mathrm{i})\,|M[(\mathrm{q}\bar{\mathrm{q}})_\mathrm{C} \to \mathrm{X}]|^2 \frac{1}{(2\pi)^3} \prod_\mathrm{f} \mathrm{d}\boldsymbol{p}. \tag{2.162}$$

On the other hand, the one-particle density is $(2\pi)^{-3}$, and hence the flux density is $|\boldsymbol{v}_1 - \boldsymbol{v}_2|/(2\pi)^6$. Therefore

$$\mathrm{d}\sigma = \frac{(2\pi)^4}{|\boldsymbol{v}_1 - \boldsymbol{v}_2|}\delta(E_\mathrm{f} - E_\mathrm{i})\delta(\boldsymbol{P}_\mathrm{f} - \boldsymbol{P}_\mathrm{i})\,|M[(\mathrm{q}\bar{\mathrm{q}})_\mathrm{C} \to \mathrm{X}]|^2 \prod_\mathrm{f} \mathrm{d}\boldsymbol{p}. \tag{2.163}$$

Now (2.158) can be rewritten as

$$\mathrm{d}\Gamma = (2\pi)^4 |\psi(0)|^2\, \delta(E_\mathrm{f} - E_\mathrm{i})\delta(\boldsymbol{P}_\mathrm{f} - \boldsymbol{P}_\mathrm{i})\,|M[(\mathrm{q}\bar{\mathrm{q}})_\mathrm{C} \to \mathrm{X}]|^2 \prod_\mathrm{f} \mathrm{d}\boldsymbol{p}, \tag{2.164}$$

Finally,

$$\Gamma = |\psi(0)|^2 |\boldsymbol{v}_1 - \boldsymbol{v}_2| \sigma[(\mathrm{q}\bar{\mathrm{q}})_\mathrm{C} \to \mathrm{X}]. \tag{2.165}$$

This formula is useful if $\sigma[(\mathrm{q}\bar{\mathrm{q}})_C \to \mathrm{X}]$ can be related to known cross-sections.

Orbital excitations require a more complicated but similar treatment (Jackson, 1977; Novikov *et al.*, 1978).

2.2.2(c) Calculation of η_c decay into hadrons

The lowest-order processes correspond to annihilation into the minimal number of gluons $c\bar{c} \to 2$ or 3 gluons, according to the C parity of the decaying state (Fig. 2.9). The $J^{PC} = 0^{-+}$ pseudoscalar η_c state will decay into two gluons, and the $J^{PC} = 1^{--}$ J/ψ state into three gluons. We can therefore apply the procedure outlined above to calculate, for instance, the decay of a *para-charmonium* state 0^{-+}, in terms of the annihilation cross-section $c\bar{c} \to 2$ gluons. Moreover, this cross-section can be simply related to the cross-section $e^+e^- \to 2\gamma$, which is well known in quantum electrodynamics.

Let us give the calculation for paracharmonium:

$$M[(c\bar{c})_{0^{-+}}\to 2\mathrm{g}]=\frac{1}{\sqrt{3}}\sum_{\substack{\alpha_1\alpha_2\\ s_1s_2}}\delta_{\alpha_1\alpha_2}\,\chi_0(s_1,s_2)\,M(\mathrm{c}_{\alpha_1 s_1}\bar{\mathrm{c}}_{\alpha_2 s_2}\to 2\mathrm{g})$$
$$=\frac{1}{\sqrt{3}}\sum_{\alpha}M[(\mathrm{c}_\alpha\bar{\mathrm{c}}_\alpha)_{S=0}\to 2\mathrm{g}]. \tag{2.166}$$

We can factorize the color quantum numbers in the amplitude. For two gluons of color states λ_1, λ_2 $(\lambda_{1,2}=1,\ldots,8)$

$$\sum_{\alpha}M[\mathrm{c}_\alpha\bar{\mathrm{c}}_\alpha)\to \mathrm{g}_{\lambda_1}\mathrm{g}_{\lambda_2}]=\sum_{a,b}\mathrm{Tr}\left(\frac{\lambda^a}{2}\frac{\lambda^b}{2}\right)e^a_{\lambda_1}e^b_{\lambda_2}\,M(\mathrm{e}^+\mathrm{e}^-\to 2\gamma)$$
$$=\tfrac{1}{2}(e_{\lambda_1}\cdot e_{\lambda_2})\,M(\mathrm{e}^+\mathrm{e}^-\to 2\gamma). \tag{2.167}$$

It is to be understood that α is substituted by α_s in $M(\mathrm{e}^+\mathrm{e}^-\to 2\gamma)$. Now, what we really have to calculate is the transition to a color-singlet state of gluons, which is

$$\omega_{2\mathrm{g}}(1,2)=\frac{1}{\sqrt{8}}\delta_{\lambda_1\lambda_2}. \tag{2.168}$$

Therefore

$$\sum_{\alpha}M[(\mathrm{c}_\alpha\bar{\mathrm{c}}_\alpha)\to 2\mathrm{g}]=\sqrt{2}\,M(\mathrm{e}^+\mathrm{e}^-\to 2\gamma). \tag{2.169}$$

Finally,

$$|M[(\mathrm{c}\bar{\mathrm{c}})_{0^{-+}}\to 2\mathrm{g}]|^2=\tfrac{2}{3}|M[(\mathrm{e}^+\mathrm{e}^-)_{S=0}\to 2\gamma]|^2 \tag{2.170}$$

and

$$\Gamma[(\mathrm{c}\bar{\mathrm{c}})_{0^{-+}}\to 2\mathrm{g}]=\tfrac{2}{3}|\psi(0)|^2\,|v_1-v_2|\,\sigma[(\mathrm{e}^+\mathrm{e}^-)_{S=0}\to 2\gamma]. \tag{2.171}$$

On the other hand, we have, for S-wave scattering $\mathrm{e}^+\mathrm{e}^-\to\gamma\gamma$, the cross-section for unpolarized electrons:

$$\sigma(\mathrm{e}^+\mathrm{e}^-\to 2\gamma)=\tfrac{1}{4}\{3\sigma[(\mathrm{e}^+\mathrm{e}^-)_{S=1}\to 2\gamma]+\sigma[(\mathrm{e}^+\mathrm{e}^-)_{S=0}\to 2\gamma]\}, \tag{2.172}$$

but since from C parity $\sigma[(\mathrm{e}^-\mathrm{e}^+)_{S=1}\to 2\gamma]=0$, one has

$$\sigma[(\mathrm{e}^+\mathrm{e}^-)_{S=0}\to 2\gamma]=4\sigma(\mathrm{e}^+\mathrm{e}^-\to 2\gamma), \tag{2.173}$$

which establishes the connection with the well known quantity,

$$\sigma(\mathrm{e}^+\mathrm{e}^-\to 2\gamma)=\frac{\pi}{|v_1-v_2|}\frac{\alpha^2}{m^2}. \tag{2.174}$$

Here m must of course be understood as the mass m_c of the c quark, and α as the strong coupling constant α_s. Then

$$\Gamma(\text{paracharmonium})=\frac{8\pi}{3}\frac{\alpha_s^2}{m_c^2}|\psi(0)|^2. \tag{2.175}$$

If we had a pure Coulombic charmonium, with a potential $-\frac{4}{3}\alpha_s/r$, we would obtain

$$|\psi(0)|^2=\frac{1}{\pi a^3}, \tag{2.176}$$

with

$$a=\frac{2}{\frac{4}{3}\alpha_s m_c}; \tag{2.177}$$

whence

$$|\psi(0)|^2=\frac{3}{8\pi}m_c^3(\tfrac{4}{3}\alpha_s)^3 \tag{2.178}$$

and

$$\Gamma=\tfrac{1}{3}\alpha_s^2(\tfrac{4}{3}\alpha_s)^3 m_c. \tag{2.179}$$

This is the result obtained by Appelquist and Politzer (1975). This result is only asymptotically correct, i.e. for a quarkonium made of quarks of very large mass, since then only pertubative gluon exchange dominates the potential. For charmonium $c\bar{c}$ and for $b\bar{b}$ systems the effective confining potential at intermediate distances is still dominant. The width (2.175) computed with *realistic* (non-Coulombic) wave functions is found to be very small, confirming the OZI rule.

2.2.3 OZI-Allowed Decays Through Pair Creation

We now turn to decays allowed by the OZI rule (Figure 2.7a). Other examples of these decays are $\Delta\rightarrow N\pi$, $N^*\rightarrow N\pi$, $\rho\rightarrow\pi\pi$, $A_2\rightarrow K\bar{K}$ and $\psi(4.03)\rightarrow D\bar{D}$, (Figure 2.10).

We shall concentrate on the case of two-body decays of hadrons. We describe a few models for these decays, with the common feature of a $q\bar{q}$ pair creation that combines with the intitial quarks to form the final hadrons. We shall begin with the Quark-Pair-Creation model (Sections 2.2.3(a)–(c)), followed by a few comments on the Cornell model (Section 2.2.3(d)) and the flux-

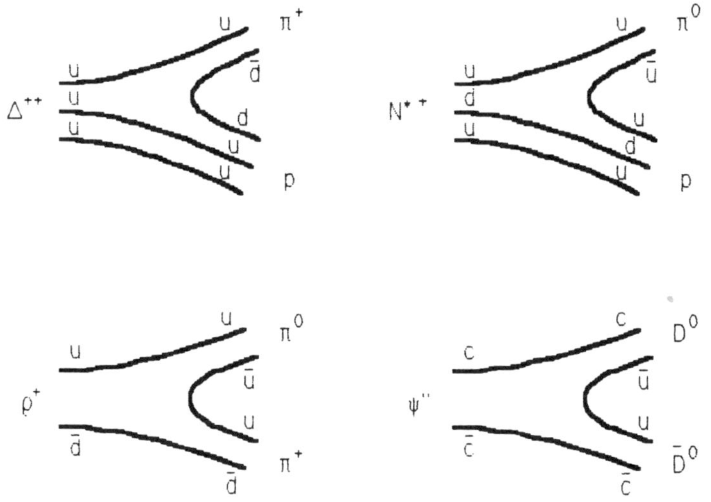

Figure 2.10 Examples of OZI-allowed decays, $\Delta^{++} \to p\pi^+$, $N^{*+} \to p\pi^0$, $\rho^+ \to \pi^0\pi^+$, $\psi'' \to D^0\bar{D}^0$.

tube model (Section 2.2.3(e)) of strong-interaction vertices. Finally in Section 2.2.5 we shall give a general angular-momentum reduction of the matrix elements for quark-pair creation.

2.2.3(a) General formulation of the Quark-Pair-Creation Model

We consider a specific model for OZI-allowed strong-interaction decays, which describes the decay by the creation of an additional quark–antiquark pair. This description is a natural consequence of the constituent-quark model of hadronic states. To further specify the model, we shall appeal to a guiding idea: in the elementary-emission model, there was an *additivity* property: the quantum is emitted by one active quark, and the others remain *spectators*, i.e. they do not play a role in the interaction.

The idea is now to generalize the additivity (Le Yaouanc *et al.*, 1973, 1975b) by postulating that, in a given process, a minimum number of quarks are active, while the others remain spectators. In hadronic decay processes this will mean that the initial quarks are spectators. A $q\bar{q}$ pair is created, but the other quarks remain unaffected (Figure 2.10). Conversely, in the absorption processes one quark and one antiquark annihilate with each other without disturbing the other quarks.

This immediately implies that the $q\bar{q}$ pair has the quantum numbers of the vacuum. With respect to the additive quantum numbers, it must be *neutral*, i.e.

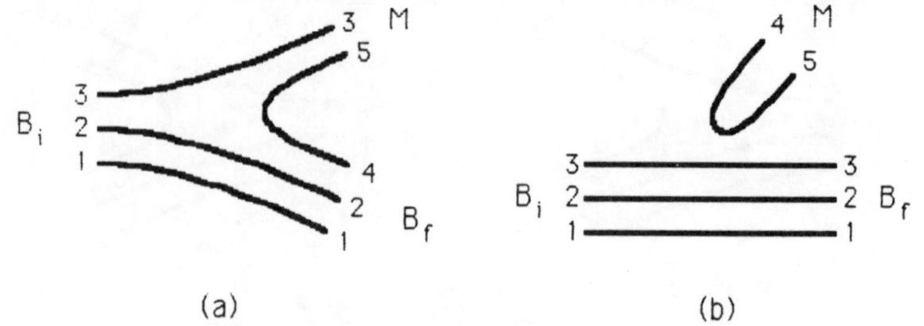

Figure 2.11 The quark-pair-creation model of hadron vertices (a); the $q\bar{q}$ pair (4,5) is created in a $^3P_0$ flavor–color singlet. The diagram of a $q\bar{q}$ scalar meson emission (b) is suppressed by the OZI rule.

singlet in color, flavor, of zero momentum and zero total angular momentum. The pair must also be of positive parity for parity conservation in the process. These requirements finally imply that the pair must be in a $^3P_0$ state (Figure 2.11). Indeed, a fermion–antifermion pair has $P=(-1)^{L+1}$, $C=(-1)^{L+S}$ (see Section 1.3.2). Vacuum quantum numbers imply $J^{PC}=0^{++}$, i.e. $L=1$ (P-wave), $S=1$ combined to give $J=0$.

Let us now express this hypothesis quantitatively. We consider the transition matrix T,

$$S = 1 - 2\pi i\, \delta(E_f - E_i)\, T, \tag{2.180}$$

and, for pair creation, we express it in the form

$$\begin{aligned} T = -\sum_{i,j} \int \mathrm{d}\boldsymbol{p}_q\, \mathrm{d}\boldsymbol{p}_{\bar{q}}\, [3\gamma\delta(\boldsymbol{p}_{\bar{q}} + \boldsymbol{p}_{\bar{q}}) \sum_m \langle 1, 1; m, -m|0, 0\rangle \\ \times \mathscr{Y}_1^m(\boldsymbol{p}_q - \boldsymbol{p}_{\bar{q}})(\chi_1^{-m}\varphi_0\omega_0)_{i,j}]\, b_i^\dagger(\boldsymbol{p}_q)\, d_j^\dagger(\boldsymbol{p}_{\bar{q}}), \end{aligned} \tag{2.181}$$

where i and j are SU(6)-color indices, φ_0 is for $SU(3)_F$ singlet ($\varphi_0 = -(u\bar{u} + d\bar{d} + s\bar{s})/\sqrt{3}$), ω_0 for color-singlet and χ_1^{-m} for triplet state of spin; $\mathscr{Y}_1^m$ is a solid harmonic polynomial reflecting the $L=1$ orbital angular momentum of the pair. This formula expresses the vacuum-like properties of the $q\bar{q}$ pair created by $b^\dagger d^\dagger$. Finally, γ is a dimensionless constant that corresponds to the strength of the transition, left undetermined by the model, which could in general depend on the momenta. The factor 3 is introduced to cancel a factor $\frac{1}{3}$ appearing after explicit calculation of color wave function overlap, to recover the definition of T when color is omitted.

Note that here we avoid the reference to a Hamiltonian operator. T is defined only for the decay process or the corresponding formation process. One might have introduced a pair-creation Hamiltonian, but this would have been another and more ambitious program.

Let us show how the matrix elements $\langle BC|T|A\rangle$ can be written in terms of wave functions. First, the states $|A\rangle$, $|B\rangle$, $|C\rangle$, are expressed as

$$\begin{aligned} |A\rangle &= \sum \Psi_{Ai_1i_2i_3} \frac{1}{\sqrt{6}} b^\dagger_{i_1} b^\dagger_{i_2} b^\dagger_{i_3} |0\rangle, \\ |B\rangle &= \sum \Psi_{Bi'_1i'_2i'_3} \frac{1}{\sqrt{6}} b^\dagger_{i'_1} b^\dagger_{i'_2} b^\dagger_{i'_3} |0\rangle, \\ |C\rangle &= \sum \Psi_{Crs} b^\dagger_r d^\dagger_s |0\rangle, \end{aligned} \tag{2.182}$$

where the indices now include all the quantum numbers, including momentum as well, and the sums extend over them.

We write T in a similar abbreviated form:

$$T = \sum P_{i_4i_5} b^\dagger_{i_4} d^\dagger_{i_5}, \tag{2.183}$$

where P stands for *pair*.

$\langle BC|\, T\,|A\rangle$ now appears as a vacuum expectation value of terms consisting of products of a certain number of creation and annihilation operators, weighted by the wave functions. To evaluate such expectation values, one has to contract in every possible way all the creation and the annihilation operators in each term. We find 18 possible ways of making the contractions—all equivalent. We thereby find the result

$$\langle BC|\, T\,|A\rangle = 3 \sum \Psi^*_{Bi_1i_2i_4} \Psi^*_{Ci_3i_5} P_{i_4i_5} \Psi_{Ai_1i_2i_3}. \tag{2.184}$$

We have discarded the type of contractions corresponding to a scalar ${}^3P_0$ particle emission with an A→B transmission (Figure 2.11(b)). By construction, since the pair carries zero total momentum, it will not contribute to the S-matrix owing to energy conservation. Moreover, this suppression is welcome, since we know that transitions with the topology of Figure 2.11(b) are OZI-forbidden.

Let us now interpret the formula (2.184). It is described by the diagram of Figure 2.11(a). The factor $3 = 18/6$ in (2.184) takes into account the number of possible rearrangements: any of the three quarks in A can go into C. It is analogous to the overall factor 3 in the elementary-emission model. For calculational purposes, the expression (2.184) can be rewritten as a scalar product:

$$\langle BC|\, T\,|A\rangle = 3\langle \Psi_B(1,2,4)\Psi_C(3,5)|\Psi_A(1,2,3)P(4,5)\rangle, \tag{2.185}$$

with $P(4,5)$ viewed as the wave function of a $q\bar{q}$ meson with quantum numbers of an SU(3)-singlet ${}^3P_0$ meson.

We can immediately work out the color part of the calculation, since color wave functions factorize out. Writing them as

$$
\left.
\begin{aligned}
\omega_{\mathrm{A}}(1,2,3) &= \frac{1}{\sqrt{6}}\varepsilon_{\alpha_1\alpha_2\alpha_3}, \\
\omega_{\mathrm{B}}(1,2,4) &= \frac{1}{\sqrt{6}}\varepsilon_{\alpha_1\alpha_2\alpha_4}, \\
\omega_{\mathrm{C}}(3,5) &= \frac{1}{\sqrt{3}}\delta_{\alpha_3\alpha_5}, \\
P(4,5) &= \frac{1}{\sqrt{3}}\delta_{\alpha_4\alpha_5}
\end{aligned}
\right\} \quad (\alpha=1,2,3), \tag{2.186}
$$

we find

$$
\langle \omega_{\mathrm{B}}(1,2,4)\omega_{\mathrm{C}}(3,5)|\omega_{\mathrm{A}}(1,2,3)\omega_{\mathrm{P}}(4,5)\rangle = \tfrac{1}{3}. \tag{2.187}
$$

Finally, we end up with an explicit expression

$$
\langle \mathrm{BC}|\,T\,|\mathrm{A}\rangle = 3\gamma\sum_m \langle 1,1;m,-m|0,0\rangle\,\langle \Phi_{\mathrm{B}}\Phi_{\mathrm{C}}|\Phi_{\mathrm{A}}\Phi_{\mathrm{vac}}^{-m}\rangle\, I_m(\mathrm{A;B,C}), \tag{2.188}
$$

where Φ are the SU(6) wave functions. In particular,

$$
\Phi_{\mathrm{vac}}^{-m} = \chi_1^{-m}\varphi_0. \tag{2.189}
$$

$I_m(\mathrm{A;B,C})$ is the spatial integral

$$
\begin{aligned}
&I_m(\mathrm{A;B,C}) \\
&= \int \prod_i \mathrm{d}\boldsymbol{p}_i\, \mathscr{Y}_1^m(\boldsymbol{p}_4-\boldsymbol{p}_5)\,\delta(\boldsymbol{p}_4+\boldsymbol{p}_5)\,\Psi_{\mathrm{B}}^{\mathrm{tot}}(\boldsymbol{p}_1,\boldsymbol{p}_2,\boldsymbol{p}_4)^*\,\Psi_{\mathrm{C}}^{\mathrm{tot}}(\boldsymbol{p}_3,\boldsymbol{p}_5)^*\,\Psi_{\mathrm{A}}^{\mathrm{tot}}(\boldsymbol{p}_1,\boldsymbol{p}_2,\boldsymbol{p}_3).
\end{aligned} \tag{2.190}
$$

Formula (2.188) is valid when the spatial functions factorize, as in the NNπ coupling. Otherwise, there will be a sum of several terms of analogous structure, as, in general, in transitions of excited states N*→Nπ.

When in (2.190) we take the additional step of going to the internal wave functions, we can easily carry out the integration over the various δ-functions describing the hadron center-of-mass motion and extract a $\delta(\boldsymbol{P}_{\mathrm{f}}-\boldsymbol{P}_{\mathrm{i}})$ reflecting as always the overall momentum conservation; this gives the amplitude M introduced in (2.27) and therefore the width according to (2.30). It is necessary to note that the color overlap happens to be the same for baryon and meson decay, $\frac{1}{3}$. The present defintion of γ agrees with that found in the literature without color for mesons and baryons (in Le Yaouanc *et al.*, (1973) the factor 3 found for baryons has been omitted; this is corrected in later papers, see Le Yaouanc *et al.* (1975b)). It is important to emphasize that the same value of γ explains the decay rates of both mesons and baryons (Ader, Bonnier and Sood, 1982).

2.2.3(b) Comparison between the QPC and elementary-emission models

The matrix element can be put into a form analogous to a model of emission of the meson C, as a matrix element of an operator between the states A and B. This operator is, in matrix form,

$$(\langle \mathrm{C}|\, T\, |0\rangle)_{i_3' i_3} = \sum_{i_5} \Psi^*_{\mathrm{C} i_3 i_5}\, P_{i_3' i_5}. \tag{2.191}$$

Here the operator is acting on the third quark, and an additional factor 3 will take into account the other possibilities. We shall write this operator explicitly in the case where C is a pion. Using matrix notation, the wave function $\Psi_{\mathrm{C} i_3 i_5}$ is considered as the element of a matrix denoted by C and P is the matrix of elements $P_{i_3' i_5}$; we can then write

$$\langle \mathrm{C}|\, T\, |0\rangle = PC^\dagger. \tag{2.192}$$

Omitting color, whose only effect is to cancel the factor 3 introduced in (2.181), we can write

$$P = \gamma \frac{I_{\text{isospin}}}{\sqrt{3}} \left(-\frac{1}{(2\pi)^{1/2}} \right) \boldsymbol{\sigma} \cdot \boldsymbol{p}\, \mathrm{i}\sigma_2 \mathscr{P}, \tag{2.193}$$

$$C = \frac{\tau}{\sqrt{2}} \frac{\mathrm{i}\sigma_2}{\sqrt{2}} \psi_\pi(2\boldsymbol{p} - \boldsymbol{k}_\pi)\, \mathrm{e}^{\mathrm{i}\boldsymbol{k}_\pi \cdot \boldsymbol{r}}\, \mathscr{P}. \tag{2.194}$$

The momentum-space matrix elements of the operator of parity $\mathscr{P}$ are

$$\langle \boldsymbol{p}'|\, \mathscr{P}\, |\boldsymbol{p}\rangle = \delta(\boldsymbol{p} + \boldsymbol{p}'), \tag{2.195}$$

and (2.193), for example, comes from considering the spatial part of P through its matrix elements:

$$\delta(\boldsymbol{p} + \boldsymbol{p}')\, \mathscr{Y}_1^m(\boldsymbol{p}' - \boldsymbol{p}) = 2\delta(\boldsymbol{p} + \boldsymbol{p}')\, \mathscr{Y}_1^m(\boldsymbol{p}').$$

The SU(6) part is easily put into matrix form with Pauli matrices, leading to the product

$$PC^\dagger = -\frac{\gamma}{2(6\pi)^{1/2}} \tau(\boldsymbol{\sigma} \cdot \boldsymbol{p})\, \psi_\pi(2\boldsymbol{p} + \boldsymbol{k}_\pi)\, \mathrm{e}^{-\mathrm{i}\boldsymbol{k}_\pi \cdot \boldsymbol{r}} \tag{2.196}$$

or

$$PC^\dagger = -\frac{\gamma}{2(6\pi)^{1/2}} \mathrm{e}^{-\mathrm{i}\boldsymbol{k}_\pi \cdot \boldsymbol{r}}\, \tau[\boldsymbol{\sigma} \cdot (\boldsymbol{p} - \boldsymbol{k}_\pi)]\, \psi_\pi(2\boldsymbol{p} - \boldsymbol{k}_\pi). \tag{2.197}$$

The internal wave function ψ_π has been defined through

$$\Psi_\pi^{\text{tot}}(\boldsymbol{p}_1, \boldsymbol{p}_2) = \delta(\boldsymbol{p}_1 + \boldsymbol{p}_2 - \boldsymbol{k}_\pi)\, \psi_\pi(\boldsymbol{p}_1 - \boldsymbol{p}_2). \tag{2.198}$$

It is interesting to compare the operator (2.197) to the one obtained in elementary emission models, for instance (2.153) or (2.154) in Section 2.2.1. In (2.197) there is a term $\boldsymbol{\sigma}\cdot\boldsymbol{p}$ in addition to the usual direct term $\boldsymbol{\sigma}\cdot\boldsymbol{k}_\pi$, which resembles the expression (2.154) with its recoil term, although the magnitude of the recoil term is different (it does not have the factor ω_π). The important difference is the wave-function factor $\psi_\pi(2\boldsymbol{p}-\boldsymbol{k}_\pi)$ coming from the composite structure of the pion, which gives a *nonlocal* character to the new emission operator. A systematic comparison of the two models as regards their phenomenological consequences will be given in Section 4.4.

2.2.3(c) Some examples of explicit calculations in the QPC model

It is instructive to perform some simple explicit calculations. In Section 2.2.5 we shall give a more general formulation of hadronic three-body vertices according to the model. We shall then see the full complexity and predictive power of the model. In Section 4.5 we will concentrate on the phenomenological applications.

Let us first consider baryon–baryon–meson vertices B_iB_fM like $NN\pi$ or $N^*\to N\pi$, $N^*\to\Delta\pi$. The internal meson wave function in momentum space is defined according to (2.198), and normalized to unity relative to the measure

$$\int d\boldsymbol{p}_1\, d\boldsymbol{p}_2\, \delta(\boldsymbol{p}_1+\boldsymbol{p}_2)\, |\psi(\boldsymbol{p}_1,\boldsymbol{p}_2)|^2=1, \tag{2.199}$$

and analogously the baryon internal wave function is defined by

$$\Psi_B^{\text{tot}}(\boldsymbol{p}_1,\boldsymbol{p}_2,\boldsymbol{p}_3)=\delta\left(\sum \boldsymbol{p}_i-\boldsymbol{p}_B\right)\psi_B(\boldsymbol{p}_1,\boldsymbol{p}_2,\boldsymbol{p}_3), \tag{2.200}$$

and normalized according to

$$\int d\boldsymbol{p}_1\, d\boldsymbol{p}_2\, d\boldsymbol{p}_3\, \delta(\boldsymbol{p}_1+\boldsymbol{p}_2+\boldsymbol{p}_3)\, |\psi_B(\{\boldsymbol{p}_i\})|^2=1. \tag{2.201}$$

The integral $I_m(B_i;B_f,M)$ of (2.190) will then be given, in the center-of-mass frame of the initial baryon B_i, by the expression

$$\begin{aligned} I_m(B_i;B_f,M)=&\frac{1}{3\sqrt{3}}\delta(\boldsymbol{p}_{B_f}+\boldsymbol{p}_M)\int d\boldsymbol{p}_\rho\, d\boldsymbol{p}_\lambda\, \mathscr{Y}_1^m\left[-2\left(\boldsymbol{p}_M+\sqrt{\tfrac{2}{3}}\boldsymbol{p}_\lambda\right)\right]\psi_{B_i}(\boldsymbol{p}_\rho,\boldsymbol{p}_\lambda)\\ &\times\psi^*_{B_f}\left(\boldsymbol{p}_\rho,\sqrt{\tfrac{2}{3}}\boldsymbol{p}_M+\boldsymbol{p}_\lambda\right)\psi^*_M\left(-\boldsymbol{p}_M-2\sqrt{\tfrac{2}{3}}\boldsymbol{p}_\lambda\right), \end{aligned} \tag{2.202}$$

where we have integrated over the δ-functions that express the hadron center-of-mass motion (2.198) and (2.200), and $\boldsymbol{p}_\rho$, $\boldsymbol{p}_\lambda$ are the relative momenta:

$$\boldsymbol{p}_\rho=\frac{1}{\sqrt{2}}(\boldsymbol{p}_1-\boldsymbol{p}_2), \qquad \boldsymbol{p}_\lambda=\frac{1}{\sqrt{6}}(\boldsymbol{p}_1+\boldsymbol{p}_2-2\boldsymbol{p}_3). \tag{2.203}$$

For the case of a three-meson decay A→B+C like ρ→2π or B→ωπ, the integral (2.190) reduces to an analogous expression, in the A center-of-mass frame:

$$I_m(\mathrm{A;B,C}) = \tfrac{1}{8}\,\delta(\boldsymbol{p}_\mathrm{B}+\boldsymbol{p}_\mathrm{C}) \int \mathrm{d}\boldsymbol{p}\, \mathscr{Y}_1^m(\boldsymbol{p}_\mathrm{B}-\boldsymbol{p})\,\psi_\mathrm{A}(\boldsymbol{p}_\mathrm{B}+\boldsymbol{p})\,\psi_\mathrm{B}^*(-\boldsymbol{p})\,\psi_\mathrm{C}^*(\boldsymbol{p}). \qquad (2.204)$$

Let us consider now vertices involving ground state hadrons, like $\mathrm{NN}\pi$ and $\rho\pi\pi$. The phenomenological coupling constants $g_{\pi\mathrm{NN}}$ and $f_{\rho\pi\pi}$ are defined by the effective Lagrangian involving hadron fields $N(x)$, $\rho_\mu(x)$, $\pi(x)$ (the arrow denotes an isospin vector),

$$\mathscr{L}_{\mathrm{NN}\pi}^{\mathrm{eff}} = \mathrm{i}g_{\pi\mathrm{NN}}(\bar{N}\gamma_5\vec{\tau}N)\cdot\vec{\pi}, \qquad (2.205)$$

$$\mathscr{L}_{\rho\pi\pi}^{\mathrm{eff}} = f_{\rho\pi\pi}(\vec{\pi}\times\partial^\mu\vec{\pi})\cdot\vec{\rho}_\mu. \qquad (2.206)$$

We shall now look for the expressions of $g_{\pi\mathrm{NN}}$ and $f_{\rho\pi\pi}$ in the quark-pair-creation model. We shall obtain $g_{\pi\mathrm{NN}}$ and $f_{\rho\pi\pi}$ in terms of the pair-creation coupling γ and a convolution integral over the internal hadron wave functions, (2.202) and (2.204) respectively. These integrals can be performed analytically in the case of the harmonic-oscillator potential, considered in Section 1.3.3 in our study of the baryon spectrum. This is a very naive but instructive calculation. The ground-state wave functions of baryons and mesons are then given by the expressions

$$\psi_\mathrm{B}(\boldsymbol{p}_1,\boldsymbol{p}_2,\boldsymbol{p}_3) = \left(\frac{3R_\mathrm{B}^2}{\pi}\right)^{3/2} \exp\left[-\tfrac{1}{6}R_\mathrm{B}^2\sum_{i<j}(\boldsymbol{p}_i-\boldsymbol{p}_j)^2\right] \qquad (2.207)$$

$$\psi_\mathrm{M}(\boldsymbol{p}_1,\boldsymbol{p}_2) = \left(\frac{R_\mathrm{M}^2}{\pi}\right)^{3/4} \exp\left[-\tfrac{1}{8}R_\mathrm{M}^2(\boldsymbol{p}_1-\boldsymbol{p}_2)^2\right], \qquad (2.208)$$

and are normalized according to (2.199) and (2.201). Then the integrals (2.202) and (2.204) are given respectively by

$$I_m(\mathrm{A;B,C}) = \left(\frac{3}{4\pi}\right)^{1/2}\frac{1}{\pi^{3/4}}(\boldsymbol{\varepsilon}_m\cdot\boldsymbol{p}_\mathrm{C})\,\tfrac{4}{3}(\tfrac{2}{3}R_\mathrm{M})^{3/2}\exp(-\tfrac{1}{12}R_\mathrm{M}^2\boldsymbol{p}_\mathrm{C}^2)\,\delta(\boldsymbol{p}_\mathrm{B}+\boldsymbol{p}_\mathrm{C}), \qquad (2.209)$$

$$I_m(\mathrm{B_i;B_f,M}) = -\left(\frac{3}{4\pi}\right)^{1/2}\frac{1}{\pi^{3/4}}(\boldsymbol{\varepsilon}_m\cdot\boldsymbol{p}_\mathrm{M})\left(\frac{3R_\mathrm{B}^2R_\mathrm{M}}{3R_\mathrm{B}^2+R_\mathrm{M}^2}\right)^{3/2}$$
$$\times\frac{4R_\mathrm{B}^2+R_\mathrm{M}^2}{3R_\mathrm{B}^2+R_\mathrm{M}^2}\exp\left(-\frac{12R_\mathrm{B}^4+5R_\mathrm{B}^2R_\mathrm{M}^2}{24(3R_\mathrm{B}^2+R_\mathrm{M}^2)}\boldsymbol{p}_\mathrm{M}^2\right)\delta(\boldsymbol{p}_\mathrm{B}+\boldsymbol{p}_\mathrm{M}), \qquad (2.210)$$

where $\boldsymbol{\varepsilon}_m (m = \pm 1, 0)$ are the vectors $\boldsymbol{\varepsilon}_{\pm 1} = (0, \pm 1/\sqrt{2}, -\mathrm{i}/\sqrt{2})$, $\boldsymbol{\varepsilon}_0 = (0, 0, 1)$.

We can now identify the T-matrix elements (2.188) with those obtained from the phenomenological Lagrangians (2.205) and (2.206). In the example of $\rho\to 2\pi$, factorizing $2\pi i\delta(E_i-E_f)$, we obtain in the ρ center-of-mass frame for the special choice $\rho^+(S_z=0)\to\pi^+\pi^0$ (Oz in the $\boldsymbol{p}_\pi$ direction):

$$(2\pi)^3\delta(\boldsymbol{p}_{\pi^+}+\boldsymbol{p}_{\pi^0})f_{\rho\pi\pi}\frac{1}{(2\pi)^{9/2}}\frac{1}{(2m_\rho)^{1/2}}\frac{1}{m_\rho}2p_\pi$$
$$=\delta(\boldsymbol{p}_{\pi^+}+\boldsymbol{p}_{\pi^0})\,\gamma\frac{1}{3\sqrt{2}}\left(\frac{3}{4\pi}\right)^{1/2}\frac{1}{\pi^{3/4}}\frac{4}{3}\left(\frac{2}{3}R_M\right)^{3/2}p_\pi\exp\left(-\tfrac{1}{12}R_M^2p_\pi^2\right), \tag{2.211}$$

where we have set $2E_\pi=m_\rho$ on the left-hand side. We choose our conventions to obtain positive values for $f_{\rho\pi\pi}$ and $g_{\pi NN}$. We find

$$f_{\rho\pi\pi}=\gamma\frac{4}{(3\pi)^{1/2}}\pi^{3/4}m_\rho^{3/2}\left(\tfrac{1}{3}R_M\right)^{3/2}\tfrac{2}{3}\exp\left(-\tfrac{1}{12}R_M^2p_\pi^2\right). \tag{2.212}$$

In the same way, we can obtain, from the identification of the calculations by the effective Lagrangian (2.206) and QPC, (2.188) with I_m given by (2.202). We find, for $g_{\pi NN}$, omitting the exponential (which corresponds to the limit $\boldsymbol{p}_\pi\to 0$),

$$\frac{g_{\pi NN}}{2m_N}=\gamma\frac{5}{3\sqrt{6}\,\pi}\pi^{3/4}m_\pi^{1/2}\left(\frac{3R_B^2R_M}{3R_B^2+R_M^2}\right)^{3/2}\left(\frac{4R_B^2+R_M^2}{3R_B^2+R_M^2}\right). \tag{2.213}$$

In these explicit calculations we have emphasized the spatial matrix element I_m, given by (2.202) or (2.204). The spin–flavor matrix elements that appear for the B_iB_fM or ABC vertices in (2.188), namely

$$\langle\Phi_{B_f}\Phi_M|\Phi_{B_i}\Phi_P^{-m}\rangle,\qquad \langle\Phi_B\Phi_C|\Phi_A\Phi_P^{-m}\rangle, \tag{2.214}$$

(where the ϕs are SU(6) flavor–spin wave functions) can be easily worked out using the wave functions of mesons and baryons of Section 1.3. For example, for the $NN\pi$ vertex, the baryon ground-state wave functions will be, from (1.91),

$$(\Phi_{N_i})_{123}=(\chi'_{123}\varphi'_{123}+\chi''_{123}\varphi''_{123})/\sqrt{2},$$
$$(\Phi_{N_f})_{124}=(\chi'_{124}\varphi'_{124}+\chi''_{124}\varphi''_{124})/\sqrt{2}, \tag{2.215}$$

and the meson and pair wave functions will be, from (1.25) and (1.67),

$$\phi_\pi=(\chi_0)_{35}(\varphi_1)_{35}$$
$$(\Phi_P^{-m})_{45}=(\chi_1^{-m})_{45}(\varphi_0)_{45}, \tag{2.216}$$

where φ_1 and φ_0 are isotriplet and flavor-singlet wave functions. In these formulae we have specified, according to Figure 2.11(a), the quark indices, so that the scalar product (2.214) is defined with the indices saturated as in (2.185). The definite symmetry of χ'_{ijk}, χ''_{ijk} (respectively antisymmetric and symmetric relative to the exchange $i\leftrightarrow j$) simplifies the calculations a great deal.

Up to now we have considered ground-state hadrons. It is very instructive to study the decay of orbitally excited mesons like $B \to \omega\pi$, $A_1 \to \rho\pi$. B and A_1 are isotriplets with $J^{PC} = 1^{+-}$ and $J^{PC} = 1^{++}$ for the neutral states B^0 and A_1^0 (see Figure 1.2). Two partial waves $l=0$, (S and D) are allowed for the relative angular momentum between the vector mesons ω or ρ and the π. Equivalently, the ω or the ρ can be emitted with transverse or longitudinal polarizations. We then have, unlike the preceding cases of $f_{\rho\pi\pi}$ and $g_{\pi NN}$, two independent amplitudes, S, D ($l=0$, 2) or equivalently $M(\pm 1), M(0)$ (transverse or longitudinal vector-meson polarization). Since the decaying mesons in this case have $q\bar{q}$ orbital angular momentum $L=1$, the integral equivalent to (2.204) will be

$$I_m = \tfrac{1}{8}\delta(\boldsymbol{p}_V + \boldsymbol{p}_\pi) \int d\boldsymbol{p}\, \mathcal{Y}_1^m(\boldsymbol{p}_\pi - \boldsymbol{p})\, \psi_1^{-m}(\boldsymbol{p} + \boldsymbol{p}_\pi)\, \psi_\pi^*(\boldsymbol{p})\, \psi_V^*(-\boldsymbol{p}), \qquad (2.217)$$

where V denotes the final vector meson and $\psi^m(\boldsymbol{k})$ is now the spatial wave function of the $L=1$ decaying meson:

$$\psi_1^m(\boldsymbol{p}) = -i\left(\frac{2}{3}\right)^{1/2} \frac{R_M^{5/2}}{\pi^{1/4}} \mathcal{Y}_1^m(\boldsymbol{p}) \exp(-\tfrac{1}{8}R_M^2 p^2). \qquad (2.218)$$

Now defining

$$I_m = \delta(\boldsymbol{p}_V + \boldsymbol{p}_\pi)\, J_m, \qquad (2.219)$$

we obtain the ratios of helicity amplitudes $M(\lambda_V)$,

$$\left[\frac{M(\pm 1)}{M(0)}\right]_{A_1 \to \rho\pi} = \frac{J_1 - J_0}{2J_1} = 1 - \frac{1}{3}R_M^2 p_\pi^2, \qquad (2.220)$$

$$\left[\frac{M(0)}{M(\pm 1)}\right]_{B \to \omega\pi} = -\frac{J_0}{J_1} = 1 - \frac{2}{3}R_M^2 p_\pi^2. \qquad (2.221)$$

The model therefore fixes the relative magnitudes of the helicity amplitudes, or equivalently the ratio of partial waves D/S:

$$\frac{D}{S} = \frac{\sqrt{2}(1-r)}{2+r} \qquad (2.222)$$

where $r = M(0)/M(\pm 1)$. We see from this formula and (2.220) and (2.221) that $D/S \simeq O(R_M^2 p_\pi^2)$, the expected behavior for a centrifugal barrier.

In Section 4.5 we shall discuss in detail a number of phenomenological applications of the $^3P_0$ quark-pair-creation model. In Section 2.2.5 we shall give a general formalism that allows us to apply the model to any three-body process like $N^* \to \rho N$ or $\Delta\pi$.

The $^3P_0$ quark-pair-creation model is found to be a very powerful and successful instrument for describing a great number of strong decays, starting with a very simple idea. On the other hand, it has a basic weakness:

there is no definite physical picture underlying the process of pair creation. Other models that we describe now try to relate the pair creation to the strong-interaction Hamiltonian. The price is paid, however, in a certain loss of simplicity. Moreover, this loss of simplicity does not yet seem to be compensated by any very significant quantitative improvements.

2.2.3(d) The Cornell model of strong decays

The Cornell group has made an ambitious proposal to describe the pair creation inherent in the usual strong decays. It is a straightforward extension of the potential model.

The assumption of the Cornell group (Eichten *et al.*, 1978) is that the decay process is governed by the same interaction Hamiltonian that describes the binding, formulated in a second-quantized form. If $V(\boldsymbol{x}-\boldsymbol{y})$ denotes the two-body confining potential, it can be extended to a four-fermion operator form:

$$H_{\mathrm{I}} = \tfrac{1}{2}\int \mathrm{d}\boldsymbol{x}\,\mathrm{d}\boldsymbol{y}\, q_a^\dagger(\boldsymbol{x})q_a(\boldsymbol{x})V(\boldsymbol{x}-\boldsymbol{y})q_b^\dagger(\boldsymbol{y})q_b(\boldsymbol{y}), \tag{2.223}$$

with a and b labelling the flavor. It is assumed moreover that $V(\boldsymbol{x}-\boldsymbol{y})$ is flavor-independent: this makes it possible to extend the same potential to any couple of flavors like $\mathrm{u\bar{u}}$, $\mathrm{c\bar{c}}$ or $\mathrm{u\bar{s}}$. Finally, color can be introduced by using the potential

$$V_{\mathrm{tot}} = \sum_{\substack{i>j\\ \alpha}} \frac{\lambda_i^\alpha}{2}\frac{\lambda_j^\alpha}{2} V_0(\boldsymbol{r}_i-\boldsymbol{r}_j), \tag{2.224}$$

to give finally

$$H_{\mathrm{I}} = \tfrac{1}{2}\int \mathrm{d}\boldsymbol{x}\,\mathrm{d}\boldsymbol{y} \sum_{a,b,\alpha} :q_a^\dagger(\boldsymbol{x})\frac{\lambda^\alpha}{2}q_a(\boldsymbol{x})\, V_0(\boldsymbol{x}-\boldsymbol{y})\, q_b^\dagger(\boldsymbol{y})\frac{\lambda^\alpha}{2}q_b(\boldsymbol{y}): \tag{2.225}$$

$V_0(\boldsymbol{x}-\boldsymbol{y})$ is related to the usual $\mathrm{q\bar{q}}$ potential within a meson $V(\boldsymbol{x}-\boldsymbol{y})$ (with color omitted) by

$$V = -\tfrac{4}{3}V_0 \tag{2.226}$$

In (2.225) the quark-field operators can be understood simply as the nonrelativistic ones, introduced in the second quantization of an N-body system with two-body forces (see e.g. Schweber, 1961) or these introduced in (1.44), which contain both creation and annihilation operators. In that case, the second-quantized version would be completely equivalent to the usual first-quantized Schrödinger equation. But we may alternatively *assume* that the quark fields are the relativistic ones, introduced in (2.3b). This is the line of thought proposed at the end of Section 1.3.5.

Then it is easy to see that the interaction (2.225) induces quark-pair creation and can therefore lead to decay. The assumption of the Cornell group is that this is the main mechanism responsible for strong-interaction decay. It is noted that (2.225) implies that the potential is considered as operating, in Dirac space, as the time component $\gamma^0\gamma^0$ of a vector-like exchange $\gamma^\mu\gamma_\mu$. Consider now the matrix elements of (2.225) between an initial state A and a final state B + C, where B and C are two hadrons. A nonrelativistic approximation is made, using the Pauli approximation (2.9). We simply quote the result given by the authors, and refer to their paper for further explanations. For the decay of a charmonium $c\bar{c}$ state ψ_n into two charmed mesons C_1 and $\bar{C}_2$, one has, in the rest frame,

$$\begin{aligned}&\langle C_1(\boldsymbol{P})\bar{C}_2(\boldsymbol{P}')|\, H_I\,|\psi_n\rangle\\ &= \frac{-i}{(2\pi)^{3/2}}\,\delta(\boldsymbol{P}+\boldsymbol{P}')\frac{1}{\sqrt{3}}\frac{1}{2m_q}\sum_s \int d\boldsymbol{x}\, d\boldsymbol{y}[\chi^\dagger(s_2')\boldsymbol{\sigma}\cdot\nabla V(\boldsymbol{x})\chi(-s_1')] \qquad (2.227)\\ &\times \varphi_1^*(\boldsymbol{x}, s_1, s_1')\,\varphi_2^*(\boldsymbol{x}-\boldsymbol{y}, s_2, s_2')\,\psi_n(\boldsymbol{y}, s_1, s_2)\, e^{-i\mu_c \boldsymbol{P}\cdot\boldsymbol{y}}\end{aligned}$$

The s_i are spin indices, φ_i the wave functions of the charmed mesons, ψ_n that of charmonium, χ the usual Pauli spinors, and $\mu_c = m_c/(m_c + m_q)$, where q is the companion of c in the charmed mesons C_1 and $\bar{C}_2$.

Formula (2.227) certainly bears a similarity with the $^3P_0$ quark-pair-creation model. In particular, the general structure of the amplitude consists in a convolution-product overlap of three hadron wave functions with a factor corresponding to the created pair. The striking point is that there is now no free parameter: we can in principle calculate the decay from knowledge of the charmonium potential.

If we want to make a more careful comparison of the Cornell model with the $^3P_0$ QPC model, we may look for the matrix element of pion emission by a quark. We find

$$\langle \pi q|\, H_I\,|q\rangle \sim \frac{1}{2m}\int d\boldsymbol{p}_3\, \psi_\pi(2\boldsymbol{p}_3 - \boldsymbol{k}_\pi)\,\tilde{V}(\boldsymbol{p}_3 + \boldsymbol{p}_4)\,\boldsymbol{\sigma}\cdot(\boldsymbol{p}_3 + \boldsymbol{p}_4), \qquad (2.228)$$

where $\boldsymbol{p}_3$ and $\boldsymbol{p}_4$ are the momenta of the two created quarks and $\tilde{V}$ is the Fourier transform of the potential (Figure 2.12). We see that we get a similar structure to the formula (2.196) of the QPC model, with a recoil term, and the Fourier transform of the pion wave function, which reflects its spatial structure.

However, there are also definite differences: in the Cornell model the created pair receives nonzero momentum from the initial quarks, and one of the initial quarks necessarily changes its momentum in the process; in contrast, in the

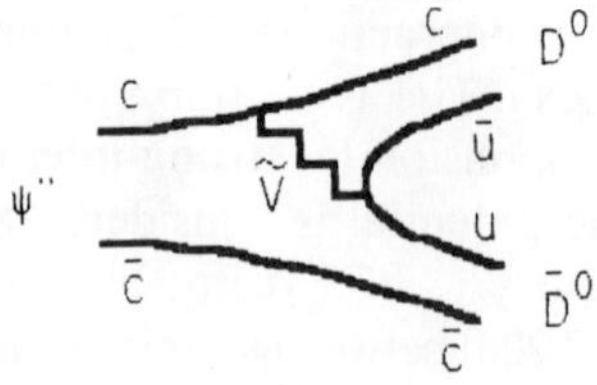

Figure 2.12 The cornell model of hadron vertices. $\tilde{V}$ is the Fourier transform of the potential responsible for the binding of quarks within hadrons.

$^3P_0$ quark-pair-creation model the pair has a zero total momentum and the initial quarks remain spectators.

Up to now, it has not been possible to clearly discriminate the two models by experimental tests. The $^3P_0$ model has been more extensively applied, probably because it is simpler. Moreover, the usual presentation of the results of the Cornell model is complicated by the introduction of the coupled-channel corrections, which will be described in Chapter 3.

2.2.3(e) Brief comments on QCD string models and further comparison between the pair-creation models

There have recently appeared new models of meson decays through quark-pair creation. We term them rather loosely QCD string models. There are essentially three of them, which we shall call respectively the Bristol model (Alcock and Cottingham, 1984), the Heidelberg model (Dosch and Gromes, 1986) and the Toronto model (Kokoski and Isgur, 1985). Generally speaking, they can be considered as attempts to give more theoretical foundation to the quark-pair-creation model of Section 2.2.3(a); they retain its main features, while possibly departing from it on some points, which we discuss below. Our comments will be very brief, since the topic is very recent and many things need to be clarified. The common idea is to give a more precise picture of the hadronic medium that is supposed to give rise to pair creation in our model; the hadronic medium is naturally described as a *gluonic* field. Moreover, this gluonic field is described by strong-coupling features akin to the idea of a *string*: the gluonic electric field is concentrated in a tube surrounding the line joining the two initial $q\bar{q}$ if the initial state is a meson. It is from this string or flux tube that the quark pair emerges. The Bristol model aims at a mostly phenomenological description of the pair-creation mechanism, so that the consequence of the string picture lies essentially in the location of the pair creation along the $q\bar{q}$ line of the initial meson, and a specific spin–orbital structure $^3S_1$ different from the popular $^3P_0$. The Heidelberg and Toronto models, on the other hand, appeal to more formal arguments of lattice QCD in the strong-coupling expansion; they connect pair creation with the bilinear quark kinetic part of the QCD Hamiltonian, which is to be treated as a pertubation in the strong-coupling expansion of the Hamiltonian or the equivalent version in Lagrangian formalism. This term is treated either in a power series (Heidelberg) or at first order (Toronto). As a consequence, in the Heidelberg model the two components of the pair can be created at distant points, while they appear at the same point in the Toronto one. In the strict strong-coupling expansion the pair must be created on the $q\bar{q}$ line (Heidelberg); but the authors abandon this aspect of the strong coupling expansion and formulate a more flexible model, in which the location of pair creation is arbitrary. As to the Toronto model, it takes into account the oscillations of the string and then concludes that creation takes place in a tube around the $q\bar{q}$ line.

Let us now compare more systematically the whole set of models, including the Orsay quark-pair-creation model and the Cornell model. First, we shall compare them from the point of view of their *phenomenological* outcome, disregarding the theoretical considerations that are at their origin. Of course, they have the obvious common feature of expressing the amplitude as an overlap of the three hadron wave functions. But they differ in three main properties of the created pair:

(i) the spin–orbital structure $^3S_1$ (Bristol) versus $^3P_0$ in the other four;

(ii) the location of the pair creation — the string or tube linking the two initial quarks (Bristol and Toronto) versus the whole space (Orsay, Cornell and Heidelberg);

(iii) the locality of the pair creation, i.e. whether the two quarks are created at the same point (Orsay, Cornell, Bristol and Toronto) or at separate points (Heidelberg).

Concerning (iii), the locality of the pair creation in the Orsay model can be exhibited by returning to configuration space in the integral (2.190) concerning baryons:

$$I_m = \int d\boldsymbol{r}_1 \dots d\boldsymbol{r}_5 \, \Psi_{\mathrm{B}}^{\mathrm{tot}}(\boldsymbol{r}_1,\boldsymbol{r}_2,\boldsymbol{r}_4)^* \, \Psi_{\mathrm{C}}^{\mathrm{tot}}(\boldsymbol{r}_3,\boldsymbol{r}_5)^* \\ \Psi_{\mathrm{A}}^{\mathrm{tot}}(\boldsymbol{r}_1,\boldsymbol{r}_2,\boldsymbol{r}_3)\, \mathscr{Y}_1^m(-2i\boldsymbol{\nabla}_4)\,\delta(\boldsymbol{r}_4-\boldsymbol{r}_5). \qquad (2.229)$$

The locality is exhibited by the δ-function, which implies $\boldsymbol{r}_4=\boldsymbol{r}_5$. On the other hand, the pair is not localized, in the sense that the pair-creation operator does not depend on the relative position with respect to the initial quarks: the dependence is only an indirect one, through the fall-off of the final hadron wave functions at large distance.

From a phenomenological point of view it is easy to test the $^3P_0$ structure, and this has been done extensively. The $^3S_1$ structure of the Bristol model seems to be discarded by the decay $B \to \omega\pi$ (Kokoski and Isgur, 1985). In contrast it is much more difficult to discriminate models on the basis of the two other properties (ii) and (iii), which lead only to differences in detail in the spatial integrals. In fact, the Toronto group admit that in practice they recover the results of the Orsay model. It therefore seems that the string structure will be difficult to exhibit experimentally. As to property (iii), the Heidelberg group finds experimental support in its formulation by considering SU(3) breaking. Two general remarks are in order in the comparison with experiment: in such models there are always additional approximations, in particular the nonrelativistic approximation, which are not derived from the general motivations of the model. It is probable that relativistic corrections will blur the details of the spatial integrals, rendering comparison more difficult; on the other hand, since there is no general consensus on the wave functions of light hadrons (see Section 4.1.2), large differences in predictions may come from such differences in the wave functions, which would have nothing to do with the decay models.

Let us now briefly compare the models on the *theoretical* side. The Orsay quark-pair-creation model of course does not claim to be more than a purely phenomenological model. The others are more ambitious. One may say that the theoretical background of the Cornell model is still simple-minded, because it is a rather straightforward extension of the potential model to pair creation. The last three models claim to appeal to more fundamental QCD ideas. On the other hand, in these string or strong-coupling models, the connection of the actual models with the fundamental ideas is rather flexible. The $^3S_1$ structure of the Bristol model does not appear to be a necessary consequence of the string idea. The authors of the Heidelberg and Toronto models recognize that the strong-coupling limit is not the relevant scale of phenomenology, so that one is compelled to make intuitive guesses in the intermediate-coupling range. Something disappointing in the present formulation of these models is that one does not find a connection between the pair-creation strength and such fundamental parameters as the string tension, after having abandoned the strict strong-coupling limit. Paradoxically, such a relation is found rather in the Cornell formulation, where the pair-creation strength is strictly related to the string tension, since the gradient of the linear confining potential, which appears in (2.227), is just the string tension κ in absolute magnitude (see (1.100)). In fact the relativistic-potential approach presented in Section 1.3.5 would have, if the Cornell approach were correct, an impressive success in relating many aspects of the quark model:

(i) the spatial dependence of the potential;

(ii) the spin dependence of the potential, mainly relevant to spectroscopy;

(iii) the strong decays.

It could also possibly deal with dynamical chiral-symmetry breaking (see the comments at the end of Section 1.3.5). All this would depend on the same basic parameter κ, the string tension. It is therefore very desirable to check precisely whether the Cornell model is able to describe such a wide domain of strong decays as the Orsay quark-pair-creation model. In fact, one may wonder why this model has been so little tested in strong decays, while on the other hand a great amount of work has been devoted to the more detailed coupled-channel effects. As to string, flux-tube or similar models, they may be an alternative to potentials if the difficulties encountered in the description of the spin structure of the confining potential, as reviewed in Section 1.3.5, persist; a major breakthrough, however, would be the prediction of the magnitude of the pair creation in these frameworks.

2.2.4 Rearrangement Models of Diquonium Decay

We end this review of decay models by some very short indications regarding a new type of situation, encountered in diquonium decays. Diquonia are states made of two quarks and two antiquarks, of the type $qq\bar{q}\bar{q}$ in a color-singlet state. In a bag model these states have exactly the same status as baryons (qqq) or mesons ($q\bar{q}$), and one can imagine in fact an infinity of possible multiquark states on an equal footing (Rosenzweig, 1976). Diquonium spectra (Jaffe and Low, 1979) have been extensively considered in the bag-model framework (Chodos *et al.*, 1974a,b; Degrand *et al.*, 1975). As to decays, in this framework hadronic strong decays always consist in the fission of the bag into two or more new bags. Although diquonia can decay into two mesons,

$$(qq\bar{q}\bar{q}) \rightarrow (q\bar{q}) + (q\bar{q}), \tag{2.230}$$

without change in the number of quarks, one needs an interaction describing the fission of the initial bag, not included in the bag Hamiltonian, at least in its usual formulation.

In the potential approach the binding Hamiltonian can in principle also describe the rearrangement decay. It is well known in atomic physics that the Coulomb potential responsible for atomic or molecular bound states may also lead to rearrangement reactions. The corresponding situation in particle physics would be that two meson states undergo a scattering at short distance owing to the same interquark forces that produce $q\bar{q}$ meson bound states. There is in fact no reason why the confining forces should not imply some *nonconfining* interaction between quarks belonging to different hadrons. It will then be possible in general that quarks in a multiquark color-singlet state (diquonium) undergo a rearrangement into a final two-meson state (Figure 2.13). Diquonia will appear as resonances in the two-meson scattering

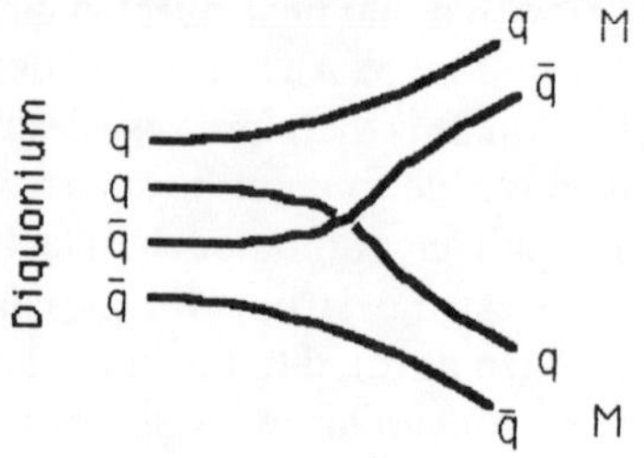

Figure 2.13 The decay of diquonium (qqq̄q̄) into two mesons by quark rearrangement.

amplitude. Discrete bound states will only be a special case that occurs when the energy level is below the two meson thresholds or when special selection rules of dynamical origin forbid the decay. In this approach, the problem of finding the diquonium states and that of finding their decay widths are both reduced to that of calculating the scattering amplitude for two mesons and finding the resonances (for instance, by looking at the phase shifts and their rate of variation with energy, one finds the resonance position and width).

The essential condition for introducing such a picture is to have a binding Hamiltonian presenting a two-meson continuum spectrum in the qqq̄q̄ color-singlet sector. The Hamiltonian must therefore include what is called *saturation of color forces*: two qq̄ color singlets must not interact too much, so that it is possible to separate them. In the potential approach, saturation can be introduced in the way suggested many years ago (Nambu, 1966; Lipkin, 1973a), by giving to the potential a color structure analogous to gluon exchange (see Chapter 1):

$$H = \sum_i \frac{p_i^2}{2m_i} + \sum_{\substack{i>j \\ \alpha}} \frac{\lambda_i^\alpha}{2} \frac{\lambda_j^\alpha}{2} V_0(\boldsymbol{r}_i - \boldsymbol{r}_j). \tag{2.231}$$

(For antiquarks, the λ_i must be understood as $\bar{\lambda}_i = -\lambda_i^* = -\lambda_i^t$). It is then easy to see that confining forces saturate in the sense that they cancel between two color-singlet states. On the other hand, the saturation is not perfect, because long-range forces, of the van der Waals type, of second order in V_0, can in principle also occur (Gavela *et al.*, 1979). We shall disregard this drawback, and present a method to get the S matrix for meson–meson scattering (Gavela *et al.*, 1978). This is an extension of the methods proposed by Mott and Massey (1949) to go beyond the Born or impulse approximations and to account for resonant scattering. The problem is here more complicated than usual because the two incident particles are composite. Although the method has not yet been extensively used and is new in the quark model, we think it is worthwhile explaining it for future applications.

The essential point is to use trial wave functions for the two-meson (q$\bar{q}$)(q$\bar{q}$) system of the type

$$\Psi = \sum_{i,j} M_i(\boldsymbol{r}'_1) M_j(\boldsymbol{r}'_2) \psi_{ij}(\boldsymbol{R}), \tag{2.232}$$

where $\boldsymbol{r}'_1$ and $\boldsymbol{r}'_2$ are the relative coordinates inside the two mesons and $\boldsymbol{R}$ is the relative coordinate between the two mesons. M_i and M_j are the sets of known wave functions of the mesons in terms of quarks, the various possible states being labelled by i and j. The coefficients $\psi_{ij}(\boldsymbol{R})$ are wave functions to be determined, which should have the asymptotic behavior of scattering states with the fundamental mesons M_0 as incident channel.

$$\psi_{ij}(\boldsymbol{R}) \underset{|\boldsymbol{R}| \to \infty}{\sim} \delta_{i0}\delta_{j0}\, e^{i\boldsymbol{k}_{00}\cdot \boldsymbol{R}} + f_{ij}(\hat{\boldsymbol{R}}) \frac{e^{ik_{ij}R}}{R}, \tag{2.233}$$

k_{ij} denoting the relative momenta in the various open channels. This formula would be exact if one could consider the whole set of mesonic excited states i, j. In practice one is compelled to truncate the series, and one is then left with a set of approximate coupled integro-differential equations for the functions ψ_{ij}. These coupled equations are obtained by the variational principle

$$\delta\left[\frac{\langle \Psi | H | \Psi \rangle}{\langle \Psi | \Psi \rangle}\right] = 0. \tag{2.234}$$

In general, the coupled equations can be solved only by numerical methods. The problem is simplified, however, if one specifies the potential by assuming the harmonic-oscillator form:

$$V_0(\boldsymbol{r}_i - \boldsymbol{r}_j) = -a(\boldsymbol{r}_i - \boldsymbol{r}_j)^2. \tag{2.235}$$

It must be emphasized that this is a much stronger assumption than the use of harmonic-oscillator trial wave functions.

With (2.235), the problem is reduced to that of a separable potential, and the coupled equations can be solved analytically. Moreover, the problem can be simplified by using a special set of relative coordinates for the qq$\bar{q}\bar{q}$ system:

$$\begin{aligned} \boldsymbol{r} &= \tfrac{1}{2}(\boldsymbol{r}_1 - \boldsymbol{r}_2 + \boldsymbol{r}_{\bar{1}} - \boldsymbol{r}_{\bar{2}}) \sim \boldsymbol{R}, \\ \boldsymbol{r}' &= \tfrac{1}{2}(\boldsymbol{r}_1 - \boldsymbol{r}_2 + \boldsymbol{r}_{\bar{2}} - \boldsymbol{r}_{\bar{1}}) \sim \boldsymbol{r}'_1 - \boldsymbol{r}'_2, \\ \boldsymbol{r}_{\mathrm{d}\bar{\mathrm{d}}} &= \tfrac{1}{2}(\boldsymbol{r}_1 - \boldsymbol{r}_{\bar{1}} + \boldsymbol{r}_2 - \boldsymbol{r}_{\bar{2}}) \sim \boldsymbol{r}'_1 + \boldsymbol{r}'_2, \end{aligned} \tag{2.236}$$

where 1 and 2 label the quarks inside mesons 1 and 2, and $\bar{1}$ and $\bar{2}$ label the antiquarks. With (2.235) and (2.236), the Hamiltonian (2.231) takes the form of a sum of squares in the color-singlet sector. However, it is *not* separable in a sum of independent harmonic oscillators, owing to color factors. But, since it is a sum of squares, the orbital angular momentum relative to each of the coordinates (2.236) is separately conserved. These orbital-angular-momentum selection rules are important in simplifying the problem. Moreover, they allow prediction of bound states for sufficiently high L_{dd}, the orbital momentum relative to $\boldsymbol{r}_{\mathrm{dd}}$. In addition, we find some resonances. If the actual potential is not too different from the harmonic oscillator, we then expect to have diquonia as resonances in meson–meson scattering.

Finally let us note that in the context of the bag model, another method, the P-matrix formalism, has been devised to study meson–meson scattering through intermediate diquonium states (Jaffe and Low, 1979). We refer to the original paper for an account of the method, which is beyond the scope of the standard potential approach in the nonrelativistic approximation.

2.2.5 Computation of Matrix Elements in the QPC Model

There is a method for handling problems involving several arbitrary angular momenta; it expresses matrix elements in terms of *reduced* matrix elements and various coupling coefficients (Messiah, 1959). In radiative transitions it is well known as the multipole expansion (Blatt and Weisskopf, 1952). We apply it to the more complicated case of strong decays. We now give the explicit formulae for the QPC model (Section 2.2.3(a)), but the same method can also be applied to other pair-creation models (Sections 2.2.3(d), (e)).

2.2.5(a) Baryon decay

The principles of the model have been described in detail in Section 2.2.3(a). The S matrix is given by (2.180) in terms of the T operator given by (2.181) in second-quantization formalism. The general structure of the T-matrix elements in terms of wave functions is indicated by (2.185), and illustrated by the diagram of Figure 2.11(a). In this section we shall give the general formulae for dealing with the angular-momenta couplings in the calculation of these matrix elements (Le Yaouanc *et al.*, 1975b).

A hadron wave function has the form $\Sigma\varphi(\chi_S\psi_L)_J$, where φ, χ_S and ψ_L are flavor, spin and space wave functions, and the quark spin S is coupled with the angular orbital momentum L to give the hadron spin J. We discard the color factor, whose effect is fully explained in Section 2.2.3(a); the total wave function for baryons is then symmetric. The sum may combine the symmetry components (in case of mixed symmetry) of χ, φ, ψ to form a symmetric total wave function. The sum will be omitted for simplicity in the following; it is assumed that a term is picked up in each hadron wave function, and a sum is to be restored in the final formula (2.247).

Let us now obtain the general form of the matrix element for $\mathrm{A} \to \mathrm{B} + \mathrm{M}$, where A and B are baryons and M is a meson. We shall use the notation J_{X}, L_{X}, S_{X} and I_{X} ($\mathrm{X} = \mathrm{A}$, B, M, P) for the spin, internal quark orbital angular momentum, total quark spin, and isospin of the hadrons (A, B, M) or of the pair (P), with the coupling $\boldsymbol{J}_{\mathrm{X}} = \boldsymbol{L}_{\mathrm{X}} + \boldsymbol{S}_{\mathrm{X}}$ (here and in the following, the order in which two coupled momenta are written is intended to give their order in the Clebsch–Gordan coefficients). The partial decay amplitudes $M(l,s)$ for the decay $\mathrm{A} \to \mathrm{B} + \mathrm{M}$ are defined with the following conventions for adding angular momenta and isospins:

$$\begin{aligned} \boldsymbol{J}_{\mathrm{B}} + \boldsymbol{J}_{\mathrm{M}} &= \boldsymbol{s}, \\ \boldsymbol{l} + \boldsymbol{s} &= \boldsymbol{J}_{\mathrm{A}}, \\ \boldsymbol{I}_{\mathrm{B}} + \boldsymbol{I}_{\mathrm{M}} &= \boldsymbol{I} = \boldsymbol{I}_{\mathrm{A}}. \end{aligned} \tag{2.237}$$

$\boldsymbol{l}$ is the orbital angular momentum between the final hadrons M and B. The total amplitude is then expressed in terms of the partial amplitudes $M(l,s)$ by the formula

$$\begin{aligned} & M(\mathrm{A} \to \mathrm{BM}) \\ & = \langle I_{\mathrm{B}} I_{\mathrm{M}} i_{\mathrm{B}} i_{\mathrm{M}} | I_{\mathrm{A}} i_{\mathrm{A}} \rangle \sum_{\substack{l,\, m \\ s,\, \mu}} M(l,s) \langle J_{\mathrm{B}} J_{\mathrm{M}} M_{\mathrm{B}} M_{\mathrm{M}} | s \mu \rangle \langle l s m \mu | J_{\mathrm{A}} M_{\mathrm{A}} \rangle\, Y_1^m(\hat{\boldsymbol{k}}_{\mathrm{M}}), \end{aligned} \tag{2.238}$$

where we have discarded a total-momentum-conserving δ-factor; A is at rest, $\boldsymbol{k}_{\mathrm{M}}$ and $-\boldsymbol{k}_{\mathrm{M}}$ are the momenta of M and B; M_{X} and i_{X} ($\mathrm{X} = \mathrm{A}$, B, M) are the spin and isospin third components of the hadrons. Our aim is now to express these partial amplitudes in terms of reduced spatial, spin and isospin matrix elements.

First let us define the reduced spatial matrix elements. In the spatial integral, given in (2.190), we introduce the internal wave functions (2.22), by $\Psi^{\mathrm{tot}} = \delta(\Sigma \boldsymbol{p} - \boldsymbol{P})\psi$, and discard a total-momentum-conserving δ-factor, obtaining

$$\begin{aligned} & I^{(L_{\mathrm{A}};L_{\mathrm{B}}L_{\mathrm{M}})}_{M_{\mathrm{A}}M_{\mathrm{P}};M_{\mathrm{B}}M_{\mathrm{M}}}(\boldsymbol{k}_{\mathrm{M}}) = 3\gamma \int \mathrm{d}\boldsymbol{p}_1 \ldots \mathrm{d}\boldsymbol{p}_5\, \psi_{\mathrm{B}}^{M_{\mathrm{B}}}(\boldsymbol{p}_1, \boldsymbol{p}_2, \boldsymbol{p}_4)^* \,\delta(\boldsymbol{p}_1 + \boldsymbol{p}_2 + \boldsymbol{p}_4 + \boldsymbol{k}_{\mathrm{M}}) \\ & \psi_{M}^{M_{\mathrm{M}}}(\boldsymbol{p}_3, \boldsymbol{p}_5)^* \,\delta(\boldsymbol{p}_3 + \boldsymbol{p}_5 - \boldsymbol{k}_{\mathrm{M}})\, \mathscr{Y}_1^{M_{\mathrm{P}}}(\boldsymbol{p}_4 - \boldsymbol{p}_5)\, \psi_{\mathrm{A}}^{M_{\mathrm{A}}}(\boldsymbol{p}_1, \boldsymbol{p}_2, \boldsymbol{p}_3)\, \delta(\boldsymbol{p}_1 + \boldsymbol{p}_2 + \boldsymbol{p}_3). \end{aligned} \tag{2.239}$$

Taking into account the transformation law of this spatial integral under rotations of $\boldsymbol{k}_{\mathrm{M}}$, one can show that it can be expressed in terms of Clebsch–Gordan coefficients by the following formula, which defines the reduced matrix elements $\mathscr{L}^{(L_{\mathrm{A}};L_{\mathrm{B}}L_{\mathrm{M}})}_{L_{\mathrm{i}}L_{\mathrm{f}};l}$:

$$I^{(L_A;L_BL_M)}_{M_AM_P;M_BM_M}(\boldsymbol{k}_M)$$
$$= \sum_{L_i,\,L_f,\,l} \mathscr{L}^{(L_A;L_BL_M)}_{L_iL_f;l}(|\boldsymbol{k}_M|)$$
$$\sum_{M_i,\,M_f,\,m} \langle L_AL_PM_AM_P|L_iM_i\rangle$$
$$\times \langle L_BL_MM_BM_M|L_fM_f\rangle\langle lL_fmM_f|L_iM_i\rangle\, Y_l^m(\hat{\boldsymbol{k}}_M). \tag{2.240}$$

This corresponds to the couplings

$$\boldsymbol{L}_A+\boldsymbol{L}_P=\boldsymbol{L}_i, \qquad \boldsymbol{L}_B+\boldsymbol{L}_M=\boldsymbol{L}_f, \qquad \boldsymbol{l}+\boldsymbol{L}_f=\boldsymbol{L}_i. \tag{2.241}$$

The spin and isospin reduced matrix elements $\mathscr{S}^{(S_A;S_BS_M)}_S$ and $\mathscr{I}^{(I_A;I_BI_M)}$ are defined by

$$\langle\chi_B^{\mu_B}(124)\,\chi_M^{\mu_M}(35)\,|\,\chi_A^{\mu_A}(123)\,\chi_P^{\mu_P}(45)\,\rangle$$
$$= \sum_{S,\,\mu} \mathscr{S}^{(S_A;S_BS_M)}_S\,\langle S_AS_P\mu_A\mu_P|S\mu\rangle\,\langle S_BS_M\mu_B\mu_M|S\mu\rangle \tag{2.242}$$

and

$$\langle\chi_B^{i_B}(124)\,\varphi_M^{i_M}(35)\,|\,\varphi_A^{i_A}(123)\,\varphi_0(45)\,\rangle$$
$$= \mathscr{I}^{(I_A;I_BI_M)}\langle I_BI_Mi_Bi_M|I_Ai_A\rangle, \tag{2.243}$$

where χ and φ are spin and isospin wave functions. This corresponds to the couplings

$$\left.\begin{array}{l} \boldsymbol{S}_A+\boldsymbol{S}_P=\boldsymbol{S}, \quad \boldsymbol{S}_B+\boldsymbol{S}_M=\boldsymbol{S} \\ \boldsymbol{I}_B+\boldsymbol{I}_M=\boldsymbol{I}_A \end{array}\right\} \tag{2.244}$$

Denoting by S_{12} the spin of the diquark (q_1q_2) (which must be the same in A and B to have a nonzero contribution), we can write an expression for the spin-matrix elements,

$$\mathscr{S}^{(S_A;S_BS_M)}_S = [(2S_B+1)(2S_M+1)(2S_A+1)(2S_P+1)]^{1/2}$$
$$\times \left\{\begin{array}{ccc} S_{12} & \frac{1}{2} & S_B \\ \frac{1}{2} & \frac{1}{2} & S_M \\ S_A & S_P & S \end{array}\right\}, \tag{2.245}$$

and for isospin, taking into account that I_P=0, we get analogously

$$\mathscr{I}^{(I_A;I_BI_M)} = (-1)^{I_{12}+I_M+I_A+I_3}[\tfrac{1}{2}(2I_M+1)(2I_B+1)]^{1/2}$$
$$\times \left\{\begin{array}{ccc} I_{12} & I_B & I_4 \\ I_M & I_3 & I_A \end{array}\right\} \tag{2.246}$$

In using (2.245) or (2.246), one must be aware that they assume spin or isospin wave functions constructed by standard Clebsch–Gordan couplings.

From (2.238), (2.240), (2.242) and (2.243), a standard (but lengthy) recoupling calculation (Messiah, 1959) gives the partial amplitudes in terms of the reduced matrix elements:

$$M(l,s) = f\mathscr{I}^{(I_A;I_BI_M)} \sum_{L_i,L_f,S} \mathscr{L}^{(L_A;L_BL_M)}_{L_iL_f;l}\mathscr{S}^{(S_A;S_BS_M)}_{S}$$
$$\times(-1)^{1+L_f+L_A+L_P}(2S+1)(2L_i+1)[\tfrac{1}{3}(2J_B+1)(2L_f+1)$$
$$\times(2J_M+1)(2s+1)]^{1/2} \tag{2.247}$$
$$\times\begin{Bmatrix} L_A & S_A & J_A \\ S & L_i & L_P \end{Bmatrix}\begin{Bmatrix} L_f & l & L_i \\ J_A & S & s \end{Bmatrix}\begin{Bmatrix} L_B & S_B & J_B \\ L_M & S_M & J_M \\ L_f & S & s \end{Bmatrix}.$$

The constant f takes the value $(\frac{2}{3})^{1/2}$ or $-(\frac{1}{3})^{1/2}$ according to the isospin $\frac{1}{2}$ or 0 of the created quarks; this is due to the assumption that the created pair is in an $SU(3)_F$-singlet state (see φ_0 in (2.181)).

2.2.5(b) Harmonic-oscillator wave functions, centrifugal barrier and anti-$SU(6)_W$ signs

We compute the reduced matrix elements $\mathscr{L}^{(L_A;L_BL_M)}_{L_iL_f;l}$ in the case $L_B=L_M=0$, which implies $L_f=0$ from (2.241). We use the harmonic-oscillator wave functions corresponding to the Hamiltonian (1.82). The baryon wave function is a sum of terms containing a spatial part proportional to $\mathscr{Y}^{m_1}_{l_1}(\boldsymbol{p}_\lambda)\,\mathscr{Y}^{m_2}_{l_2}(\boldsymbol{p}_\rho)$. Since $L_B=0$, it is obvious that only $\mathscr{Y}^{M_A}_{L_A}(\boldsymbol{p}_\lambda)$ gives a nonvanishing overlap for the ρ variable in (2.239). We thus write

$$\psi^{M_A}_{L_A}=(-i)^{L_A}N_A\,\mathscr{Y}^{M_A}_{L_A}(\boldsymbol{p}_\lambda)\exp[-\tfrac{1}{2}R_A^2(\boldsymbol{p}_\lambda^2+\boldsymbol{p}_\rho^2)] \tag{2.248}$$

$$N_A=(3\sqrt{3})^{1/2}\frac{R_A^{3+L_A}}{\pi^{3/4}}\left(\frac{2^{L_A+2}}{(2L_A+1)!!}\right)^{1/2}. \tag{2.249}$$

Using the relation

$$\mathscr{Y}^m_l(\boldsymbol{a}+\boldsymbol{b}) = \sum_{\substack{l_1+l_2=l\\ m_1+m_2=m}} \mathscr{Y}^{m_1}_{l_1}(\boldsymbol{a})\,\mathscr{Y}^{m_2}_{l_2}(\boldsymbol{b})\left(\frac{4\pi(2l+1)}{(2l_1+1)(2l_2+1)}\right)^{1/2}$$
$$\times\left(\frac{(l+m)!(l-m)!}{(l_1+m_1)!(l_1-m_1)!(l_2+m_2)!(l_2-m_2)!}\right)^{1/2}, \tag{2.250}$$

we find from (2.239 and (2.240) that

$$\mathscr{L}^{(L_A;0,0)}_{L_A+1,0;L_A+1} = C(-\mathrm{i}\sqrt{\tfrac{3}{2}})^{L_A}\left(\frac{3(L_A+1)}{2L_A+1}\right)^{1/2}(-xy^{L_A})|\boldsymbol{k}_M|^{L_A+1},$$

$$\mathscr{L}^{(L_A;0,0)}_{L_A-1,0;L_A-1} = -C(-\mathrm{i}\sqrt{\tfrac{3}{2}})^{L_A}\left(\frac{3L_A}{2L_A-1}\right)^{1/2}(y|\boldsymbol{k}_M|)^{L_A-1}\left(\frac{2L_A+1}{\rho^2} - xy\boldsymbol{k}_M^2\right), \tag{2.251}$$

with of course $L_P=1$, and the constants ρ^2, C, x and y are given by

$$\rho^2 = 3R_A^2 + R_M^2,$$
$$C = 3\gamma N_A N_B N_M \pi(\pi/\rho^2)^{3/2},$$
$$x = \frac{4R_A^2+R_M^2}{2\rho^2}, \qquad y = \frac{2R_A^2+R_M^2}{2\rho^2},$$
$$N_B = \frac{(3\sqrt{3})^{1/2}R_A^3}{\pi^{3/2}}, \qquad N_M = \left(\frac{R_M}{\pi^{1/2}}\right)^{3/2}, \tag{2.252}$$

where we have assumed $R_A = R_B$.

In order to underline some physical consequences of these formulae, let us define

$$\mathscr{L}^+ = \left(\frac{2L_A+1}{3(L_A+1)}\right)^{1/2}\mathscr{L}^{(L_A;0,0)}_{L_A+1,0;L_A+1},$$
$$\mathscr{L}^- = -\left(\frac{2L_A-1}{3L_A}\right)^{1/2}\mathscr{L}^{(L_A;0,0)}_{L_A-1,0;L_A-1}. \tag{2.253}$$

We easily obtain

$$\frac{\mathscr{L}^+}{\mathscr{L}^-} = \frac{-xy\boldsymbol{k}_M^2}{(2L_A+1)/\rho^2 - xy\boldsymbol{k}_M^2}. \tag{2.254}$$

The vertex symmetry $SU(6)_W$ predicts $\mathscr{L}^+/\mathscr{L}^- = 1$. This is easy to understand, since $SU(6)_W$ symmetry (see Section 4.1.4 and Lipkin, 1967) is equivalent to neglecting the quark motion in the direction transversal to the momentum $\boldsymbol{k}_M$. This is the case when the radii of the wave functions go to infinity (which, by the uncertainty relation, implies no internal momentum for the quarks); then the first term in the denominator of (2.254) vanishes.

Now we see that the model predicts the anti-$SU(6)_W$ signs (i.e. $\mathscr{L}^+/\mathscr{L}^-$ negative) if

$$\mathbf{k}_M^2 < \frac{2L_A+1}{xy\rho^2}. \tag{2.255}$$

In fact if we take realistic radii, that is $R_B^2 = R_A^2 = 6\,\mathrm{GeV}^{-2}$, $R_M^2 = 8\,\mathrm{GeV}^{-2}$, then

$$\frac{1}{xy\rho^2} = 0.163\,\text{GeV}^{-2}. \tag{2.256}$$

It happens that in all reactions experimentally studied (Le Yaouanc *et al.*, 1975b) we are indeed *in the case of anti*-SU(6)$_W$ *relative signs.*

From (2.251) and the values of the radii we have adopted we can also easily deduce the *centrifugal barrier effect*, since when $\boldsymbol{k}_M^2$ is small

$$\mathscr{L}^+/\mathscr{L}^- \sim (r\boldsymbol{k}_M)^2, \tag{2.257}$$

with $r^2 = xy\rho^2/(2L_A+1)$. Finally we note that the relations (2.251) also imply a cancellation of the lowest partial wave when

$$\boldsymbol{k}_M^2 = \frac{2L_A+1}{xy\rho^2}. \tag{2.258}$$

2.2.5(c) Meson decay

Let us now consider the case where A and B are mesons. The formulae of Section 2.2.5(a) apply, with the following changes. The quark 1 (and momentum $\boldsymbol{p}_1$) have to be suppressed in (2.239), (2.242) and (2.243). The variables 2 and 5 correspond to quarks, while 3 and 4 correspond to antiquarks. In (2.245), S_{12} has to be replaced by $\frac{1}{2}$ (the spin of the quark 2), and in (2.246) I_{12} has to be replaced by the isospin of the quark 2 ($\frac{1}{2}$ for u and d and 0 for s, c etc.).

The factor 3 (sum on the 3 quarks) must be suppressed in (2.239). With these changes, (2.247) gives the contribution of the diagram where the antiquark 3 goes into the final meson M. We must also add the diagram with the quark 2 ending in the meson M. The corresponding changes in the formulae of Section 2.2.5(a) are easy to deduce. It is essential to take care with the signs when exchanging variables and to remember our conventions: we have always written the meson wave functions with the quark first, and the formulae in Section2.2.5(a) assume that the final hadron B is written before the meson M (the corresponding signs have been taken for the Clebsch–Gordon coefficients). It may happen that one of the diagrams is forbidden. For instance in the case of $\phi \to K\bar{K}$, if M is taken to be the final K then it is obvious that the only contribution comes from the diagram with the antiquark 3 ending in M. When both diagrams contribute, they generally give the same absolute value, and the relative sign is minus for decays that are fobidden by C, P or G-parity conservation, and plus for allowed decays (giving a factor 2 in (2.247)). When the two mesons are identical, one distinguishes them by choosing M as the one that goes, say, in the forward hemisphere. If the decay is allowed, the two diagrams give the same contribution, but the corresponding factor 2 is tempered by a reduction factor of $1/\sqrt{2}$ in the amplitude ($\frac{1}{2}$ in width), owing to the restriction of the final phase space to one hemisphere.

Finally, it is useful to give the decay widths in terms of the amplitudes $M(l,s)$ (with the final states expanded in partial waves instead of plane waves as in (2.30) and (2.31a)):

$$\Gamma^{(l,s)}_{A\to BM} = 2\pi \frac{E_B E_M}{m_A} k_M |M(l,s)|^2. \tag{2.259}$$

CHAPTER 3

The quark model beyond the nonrelativistic approximation

Though this be madness, yet there is method in't.

W. SHAKESPEARE, *Hamlet*, II.2.

Zanetto, lascia le donne, e studia la matamatica (sic).

J. J. ROUSSEAU, *Les Confessions, livre VII.*

... sans difficile pas de facile.

MAO TSE TOUNG, *De la Contradiction.*

3.1 INTRODUCTION

As underlined in Chapter 1, the nonrelativistic approximation is justified by simplicity rather than by serious theoretical arguments. It gives in a straightforward way the qualitative features of transition amplitudes and crude quantitative estimates. As we shall see in Chapter 4, these are surprisingly good when compared with experiment. On the whole, this naive model is qualitatively correct, and the relativistic and other higher-order corrections to the naive model are perhaps not essential to get a first view of the quark model. The student may therefore leave the present chapter for a second reading. This is also the reason why we have refrained from making detailed calculations except on the v/c expansion of the Dirac equation. However, it is very important to be aware of what is neglected in the naive approach if one wants to have a better understanding of the status of this approach. It is also important for the future progress of phenomenology, which may become more and more accurate, to know the methods allowing to go beyond the naive model. This is of course also crucial when the naive model fails too badly.

A completely relativistic treatment is in principle contained in the field theory of strong interactions, the local QCD Lagrangian. But there is as yet no systematic treatment along these lines for low- and medium-energy hadronic transitions. This situation is in contrast with that concerning the so-called *hard phenomena*, which can be described by perturbative QCD. Moreover, this procedure would require that the gluon degrees of freedom be taken fully into account. This goes far beyond the simplicity of the naive quark model. Our aim, while being conscious of the ultimate necessity of considering such gluon degrees of freedom, is less ambitious. We shall present methods for introducing relativistic and other corrections to the naive quark model, but we shall retain the picture of dynamics at the level of forces being described phenomenologically by a potential or an effective Bethe–Salpeter kernel. One may question the validity of such intermediate schemes that lack a complete and rigorous field-theoretical basis. Since the applications have not been developed to the same extent as for the naive quark model, we cannot support these relativistic approaches by any striking agreement with experiment except for charmonium E_1 radiative transition (see Section 4.3.2). These approaches are only tentative at present, and have yet to be confirmed. Nevertheless, they can indicate corrections that may be reasonably expected, and this is one of the aims of the present discussion.

Before proceeding with technical developments, it is useful to consider qualitatively some possible departures that can be expected from the naive nonrelativistic model. First, quarks have large velocities when the velocity of the hadron itself is large. Of course, this is not a frame-independent statement, and hadrons may move slowly in one frame while moving fast in another. However, it is clear that in many cases some large hadron velocities are unavoidable, for instance in transitions with large mass differences, as in $\omega \to \pi^0\gamma$, or when the momentum transfer is large, as in large-angle electron–nucleon scattering (nucleon form factors, electroproduction). In low-q^2 transitions like $\omega \to \pi^0\gamma$ one can still use nonrelativistic calculations with a suitable choice of reference frame (see the discussion of Section 4.1.3, and for electroproduction Le Yaouanc *et al.*, 1973). For large q^2 it is essential to have a relativistic treatment of the center-of-mass motion (see Section 3.4.7).

The other aspect of relativistic motion, which is more important for the developments in this chapter, is the large *internal* quark velocity. Even at $\boldsymbol{P}=0$, the rest frame of the hadron, the quarks (especially the light ones) have on average large velocities. We have said in Chapter 1 that the nonrelativistic expansion for light quarks is only formal. Let us now show this in more detail. Our first remark is that the situation is not one of *weak binding*. Indeed in such a case the excitation levels would be close to the ground state, and the interval ω between two levels would be small with respect to the mass of the ground state. This is certainly not the actual case. If we consider for instance the nucleon or the $\Delta(1236)$ and the orbital excitation of the nucleon $N^*(1520)$, we obtain $\omega = 582$ or 284 MeV, of the order of the nucleon mass. In order to understand the impressive successes of SU(6) × O(3) (Dalitz, 1965, 1966, 1967a,b; Dalitz and Sutherland, 1966) it was proposed by Morpurgo (1965) that there is very *strong binding* with very large quark masses ($m_q \gg m_N$) cancelled partly by the very strong potential; there could then be slow quark motion. This was the direction taken later by Böhm, Joos and Krammer (1973b). In that case it can easily be seen that the formulation is very different from that involving the nonrelativistic Schrödinger equation, and the connection with this latter successful approach is very indirect. A simpler and successful attitude, which we adopt throughout this book, is to stick in a first step to the *Schrödinger equation.* Assuming moreover a harmonic-oscillator potential, it has been possible (Faiman and Hendry, 1968; Copley, Karl and Obryk, 1969) to describe the hadron spectrum and hadronic transitions in the framework of Chapter 2. The two essential parameters are the quark mass m and the oscillator radius R. They are sufficiently constrained by the data (for further discussion see Section 4.1.2) so as to imply

$$\frac{v}{c} = \frac{p}{E} \sim \frac{1}{mR} \sim 1. \tag{3.1}$$

Indeed, if one wants to retain the nonrelativistic formula for the nucleon magnetic moment $\mu_p = e/2m$, one sees that m is around 300 MeV, i.e. roughly equal to ω; then (3.1) comes from $\omega \sim m$ and the oscillator formula $\omega \sim 1/mR^2$. This set of parameters is further supported by study of the decay of excitations. We then reach a paradoxical conclusion: the very success of the nonrelativistic quark model for light quarks implies that the parameter v/c is not small. This is why we stress that the v/c expansion is *a priori* only formal in the case of light quarks. We do not offer any explanation for the success of the lowest-order approximation developed in Chapter 2. But we still think that *it is successful* and that the higher orders nevertheless make sense in analysing *the trend* of relativistic effects.

We must be aware that apart from relativistic effects connected with fast spatial motions, there are other types of corrections to the naive quark model. In fact, pair creation by itself is necessarily a relativistic phenomenon: first, pair creation needs kinematically an energy of order $2m$, which would not be available in a weak-binding situation; secondly, the concept of pair creation derives from relativistic fields. And indeed the models that will be explained in Sections 3.3 and 3.4 describe the pair creation present in strong decays as a relativistic effect. We call all these effects higher-order effects, since they appear in an *expansion in powers of the interaction* between quarks.

Since the discovery of charmonium, the hope has arisen that we have finally met a situation with great theoretical simplifications due to the large mass of the charmed quark. Indeed, heavy quarkonia $c\bar{c}$ and $b\bar{b}$ systems appear to be nonrelativistic to a much better degree of approximation than light-quark systems like $d\bar{d}$, $u\bar{u}$ and $s\bar{s}$. There is then the exciting hope of testing the quark model in the nonrelativistic limit, in contrast with the complex situation prevailing in the domain of light quarks. It must not be forgotten, however, that many relativistic aspects remain even in heavy-quarkonium decay, in particular in exclusive channels containing fast-moving light hadrons, or in the physics of the D mesons which contain a light quark. In addition, the discussion of the ψ and Υ systems ($c\bar{c}$ and $b\bar{b}$) is becoming so detailed that even here one can consider relativistic corrections, due to non-negligible v/c; there are even cases for charmonium where the relativistic corrections are not small (E_1 transitions).

Anyway, light-quark physics is unavoidably relativistic—whence the interest of discussing the approaches described below.

3.2 CURRENT-MATRIX ELEMENTS IN THE DIRAC EQUATION

3.2.1 Introduction

The Dirac equation in a central potential has an appealing simplicity, while at the same time it contains a whole set of relativistic effects. Of course, the approximation of neglecting the recoil is a crude one. It prevents us from discussing any effect connected with the overall hadron center-of-mass motion. But this may also be considered as an advantage, because then the Dirac equation isolates the effect of quark internal velocities, which deserve special study. One must be aware that the usual static-potential approximation also leaves aside retardation effects and the important two-body spin–spin forces, which come from transverse gluon exchange (De Rújula *et al.*, 1975). This latter effect is qualitatively important because it explains the hyperfine mass differences, for example between N and Δ. Therefore we cannot handle such mass differences with the Dirac equation in an external static potential.

On the other hand, the Dirac equation represents a rather good approximation, compared with the Bethe–Salpeter (1951) equation in the ladder approximation, since it contains the effect of crossed diagrams. Therefore at least certain effects of pair contributions—in the language of old-fashioned perturbation theory—are included that do not appear in the latter treatment (Figure 3.1).

In Figure 3.1 it is understood that we work in second-quantized perturbation theory, the unperturbed part being the Hamiltonian for free Dirac fields. The first diagram in Figure 3.1 is the crossed diagram in Feynman perturbation theory. The second is one of the contributions to this diagram in time-ordered perturbation theory, considered in an instantaneous approximation. Finally, if one performs a static approximation on the quark in the bottom line, one obtains diagram (c), included in a perturbation expansion of

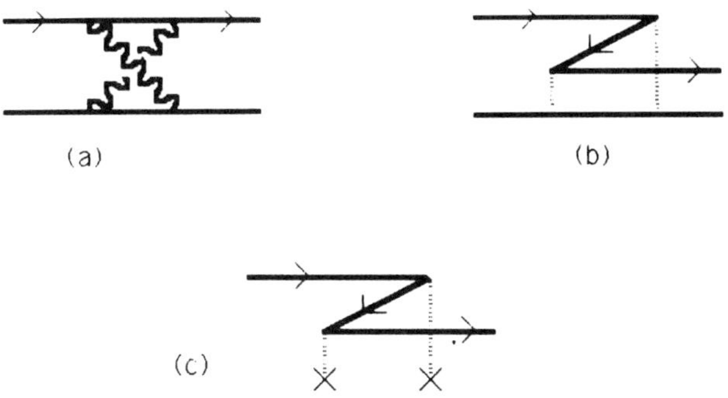

Figure 3.1 A Feynman diagram (a) beyond the ladder approximation in QED, which contains diagram (b) of usual perturbation theory, a Z diagram with an intermediate state containing an extra quark pair; diagram (c) is the external-field analogue of diagram (b) and is fully taken into account by the Dirac equation. Wavy lines correspond to photon propagators, and dotted lines to the instantaneous interaction.

the second-quantized Dirac equation, the perturbation being the potential term.

Let us now make explicit the wave equation for a system of N quarks. We write

$$H_D = \sum_{i=1}^{N} H_D(i) = \sum_{i=1}^{N} [m\beta(i) + \boldsymbol{\alpha}(i)\cdot\boldsymbol{p}_i + V_V(\boldsymbol{r}_i) + \beta(i)V_S(\boldsymbol{r}_i)], \qquad (3.2)$$

$$H_D\Psi = E\Psi \qquad (3.3)$$

for a hadron state of energy E, to be identified with the mass since there is no center-of-mass momentum. On the other hand, we shall see that some states represent *spurious* motions of the center of mass not to be identified with hadrons. The potential contains two main pieces:

$$V(\boldsymbol{r}_i) \equiv V_V(\boldsymbol{r}_i) + \beta V_S(\boldsymbol{r}_i)$$

The *Lorentz-vector* part $V_V(\boldsymbol{r}_i)$ contains a short-distance piece, and $V_S(\boldsymbol{r}_i)$ represents the possible confining *Lorentz-scalar* part, which corresponds to the usual two-body potentials considered in spectroscopy (see Section 1.3.5). The Lorentz structure of the potential is important for the calculations of transitions, since V_V and V_S will enter in different ways in the final expressions.

3.2.2 The Bag Model and the Question of Quark Masses

The MIT bag model (Chodos *et al.*, 1974a) has been extensively used in recent years. Leaving apart the theoretical motivations and the refinements concerning the center-of-mass motion of the bag, the bag model consists essentially in the use of the Dirac equation in a domain limited by an infinite potential barrier. At least, this is what appears to be relevant for the calculations of transitions. The specific feature of the bag model is that we deal with free-quark Dirac spinors subject to certain boundary conditions. This greatly simplifies the calculations and in principle allows simple explicit results to be obtained in the static spherical cavity approximation.

But, apart from the question of simplicity, the bag model raises an important issue, which deserves discussion. At least for light quarks, it is claimed (Chodos *et al.*, 1974a) that the masses to be inserted in the Dirac equation are small or zero masses, the so-called *current masses* that parametrize the departure from exact chiral symmetry in the QCD Lagrangian. This argument is not very compelling, since either we are dealing with effective quarks and we should give them the constituent masses, or, if we insist on starting with current masses, we should consider the rather complex phenomenon of dynamical chiral symmetry breaking (Nambu and Jona-Lasinio, 1961a,b), that is certainly omitted in the Dirac equation (see Section 1.2.4). To take this phenomenon into account has required a substantial complication of the bag picture, which is outside our scope.

Leaving aside the theoretical arguments, one can consider the mass as a parameter in the Dirac equation and look for the best agreement with experiment. In fact, the value cannot be firmly estimated owing to the scarcity and the ambiguities of the relativistic calculations done so far. It must not be forgotten that the potential itself can be chosen rather freely except for the general confinement property at long distances and asymptotic freedom at very short distances. The calculations, done essentially for the ground state, do not seem able to discriminate between small masses of a few MeV, as used in the bag, and masses of a few hundred MeV, which have been termed *constituent masses.* The calculation of static quantities like magnetic moments does not yield a decisive test, since what in fact enters is the mean quark energy of about 300 MeV. The role of the *mass* in nonrelativistic expressions is often played by the purely *kinetic energy* of the quarks, which is of the same order as the mass of the nonrelativistic quark model. Our personal feeling is that the relativistic corrections introduced by the bag model are often too large (e.g. for the magnetic moments), but other possible effects must be kept in mind. Something at least should be clear: owing to the large size of the relativistic corrections, the Dirac quark mass cannot be the same as the mass used in the nonrelativistic quark model.

3.2.3 Decays by Quantum Emission

It is very simple, with the Dirac equation, to treat current-matrix elements governing the electromagnetic and weak decays. Also, for the description of strong decays, the only possible framework offered by this equation is the model of emission of an elementary quantum coupled to quark bilinear currents. We therefore have to concentrate on the problem of the matrix elements of currents.

The interaction Hamiltonian and therefore the current operators will be exactly the same as in the naive quark model, without the nonrelativistic approximation. For electromagnetism we shall have, for example,

$$H_{\mathrm{I}} = e\int \mathrm{d}\boldsymbol{x}\, j^{\mu}(\boldsymbol{x}) A_{\mu}(\boldsymbol{x}), \tag{3.4}$$

$$j^{\mu}(\boldsymbol{x}) = \bar{q}(\boldsymbol{x})\gamma^{\mu} Q q(\boldsymbol{x}). \tag{3.5}$$

We could expand the quark field q over any basis of states. In particular, we could still choose the plane-wave solutions of the free Dirac equation, as in Chapter 2. However, in such a case the link with the wave equation (3.3) would be rather indirect. In fact, an eigenstate of (3.3) contains negative-energy components of free quarks, or equivalently Z-graph contributions of the type of Figure 3.1(c). It is more direct to expand the fields in terms of the eigenstates of $H_{\mathrm{D}}(i)$, the Dirac Hamiltonian for one quark in the static potential. The nice feature of the Dirac equation (3.3) is that the quarks remain independent, allowing for a treatment analogous to that of free quarks.

In terms of the eigenstates $\psi_{\alpha_i}(i)$ of $H_{\mathrm{D}}(i)$, the eigenstates of (3.3) can be written as

$$\Psi = \sum_{\{\alpha_i\}} c(\alpha_i) \prod_{i=1}^{N} \psi_{\alpha_i}(i), \tag{3.6}$$

where α_i labels the eigenstates of $H_{\mathrm{D}}(i)$.

For instance, the ground state is described by the product of individual ground states $\alpha_i = 0$:

$$\Psi_0 = \prod_{i=1}^{N} \psi_0(i). \tag{3.7}$$

It must not be forgotten that the complete set of eigenstates of $H_{\mathrm{D}}(i)$ includes negative-energy eigenvalues. Although they are not retained in the composition of physical states, they naturally appear through the action of operators like (3.5). They are then easily reinterpreted in terms of creation or absorption of an antiquark; this is automatically done by writing

$$q(\boldsymbol{x}) = \sum_{\alpha} b_{\alpha} u_{\alpha}(\boldsymbol{x}) + \sum_{\alpha} d_{\alpha}^{\dagger} v_{\alpha}(\boldsymbol{x}). \tag{3.8}$$

Here $\{u_{\alpha}(\boldsymbol{x})\}$ and $\{v_{\alpha}(\boldsymbol{x})\}$ are complete orthonormal sets of positive- and negative-energy eigenstates of $H_{\mathrm{D}}(i)$, and b_{α} and $d_{\alpha}^{\dagger}$ are respectively annihilation and creation operators of quarks and antiquarks. As a consequence, the wave function of antiquarks is $v_{\alpha}^{\dagger}(\boldsymbol{x})$. The presence of antiquarks will manifest itself in current-matrix elements by the presence of Z-graphs of a new type, induced by the quantized electromagnetic field as we see in Figure

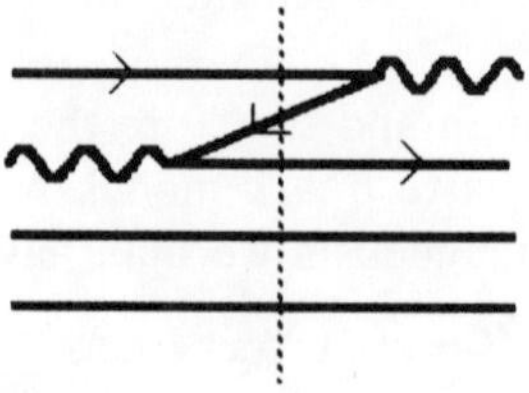

Figure 3.2 A Z-graph occurring in photon scattering, with quark and antiquark states in an external potential.

3.2 for a graph describing photon scattering. Note that Z-graphs of Figure 3.1(c) are no longer to be considered in the formalism under discussion, since the straight lines in Figure 3.2 now represent solutions of the full equation (3.3), and we are no longer perturbing with respect to the binding potential V.

An important remark concerning the Dirac wave functions is that eigenstates of (3.3) may be eigenstates of the total one-quark angular momentum $\boldsymbol{j}_i$,

$$\boldsymbol{j}_i = \boldsymbol{r}_i \times \boldsymbol{p}_i + \tfrac{1}{2}\boldsymbol{\sigma}_i, \tag{3.9}$$

but not of $\boldsymbol{r}_i \times \boldsymbol{p}_i$ and $\boldsymbol{\sigma}_i$ separately. Therefore it is necessary to directly couple states of definite $\boldsymbol{j}_i$ to constitute an N quark state. This is the so-called j–j coupling, distinct from the $L-S$ coupling familiar in the nonrelativistic quark model.

Finally, what we have described is just the spin–space part of the wave function. To obtain the complete wave function, one has still to multiply by color and flavor wave functions. Note that the factorization of flavor wave functions is only justified if one assumes the N quarks to have the same mass. In the following, for the sake of simplicity, we omit color and flavor, which can be trivially accounted for.

The effect of second-quantization current operators like $\bar{q}\gamma_\mu Qq$ taken in matrix elements between states with the same quark number N is simply equivalent to the first-quantization operator

$$\sum_{i=1}^{N} \gamma_0(i)\gamma_\mu(i)\delta(\boldsymbol{x}-\boldsymbol{r}_i)Q(i) \tag{3.10}$$

acting between wave functions like (3.6) in the configuration representation. Still, for simplicity, we shall omit the summation over i and consider only one-quark states. In fact the additivity, manifest in (3.10), allows reduction of the calculation to one-quark matrix elements. Furthermore, omitting the charge operator, one is left with a matrix element

$$\langle\beta|\, \gamma_0\gamma_\mu\delta(\boldsymbol{x}-\boldsymbol{r})\,|\alpha\rangle. \tag{3.11}$$

The matrix element of H_{I} corresponding to (3.11) is

$$\langle\beta|\mathrm{H}_I|\alpha;\boldsymbol{k},\varepsilon\rangle = e\frac{1}{(2\pi)^{3/2}}\frac{1}{(2k^0)^{1/2}}\langle\beta|\,\gamma_0\gamma_\mu\varepsilon^\mu\,\mathrm{e}^{\mathrm{i}\boldsymbol{k}\cdot\boldsymbol{r}}|\alpha\rangle \tag{3.12}$$

for the absorption of a photon. This formula becomes, in the radiation gauge,

$$\langle\beta|H_{\mathrm{I}}|\alpha;\boldsymbol{k},\varepsilon\rangle = e\frac{1}{(2\pi)^{3/2}}\frac{1}{(2k^0)^{1/2}}\langle\beta|\,\boldsymbol{\alpha}\cdot\boldsymbol{\varepsilon}\,\mathrm{e}^{\mathrm{i}\boldsymbol{k}\cdot\boldsymbol{r}}|\alpha\rangle. \tag{3.13}$$

From now on, we consider only the matrix element

$$\langle\beta|\,\boldsymbol{\alpha}\cdot\boldsymbol{\varepsilon}\,\mathrm{e}^{\mathrm{i}\boldsymbol{k}\cdot\boldsymbol{r}}|\alpha\rangle, \tag{3.14}$$

which straightforwardly gives the S-matrix and transition rates. If we desire to calculate these observable quantities from (3.14), we have to perform the following steps. First the S-matrix is still given by

$$S = 1 - 2\pi\mathrm{i}\,\delta(E_{\mathrm{f}} - E_{\mathrm{i}})\,\langle\mathrm{f}\,|H_{\mathrm{I}}|\,\mathrm{i}\rangle, \tag{3.15}$$

and the transition rates by

$$W = 2\pi\mathrm{i}\,\delta(E_{\mathrm{f}} - E_{\mathrm{i}})\,|\langle\mathrm{f}\,|\,H_{\mathrm{I}}\,|\mathrm{i}\rangle|^2. \tag{3.16}$$

An essential difference with Chapter 2 is that we no longer have translational invariance, and therefore no momentum conservation and no momentum label attached to the discrete hadron states, which have total energies E_α and E_β, given directly by the discrete spectrum of (3.3). This leads to a different phase space, with the photon momentum as the only degree of freedom. Moreover, the transition rate for one particle is now W itself, whence $\Gamma = \int \mathrm{d}\boldsymbol{k}\, W$ instead of $\int \mathrm{d}\boldsymbol{p}_{\mathrm{B}}\,\mathrm{d}\boldsymbol{p}_{\mathrm{C}}\, W$. Then, with

$$\rho_{\mathrm{f}} = \int \mathrm{d}\boldsymbol{k}\,\delta(E_\beta + k - E_\alpha) = 4\pi k^2, \tag{3.17}$$

where $k = E_\alpha - E_\beta$ is the actual photon momentum,

$$\Gamma = 2\pi\rho_{\mathrm{f}}|\langle\beta|\,H_{\mathrm{I}}\,|\alpha\rangle|^2 = \frac{e^2}{2\pi}k|\langle\beta|\,\boldsymbol{\alpha}\cdot\boldsymbol{\varepsilon}\,\mathrm{e}^{-\mathrm{i}\boldsymbol{k}\cdot\boldsymbol{r}}\,|\alpha\rangle|^2. \tag{3.18}$$

Let us recall that in (3.18) we have omitted, for the sake of simplicity, the summation over quarks, flavor and color matrix elements, the coefficients $c(\alpha)$ in (3.6) and the sum and average over polarizations. The sign of $\boldsymbol{k}$ has been changed for emission.

3.2.4 General Remarks on the v/c Expansion

Of course, the most direct way to estimate for example the matrix element (3.14) would be to find, analytically or numerically, the exact solutions of the one-quark Dirac equation

$$H_D\psi = [m\beta + \boldsymbol{\alpha}\cdot\boldsymbol{p} + V(\boldsymbol{r})]\psi = E\psi \qquad (3.19)$$

As we have pointed out, this approach of exactly solving the Dirac equation has been developed in the framework of the bag model (Chodos *et al.*, 1974a,b; De Grand *et al.*, 1975). It has also been used with various potentials by Bando and coworkers (see e.g. Abe *et al.*, 1980). This procedure seems to have been followed particularly for ground-state matrix elements. For baryons the ground state is of course understood as including the Δ together with the N, since both have the same spatial wave function. The matrix element for decays like $\Sigma^0\to\Lambda\gamma$ or $\Delta^+\to p\gamma$ will be proportional to the one giving the nucleon magnetic moment, apart from different Clebsch–Gordan coefficients.

As an instructive example, let us recall that in the Dirac approach initiated by Bogoliubov (1968) one finds for the proton magnetic moment

$$\mu_p = \frac{1}{2E_0}(1-\delta), \qquad \text{with} \qquad \delta = \frac{2}{3}\frac{\int \mathrm{d}\boldsymbol{r}\,|\psi_2|^2}{\int \mathrm{d}\boldsymbol{r}\,(|\psi_1|^2+|\psi_2|^2)} \qquad (3.20)$$

where E_0 is the quark energy level ($m_p = 3E_0$), and ψ_1 and ψ_2 are the large and small components of the quark Dirac spinor. The relation (3.20) exhibits the fact that in the relativistic calculation, there is no longer the simple $1/m_q$ dependence on the quark mass of the nonrelativistic expression $\mu_p = 1/2m_q$ (for the nonrelativistic calculation see Section 4.3.1). The small components of the quark Dirac spinors provide sizeable corrections.

A more general and instructive method for analysing the output of the Dirac treatment of transitions is the v/c expansion. It permits a transparent comparison with the nonrelativistic model, which consists in retaining only the lower orders in the expansion. The great interest of the Dirac equation, in contrast with more complete treatments, is that it allows this v/c expansion to be performed rather simply. This is the reason why we shall give in detail the v/c expansion of decay-matrix elements in the Dirac formalism, while being much more concise in our treatment of the Hamiltonian and Bethe–Salpeter approaches. In fact, it will give us a good idea of the qualitative effects to be expected from quark internal velocities. We must not forget that v/c is actually not small for light quarks, so that we cannot claim to obtain realistic estimates, but only a general trend of the effects.

Of course, the v/c expansion of the Dirac equation is something that is standard in quantum mechanics (Messiah, 1959), but it has to be adapted to the concrete problems of the quark model. Indeed, the basic methods are usually presented in the context of purely electromagnetic interactions, and, as we shall see in the next section, care must be exerted in translating them to the case of a combination of a strong-interaction binding potential and fields representing emitted or absorbed quanta. Also, one must realize the specificity of a bound-state problem in contrast with a scattering problem; the quark velocity is no longer an independent parameter, but is on average determined by the strength of the potential: $p^2/2m \sim V/m$; as we shall see precisely, this implies that the v/c expansion is in fact an expansion in powers of the potential. Finally, a new specificity comes from the fact that we want to treat transitions; therefore what we have to expand is not only a Hamiltonian determining the binding energies, but a bilinear combination of the Dirac wave functions; in addition, most often we want to treat real transitions, and this has consequences for the external momentum k that enters the transition amplitude; for instance, for a real photon, $k = E_\beta - E_\alpha$ is of the order of the energy levels $mO(v^2/c^2)$, and this must be taken into account in the v/c expansion.

To define the v/c expansion systematically, we first have to define the lowest-order approximation, with respect to which we shall calculate corrections of relative order v^n/c^n. This lowest-order approximation is defined by taking Dirac spinors with only upper components, which satisfy the one-body Schrödinger equation with potential $V = V_S + V_V$; the zeroth order for energy will be defined as m, and the energy levels of the Schrödinger equation are then of order $mO(v^2/c^2)$ as stated above.

Let us then consider vector or axial-vector current-matrix elements, which are $\int \mathrm{d}\boldsymbol{r}\, \bar{\psi}\gamma_\mu\psi \mathrm{e}^{\mathrm{i}\boldsymbol{k}\cdot\boldsymbol{r}}$ or $\int \mathrm{d}\boldsymbol{r}\, \bar{\psi}\gamma_\mu\gamma_5\psi \mathrm{e}^{\mathrm{i}\boldsymbol{k}\cdot\boldsymbol{r}}$. Only the time component of the vector and the space component of the axial vector will remain at this order:

$$\left.\begin{array}{l}\int \mathrm{d}\boldsymbol{r}\,\psi_1^\dagger\psi_1 \mathrm{e}^{\mathrm{i}\boldsymbol{k}\cdot\boldsymbol{r}} \\ \int \mathrm{d}\boldsymbol{r}\,\psi_1^\dagger\boldsymbol{\sigma}\psi_1 \mathrm{e}^{\mathrm{i}\boldsymbol{k}\cdot\boldsymbol{r}}\end{array}\right\}. \tag{3.21}$$

This is a very crude approximation, since it gives zero for radiative transitions, for the pion pseudoscalar coupling and for the parity-violating nonleptonic weak matrix element. This is the same difficulty that we have analysed in Section 2.1.1(b): the use of the nonrelativistic field (1.44) in transition operators gives a too drastic approximation. Therefore, as in Chapter 2, we usually take as the nonrelativistic approximation the one that retains in the transition-matrix elements v/c corrections to the lowest-order estimate of Dirac spinors; these corrections correspond to the Pauli approximation:

$$\psi_2 = -\mathrm{i}\frac{\boldsymbol{\sigma}\cdot\boldsymbol{\nabla}}{2m}\psi_1. \tag{3.22}$$

ψ_1 still obeys the Schrödinger equation, since we consistently neglect the corrections of order v^2/c^2. With such an approximation, we have nonzero radiative transitions, pion pseudoscalar coupling and parity-violating matrix elements, as found in Chapter 2.

It must be noted that the same degree of approximation in v/c for the fields may give the transition-matrix elements up to different orders in v/c for different couplings, even if these couplings are related by identities. An important example is offered by the pion pseudoscalar and pseudovector couplings of Section 2.2.1. From

$$\left.\begin{aligned} i\gamma^{\mu}\partial_{\mu}q &= (m+V_S)q + V_V\gamma^0 q \\ -i\partial_{\mu}\bar{q}\gamma^{\mu} &= (m+V_S)\bar{q} + V_V\bar{q}\gamma^0 \end{aligned}\right\} \tag{3.23}$$

one deduces the identity

$$-i\partial_{\mu}(\bar{q}\gamma^{\mu}\gamma_5 q) = 2(m+V_S)\bar{q}\gamma_5 q. \tag{3.24}$$

To get transition-matrix elements, the fields are substituted by ψs and the Fourier transform obtained by integrating both sides of (3.24) with $d\boldsymbol{r}\, e^{i\boldsymbol{k}\cdot\boldsymbol{r}}$:

$$k^{\mu}\int d\boldsymbol{r}\bar{\psi}\gamma_{\mu}\gamma_5\psi e^{i\boldsymbol{k}\cdot\boldsymbol{r}} = 2\int d\boldsymbol{r}(m+V_S)\bar{\psi}\gamma_5\psi e^{i\boldsymbol{k}\cdot\boldsymbol{r}} \tag{3.25}$$

Both sides will be equal order by order in v/c. The left-hand side of (3.25) contains an external factor k_{μ}; if it is assumed that $m_{\pi}=0$ and that the pion is real, $k=k^0=mO(v^2/c^2)$. Then an expansion of the field up to order $(v/c)^n$ inserted in the left-hand side gives order $(v/c)^{n+2}$, while on the right-hand side it gives the expansion only up to $(v/c)^n$. The zeroth order in the fields, used on the right-hand side, will give the order zero, which vanishes, while, on the left-hand side, it gives order $(v/c)^2$, which does not vanish and is the $\boldsymbol{\sigma}\cdot\boldsymbol{k}$ coupling. The Pauli approximation (3.22) gives a correct result to $O(v^3/c^3)$ on the left-hand side which is only possible to obtain on the right-hand side through higher orders of the field expansion and cumbersome calculations. This $O(v^3/c^3)$ contains the recoil term (from the $\gamma^0\gamma^5$ term), which shows that, as pointed out in Section 2.2.1, *the pseudoscalar coupling also contains a recoil term.* But it is clear that it is *easier to expand the pseudovector coupling.*

Anyway, the real difficulty is that we must be able to go beyond the approximation (3.22) to get arbitrarily high-order corrections — and this is not easy. We should like to eliminate the small components by expressing them in terms of the large ones; but the problem is that this is much more difficult to perform at higher orders than it was in (3.22), because the expression of ψ_2 in terms of ψ_1 depends on the energy itself, so that we would have to simultaneously make explicit the expansion of E. The simpler method of Foldy and Wouthuysen will now be explained.

3.2.5 Systematic v/c Expansion

There is a clear need for an algorithm yielding a systematic v/c expansion of matrix elements such as (3.14), beyond the approximations obtained in Chapter 2. Indeed, we see that we can improve our results within the Pauli approximation by a clever choice of couplings and the use of certain identities, but this is not a systematic method. The desired method is supplied by the Foldy–Wouthuysen (1950) transformation. This transformation allows the reduction of Dirac operators and wave functions to two-component forms instead of four-dimensional spinors, the reduction being performed in successive v/c orders.

In practice, it consists in eliminating in H the so-called *odd operators* $\mathcal{O}$, which anticommute with $\beta=\gamma^0$, up to a given order $(v/c)^{2n}$, with the help of a unitary transformation. The remaining *even operators* commuting with β, denoted by $\mathcal{E}$, will act separately on the subspaces $\gamma^0\psi=\pm\psi$, which are two-dimensional. Therefore the action of these even operators is effectively reduced to a two-component theory in the sector of positive eigenvalues of H. At the same time we have therefore discarded the negative-energy solutions, which were to be eliminated from the physical states, with the exception of the explicit introduction of valence antiquarks in the hadron states. As a consequence, if the total Hamiltonian H_D+H_I, where H_I induces a transition, has been reduced in such a way, there will be no transition from positive- to negative-energy states of H_D, or equivalently no creation of quark–antiquark pairs by H_I at the considered order. This indicates that pair creation by H_I vanishes faster than any power of $v/c\to 0$. This phenomenon is analysed in Gavela *et al.* (1982c).

In fact, as the reader may have noticed, and as will be confirmed below, the first steps are rather similar to the steps performed at the end of Section1.3.5 and in Section 2.1.1(b). Indeed, we performed there a canonical transformation on a relativistic Hamiltonian (1.112) in order to put it into the form (1.43), i.e. a two-component theory, or more exactly a theory where upper components representing quarks and lower components representing antiquarks are completely separated. In fact, (1.113) represents the first Foldy–Wouthuysen transformation S_1, stopped at order v^2/c^2 (see below). However, such a transformation based on the expansion of free-quark spinors is unable to eliminate pair creation and annihilation terms from the Hamiltonian at higher orders. In contrast it is possible to reduce the Dirac Hamiltonian to a two-component form at any order by using new transformations that involve the potential.

The general algorithm for reducing the Hamiltonian is as follows. Suppose we have an expansion in powers of m (m, m^0, $1/m$, $1/m^2, \ldots$):

$$H=\beta m+\mathcal{E}+\mathcal{O}, \tag{3.26}$$

where $\mathscr{E}$ and $\mathscr{O}$ are of order m^0 at most and $[\mathscr{E}, \beta]=0$, $\{\mathscr{O}, \beta\}=0$.

The unitary transformation

$$\left.\begin{aligned} \psi \to \psi' &= e^{iS}\psi \\ H \to H' &= e^{iS} H e^{-iS} - ie^{iS}\frac{\partial}{\partial t}e^{-iS} \end{aligned}\right\}, \tag{3.27}$$

with

$$S = -\frac{i}{2m}\beta\mathscr{O}, \tag{3.28}$$

transforms H into

$$H' = \beta m + \mathscr{E}' + \mathscr{O}', \tag{3.29}$$

where $\mathscr{O}'$ is now of order $1/m$ at most. e^{iS} is unitary because S is Hermitian ($\mathscr{O}$ is Hermitian and $\{\mathscr{O}, \beta\}=0$).

In the same way,

$$S' = -\frac{i}{2m}\beta\mathscr{O}' \tag{3.30}$$

reduces H' to an expression

$$H'' = \beta m + \mathscr{E}'' + \mathscr{O}'', \tag{3.31}$$

where $\mathscr{O}''$ is of order $1/m^2$, and so on.

This procedure may be applied to H_D, in which case

$$\mathscr{E} = V_V + \beta V_S = V_D, \tag{3.32}$$

$$\mathscr{O} = \boldsymbol{\alpha}\cdot\boldsymbol{p}, \tag{3.33}$$

and we have to find the orders of $\mathscr{E}$ and $\mathscr{O}$ in powers of m. We count p as an independent parameter; this leads to the identification of $\mathscr{O}$ as indeed being of order m^0. It also means that the limit $m\to\infty$ is the nonrelativistic limit $p/m\to 0$. Finally, it is easily shown that the whole $1/m$ expansion is in fact a p/m expansion. First, inside a bound state, with Schrödinger equation, the potential V is of order $p^2/2m$. Then the three terms in H (3.26), one proportional to m, one to $\boldsymbol{p}$ and one to V_D, are all of the form $m(p/m)^n$. Each step of the Foldy–Wouthuysen procedure will increase the power in $1/m$ of the odd part, and gives a series in $1/m$ for the even part. But since each term of H is of the form $m(p/m)^n$ and S is a power $(p/m)^{n'}$, it is clear that, at each step, the powers in $1/m$ in the new Hamiltonian must be compensated by powers of p to give the desired dimension of a mass: the series in $1/m$ is also a series in p/m times m. What gives the scale of p/m? As we have seen, $(p/m)^2$ is of order V_D/m, and, as an expansion parameter, p/m can be replaced by V_D/m.

This is the treatment of the Dirac equation describing the bound state with the central potential (3.32). However, what is also of interest to us is to expand *transition*-matrix elements like (3.14). A first method to do this would be to apply the above procedure to the full Hamiltonian $H_D + H_I$ (Friar, 1974, 1975). However, there is a much more economical way, since we are only interested in first-order perturbation theory in H_I. We consider the matrix element

$$\langle\beta| H_I |\alpha\rangle, \tag{3.34}$$

and we apply to both the wave functions α and β and the operator H_I the successive unitary transformations $S_1, S_2, \ldots$ devised to diagonalize H_D. It is easily seen that this gives us the desired v/c expansion of (3.34).

Indeed, suppose that we have transformed H_D into an even form up to order $m(v/c)^{2n}$, with a corresponding odd form reduced to order $m(v/c)^{2n+1}$. (It is found that in the present case each transformation increases the power by two units, thanks to the fact that V is $mO(v^2/c^2)$, whence the power $2n$). Then, as we show below, the transformed Dirac wave functions have small components of order $(v/c)^{2n+1}$ at least, with respect to the large components. The negative-energy solutions will have small (upper) components of the same order. We can say that the positive- and negative- energy solutions correspond respectively to the two-dimensional subspaces of upper or lower components with an accuracy $(v/c)^{2n}$. Finally, as we also show, one can use perturbation theory in these two-dimensional subspaces to calculate corrections to the Schrödinger wave functions up to order $(v/c)^{2n-2}$.

For example, for $n=1$, H_D is transformed into the Schrödinger form

$$H' = \beta m + H^{(2)} + mO(v^3/c^3), \tag{3.35}$$

where

$$H^{(2)} = \frac{\beta p^2}{2m} + V_V + \beta V_S. \tag{3.36}$$

Then the small components are of order 3 with respect to the large components, which are approximated by the Schrödinger wave function, an eigenstate of (3.36), conventionally defined as the *order* 0. At the next step $n=2$,

$$H'' = \beta m + H^{(2)} + H^{(4)} + mO(v^5/c^5), \tag{3.37}$$

where we denote by $H^{(2n)}$ the additional even term obtained at each step. The small components are now of order v^5/c^5. $H^{(4)}$ contains in the vector case the well-known spin–orbit and Darwin terms. Including a scalar part, we find

$$\begin{aligned} H^{(4)} = & -\beta\frac{\boldsymbol{p}^4}{8m^3}+\frac{1}{8m^2}\nabla^2 V_{\mathrm{V}} \\ & -\frac{1}{8m^2}\{\boldsymbol{p},\{\boldsymbol{p},\beta V_{\mathrm{S}}\}\}+\frac{1}{4m^2}\boldsymbol{\sigma}\cdot[\nabla(V_{\mathrm{V}}-\beta V_{\mathrm{S}})\times\boldsymbol{p}]. \end{aligned} \tag{3.38}$$

Note that $\boldsymbol{\sigma}$ is to be understood as a vector of 4×4 matrices having the usual Pauli matrices as diagonal blocks. In the last term in (3.38) the spin–orbit potential is of opposite sign for a given Schrödinger potential (3.36) depending on its underlying Lorentz scalar or vector structure.

We now derive the order of magnitude of the small components by perturbation theory; the unperturbed Hamiltonian is the even part in lowest nontrivial order, $\beta m + H^{(2)}$, and the perturbation responsible for the appearance of the small components is the odd part, of order $m(v/c)^{2n+1}$. For instance, at $n=2$,

$$\frac{1-\beta}{2}\psi''_\alpha = \sum_\gamma \frac{\left(\begin{pmatrix} 0 \\ \varphi_\gamma^- \end{pmatrix}, H''_{\mathrm{odd}}\begin{pmatrix} \varphi_\alpha^+ \\ 0 \end{pmatrix}\right)}{E_\alpha^+ - E_\gamma^- + 2m}\begin{pmatrix} 0 \\ \varphi_\gamma^- \end{pmatrix}. \tag{3.39}$$

Here $\varphi_\gamma^\pm$ and $E_\gamma^\pm$ represent respectively the upper or lower components of the eigenstates, and the eigenvalues of $H^{(2)}$ in the sectors $\beta\psi = \pm\psi$. The perturbation H''_{odd} has only matrix elements between these two sectors. Since $E_\gamma^\pm$ is of order $m(v^2/c^2)$, the denominator can be approximated by $2m$, and the final result is $O(v^5/c^5)$.

As to large components, at each new step $n>1$, we can calculate corrections up to order $(v/c)^{2n-2}$ by perturbation theory with respect to the unperturbed Hamiltonian $\beta m + H^{(2)}$ separately in the two two-dimensional subspaces, the perturbation being $H^{(4)}+\ldots+H^{(2n)}$. For instance, at $n=2$, the upper components of the transformed wave function are given by

$$\frac{1+\beta}{2}\psi''_\alpha = \begin{pmatrix} \varphi_\alpha^+ \\ 0 \end{pmatrix} + \sum_\gamma \frac{\left(\begin{pmatrix} \varphi_\gamma^+ \\ 0 \end{pmatrix}, H^{(4)}\begin{pmatrix} \varphi_\alpha^+ \\ 0 \end{pmatrix}\right)}{E_\alpha^+ - E_\gamma^+}\begin{pmatrix} \varphi_\gamma^+ \\ 0 \end{pmatrix}. \tag{3.40}$$

The correction is of order v^2/c^2 because this time the denominator is of order mv^2/c^2.

We emphasize that we are not performing the v/c expansion of the initial quantities H and ψ. We are introducing at each step a new basis. At each new step the small components are reduced by an additional factor $O(v^2/c^2)$. The matrix element (3.34) is of course invariant under these changes of basis.

We can generalize the above formulae to an arbitrary number of successive Foldy–Wouthuysen transformations. The general lesson is clear: after n steps the upper components can be computed to an accuracy $O((v/c)^{2n-2})$ and can be used to compute matrix elements to the same accuracy; their v/c expansion is given by the standard perturbation theory on the eigenstates of the usual nonrelativistic Schrödinger equation, the perturbation being given by $H^{(4)} + H^{(6)} + \ldots$. The small components are reduced to order $(v/c)^{2n+1}$, and can be simply omitted.

We have still to express the transformed transition operator

$$\mathrm{e}^{\mathrm{i}S_n} \ldots \mathrm{e}^{\mathrm{i}S_1} H_\mathrm{I} \mathrm{e}^{-\mathrm{i}S_1} \ldots \mathrm{e}^{-\mathrm{i}S_n} \tag{3.41}$$

in terms of the two-component spinor formalism. It is very simple to get the v/c expansion of (3.41) at order $(v/c)^{2n-2}$. Apart from explicit space or momentum-operator dependence, H_I contains essentially the factor $\mathrm{e}^{ik\cdot r} = \mathrm{e}^{ikz}$ describing the momentum transfer. The expansion of this factor is very simple. We have, for absorption,

$$k = E_\beta - E_\alpha. \tag{3.42}$$

k is then of order $mO(v^2/c^2)$, while, on average, owing to the uncertainty relation $z = O(1/p)$,

$$z = \frac{1}{m} O\left(\left(\frac{v}{c}\right)^{-1}\right). \tag{3.43}$$

Therefore

$$kz = O(v/c), \tag{3.44}$$

and e^{ikz} can be consistently expanded in a power series $\mathrm{e}^{ikz} = \Sigma(ikz)^m/m!$, which gives the v/c expansion at the same time. It must be noted that it is not significant to retain the full e^{ikz} factor in the nonrelativistic quark model, since one is proceeding to a low-order approximation in v/c. For instance, the operator e^{ikz} taken betwen Gaussian wave functions generates a Gaussian factor in the amplitude of the form e^{-Ak^2}, which is not significant and should be expanded. The expansion of the exponential factors $\mathrm{e}^{\mathrm{i}S_j}$ in (3.41) is also straightforward because S_j is of order $(v/c)^{2j-1}$, so that we just have to write $\mathrm{e}^{\mathrm{i}S_j} = \Sigma(\mathrm{i}S_j)^m/m!$, and when j increases, we have to retain fewer terms in the series. We shall not even need S_n, because it is already of order $2n-1$, which we neglect. Finally, the result is much simplified by the fact that the S_j are odd matrices; we can drop all the odd matrices in the final result since the lower components are negligible.

Let us write the first two operators S_j:

$$S_1 = -\frac{\mathrm{i}}{2m}\beta\boldsymbol{\alpha}\cdot\boldsymbol{p}, \tag{3.45}$$

$$S_2 = \frac{\mathrm{i}}{6m^3}\beta\boldsymbol{p}^2\boldsymbol{\alpha}\cdot\boldsymbol{p} - \frac{\mathrm{i}}{4m^2}[\boldsymbol{\alpha}\cdot\boldsymbol{p}, V_{\mathrm{D}}], \tag{3.46}$$

V_{D} being defined in (3.32).

If we stop our expansion at second order v^2/c^2, we need only S_1: the relativistic expansion of (3.41) is the same as it would have been calculated with free Dirac spinors, since S_1 is, to this order, the Foldy–Wouthuysen transformation that maps the free Dirac spinors (normalized to $u^\dagger u = 1$) onto the Pauli spinors (with $\chi^\dagger\chi = 1$). This is no longer true at third order, since (3.46) contains the potential.

Let us now examine two simple cases: pion and photon transitions.

3.2.6 Pion Transitions

We shall write the formulae for pion absorption. We use the axial coupling (2.151). Proceeding analogously to what was done for the electromagnetic interaction, from (3.10) to (3.13), we pass to first-quantization operators and take the matrix element with an initial pion. Instead of the nonrelativistic formula (2.154), we get

$$H_{\mathrm{I}} = \frac{\mathrm{i}}{(2\pi)^{3/2}}\frac{1}{(2\omega_\pi)^{1/2}}\frac{g_{\pi qq}}{2m}\{\boldsymbol{\sigma}\cdot\boldsymbol{k}\,\mathrm{e}^{\mathrm{i}\boldsymbol{k}\cdot\boldsymbol{r}} - \omega_\pi\gamma_5\,\mathrm{e}^{\mathrm{i}\boldsymbol{k}\cdot\boldsymbol{r}}\}. \tag{3.47}$$

If $m_\pi = 0$, $|k| = \omega_\pi$, and assuming that the pion moves towards $z > 0$ along the z axis,

$$H_{\mathrm{I}} = \frac{\mathrm{i}}{(2\pi)^{3/2}}\frac{1}{(2\omega_\pi)^{1/2}}\frac{g_{\pi qq}}{2m}|k|\{\sigma_z\mathrm{e}^{\mathrm{i}kz} - \gamma_5\,\mathrm{e}^{\mathrm{i}kz}\}. \tag{3.48}$$

It is now straightforward to use the method of the preceding section to get the v/c expansion of the matrix element (Gavela *et al.*, 1982c):

$$\langle\beta|\,\sigma_z\mathrm{e}^{\mathrm{i}kz} - \gamma_5\mathrm{e}^{\mathrm{i}kz}\,|\alpha\rangle, \tag{3.49}$$

which we write as

$$(\beta|\,X\,|\alpha), \tag{3.50}$$

where $|\ldots)$ denotes the new states $(\psi', \psi''\ldots)$ obtained by the sequence of unitary transformations, and X is the new operator. To second order, we find

$$X = \sigma_z e^{ikz} - \frac{1}{2m}\{\boldsymbol{\sigma}\cdot\boldsymbol{p}, e^{ikz}\}$$

$$- \frac{1}{8m^2}(2p^2\sigma_z - 2\boldsymbol{\sigma}\cdot\boldsymbol{p}\,\sigma_z\boldsymbol{\sigma}\cdot\boldsymbol{p}). \tag{3.51}$$

In the last term we have consistently set $e^{ikz} \sim 1$. In the first two terms we should set $e^{ikz} = 1 + ikz + \frac{1}{2}(ikz)^2$ or $1 + ikz$ respectively. Concerning the wave function, we have to consider the second-order corrections to the Schrödinger wave function. However, in fact they contribute only to the matrix element of the first term, because for the last two terms, they will give $O(v^3/c^3)$. We therefore consider

$$(\beta|\,\sigma_z e^{ikz}\,|\alpha) = (\beta|\,\sigma_z\,|\alpha) + (\beta|\,\sigma_z ikz\,|\alpha) + (\beta|\,\sigma_z \tfrac{1}{2}(ikz)^2\,|\alpha). \tag{3.52}$$

In the last two terms we can still approximate $|\alpha)$ and $|\beta)$ by the Schrödinger wave function. There remains a possible contribution from wave-function corrections in $(\beta|\sigma_z|\alpha)$. If, furthermore, one of the states, α for example, is the ground state, then

$$\sigma_z|\alpha)_{\text{SCHRO}} = \pm|\alpha)_{\text{SCHRO}}. \tag{3.53}$$

The transformed state $|\alpha)$ under consideration is given by (3.39) and (3.40). The small components (3.39) are negligible. As to (3.40), since the only spin-dependent part in $H^{(4)}$ is the spin–orbit coupling, it is zero when acting on the unperturbed ground state, and the left-hand side of (3.40) is still an eigenstate of σ_z; therefore (3.53) is also valid for $|\alpha)$. $(\beta|\sigma_z|\alpha)$ is 0 or ± 1, and there is finally no need to consider wave-function corrections in that particular case. We can set

$$|\alpha), |\beta) \approx |\alpha), |\beta)_{\text{SCHRO}}. \tag{3.54}$$

Let us compare this with the nonrelativistic calculation (2.154). We assume that e^{ikz} is also consistently expanded in the nonrelativistic calculation, as it is here. The only difference comes finally from the presence of the third term in (3.51), which can be further reduced to

$$X^{(2)} = -\frac{1}{2m^2}(p_T^2\sigma_z - p_z\boldsymbol{\sigma}\cdot\boldsymbol{p}_T). \tag{3.55}$$

This second-order correction is especially crucial for a transition from a *positive-parity* excited state to the ground state, because there is no zeroth-order term, since $(\beta|\,\sigma_z\,|\alpha) = 0$, and no first-order term, by parity conservation. Then the whole matrix element is of *second order*, and the correction is of the same order.

A new feature characterizing the difference between the nonrelativistic calculation and the result from the Dirac equation appears if we proceed to a further reduction of the powers of e^{ikz} by the following device. Suppose we have a term like

$$(\beta|(k)^n \mathcal{O}|\alpha), \tag{3.56}$$

where $\mathcal{O}$ is any k-independent operator. We may eliminate $(k)^n$ by rewriting (3.56) as the matrix element of the n-multiple commutator of $\mathcal{O}$ by H:

$$(\beta|\underbrace{[H,[H,[\ldots,[H,\mathcal{O}]\ldots]]]}_{n \text{ times}}|\alpha) \tag{3.57}$$

If we apply this device to the matrix element $(\beta|X|\alpha)$, we find

$$(\beta|X|\alpha) = (\beta|\sigma_z - \frac{1}{m}\boldsymbol{\sigma}\cdot\boldsymbol{p}_{\mathrm{T}} - \frac{1}{2m^2}\boldsymbol{p}_{\mathrm{T}}^2\sigma_z - \frac{1}{2m^2}p_z\boldsymbol{\sigma}\cdot\boldsymbol{p}_{\mathrm{T}} + \frac{1}{m}z\boldsymbol{\sigma}\cdot\boldsymbol{\nabla}_{\mathrm{T}}V|\alpha). \tag{3.58}$$

If we had not taken the contribution $X^{(2)}$ (3.55) into account, which would correspond to the nonrelativistic calculation of Section 2.2.1, we should have found instead

$$(\beta|\sigma_z - \frac{1}{m}\boldsymbol{\sigma}\cdot\boldsymbol{p}_{\mathrm{T}} - \frac{1}{m^2}p_z\boldsymbol{\sigma}\cdot\boldsymbol{p}_{\mathrm{T}} + \frac{1}{m}z\boldsymbol{\sigma}\cdot\boldsymbol{\nabla}_{\mathrm{T}}V|\alpha). \tag{3.59}$$

One important feature distinguishing (3.58) and (3.59) is the following. The analysis of strong decays emphasizes the importance of the behavior of the transition operator under separate spin and space rotations. The behavior under the space rotations around the z axis is labelled by the usual quantum number denoting the irreducible representation to which the operator belongs. For instance, $\Delta L_z = \pm 1$ denotes a behavior similar to that of the transverse coordinates p_x and p_y. Of course, to identify the transition operator unambiguously, we must make precise the wave functions on which it operates. Since we want to compare with the naive quark model, we agree to define the transition operator as acting on the nonrelativistic wave functions. Therefore all the differences with the naive quark model are transferred to the transition operator. Indeed, for (3.58) the β and α wave functions can in fact be replaced by the nonrelativistic wave functions (3.54). Remember also that σ_z does not contribute to transitions between the ground state and excited states (see the discussion preceeding (3.54)). Therefore we see that in the nonrelativistic expression (3.59) these transitions will be characterized by $\Delta L_z = \pm 1$ and odd-parity behavior. In contrast, (3.58) contains the additional term

$$-\frac{1}{2m^2}\boldsymbol{p}_{\mathrm{T}}^2\sigma_z, \tag{3.60}$$

which has the structure $\Delta L_z = 0$ and even parity, invariant under spatial rotations around the z axis. Therefore we see that the relativistic corrections drastically modify the ΔL_z structure of the transition for decays of positive-parity excitations (remember that they are of the same order as the nonrelativistic terms contributing to the same transitions), while they do not contribute to the decays of the negative-parity hadrons. The term (3.60) also gives a correction to the transitions within the ground state. However, this term is not so typical in this case, since it will not change the ratio of two amplitudes, like for example the ratio of coupling constants $g_{\Delta N\pi}/g_{NN\pi}$. This correction will of course also be present in the matrix elements of the axial current within the ground state, which govern for instance β decay $n \to pe^- \bar{\nu}_e$ and the corresponding hyperon decays. The correction (3.58) will reduce the seemingly too large value (Section 4.2.1) found in the nonrelativistic approximation, an effect analogous to (3.20) for the magnetic moment. Finally, we must not forget that the quark axial current

$$\bar{q}\gamma^\mu \gamma_5 q \tag{3.61}$$

is subject to renormalization effects due to the radiative corrections of strong interactions (Figure 3.3 and (4.47)) (Gavela *et al.*, 1980a,b), and we should have to combine the various effects before comparing with experiment (see further comments in Section 3.4.8).

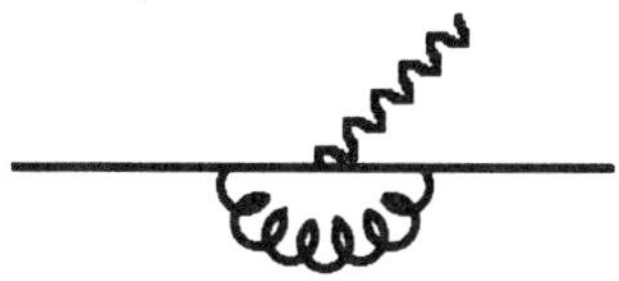

Figure 3.3 A vertex radiative correction in QCD. The wavy line represents a vector boson and the curly line a gluon.

3.2.7 Radiative Transitions

Photon decays are a little more complicated, because the lowest order is already v/c, and the next order beyond this nonrelativistic approximation is v^3/c^3. Therefore one also needs the second Foldy–Wouthuysen transformation (3.46), and the potential V_D will immediately enter the expression for the transition operator: *it is not possible to evaluate it with free Dirac spinors.* Moreover, one cannot avoid the evaluation of the corrections to the Schrödinger wave functions in addition to those concerning the transition operator. In the Foldy–Wouthuysen representation, (3.14) is found to be, up to v^3/c^3 terms,

$$(\beta|\left\{\frac{\boldsymbol{p}\cdot\boldsymbol{\varepsilon}}{2m}, \mathrm{e}^{\mathrm{i}kz}\right\}+\mathrm{i}\frac{\boldsymbol{\sigma}\cdot(\boldsymbol{k}\times\boldsymbol{\varepsilon})}{2m}\mathrm{e}^{\mathrm{i}kz}$$
$$-\frac{1}{2m^3}\boldsymbol{p}^2(\boldsymbol{p}\cdot\boldsymbol{\varepsilon})+\frac{1}{2m^2}[\boldsymbol{\sigma}\cdot(\boldsymbol{\nabla}V_{\mathrm{V}}\times\boldsymbol{\varepsilon})-\{\boldsymbol{p}\cdot\boldsymbol{\varepsilon}, V_{\mathrm{S}}\}]\,|\alpha). \tag{3.62}$$

The first two terms come from the naive calculation. In the terms of order v^3/c^3, which are new, we have set $\mathrm{e}^{\mathrm{i}kz}=1$; they are dependent on the binding potential and moreover on the precise Lorentz structure of the potential. These terms can be written as $(\beta|\boldsymbol{j}^{(3)}\cdot\boldsymbol{\varepsilon}|\alpha)$, where

$$\boldsymbol{j}^{(3)}=-\frac{1}{2m^3}\boldsymbol{p}^2\boldsymbol{p}+\frac{1}{2m^2}[\boldsymbol{\sigma}\times\boldsymbol{\nabla}V_{\mathrm{V}}-\{\boldsymbol{p}, V_s\}]. \tag{3.63}$$

The superscript (3) means order $(v/c)^3$. If we want to classify the terms according to their rotational behavior, we notice that the first and last are of the type spin scalar by space vector, already present in the naive expression. On the other hand, the second term is of the type spin vector by space vector. It does not have the same structure as the magnetic term, since $\boldsymbol{k}$ is along the z direction, and we therefore have for the magnetic term $\Delta L_z=0$, while here we may also have $\Delta L_z=\pm 1$ contributions. This term is called a *spin–orbit contribution*—not to be confused with the spin–orbit contribution to the binding Hamiltonian, present in (3.38).

Usually, the spin–orbit term, when taken into account in the quark-model calculations, is taken from the original paper of Foldy and Wouthuysen, which considers only electromagnetic fields. We want to make here a remark that illustrates the particular situation we are dealing with in the quark model, and which has been overlooked, leading to theoretically uncorrect results (exceptions are the works of Hardekopf and Sucher 1982 and of MacClary and Byers, 1984). In radiative transitions of hadrons we have two different Lorentz-vector fields to consider: the vector binding potential V_{V}, whose origin is in principle QCD, and the radiation electromagnetic field $\boldsymbol{A}(\boldsymbol{r},t)$. In atoms both fields have the same dynamical origin—the electromagnetic field. In the original deduction of Foldy and Wouthuysen (1950) there is indeed a spin–orbit term of the form

$$-\frac{\mathrm{e}}{4m^2}\boldsymbol{\sigma}\cdot[\boldsymbol{E}\times(\boldsymbol{p}-e\boldsymbol{A})], \tag{3.64}$$

where $\boldsymbol{E}$ is the electric field,

$$\boldsymbol{E}=-\frac{1}{e}\boldsymbol{\nabla}V-\frac{\partial\boldsymbol{A}}{\partial t}, \tag{3.65}$$

$V=-Ze^2/r$ being the Coulomb potential in the atom and $\boldsymbol{A}$ the radiation field;

e is here the *algebraic* charge of the electron. The first term in (3.65) comes from the fact that $V = e\varphi$.

In the quark model, (3.64) is still valid if, instead of (3.65), where both terms are of electromagnetic origin, we put

$$\boldsymbol{E} = -\frac{1}{e}\boldsymbol{\nabla} V_{\mathrm{V}} - \frac{\partial \boldsymbol{A}}{\partial t}, \tag{3.66}$$

where now V_{V} is the vector binding potential of strong-interaction origin, and $\boldsymbol{A}$ is the radiation electromagnetic field, and the scalar potential V_{S} is absent. This field $\boldsymbol{E}$ is a mixture of a piece of strong origin and a piece of electromagnetic origin. Let us show that if we substitute (3.66) into the Foldy–Wouthuysen expression (3.64) we do indeed recover our result for the spin–orbit current, the second piece of (3.63):

$$\boldsymbol{j}^{\mathrm{so}} = \frac{1}{2m^2}\boldsymbol{\sigma} \times \boldsymbol{\nabla} V_{\mathrm{V}}. \tag{3.67}$$

Since the radiation field is

$$\boldsymbol{A}(\boldsymbol{r}, t) = \frac{1}{(2\pi)^{3/2}}\frac{1}{(2k)^{1/2}}\boldsymbol{\varepsilon}\,\mathrm{e}^{\mathrm{i}(\boldsymbol{k}\cdot\boldsymbol{r} - \omega t)} \tag{3.68}$$

we see that (3.66) becomes

$$\boldsymbol{E} = -\frac{1}{e}\boldsymbol{\nabla} V_{\mathrm{V}} + \mathrm{i}\omega \boldsymbol{A}, \tag{3.69}$$

and (3.64) therefore becomes

$$-\frac{e}{4m^2}\boldsymbol{\sigma}\cdot\left[\left(-\frac{1}{e}\boldsymbol{\nabla} V_{\mathrm{V}} + \mathrm{i}\omega \boldsymbol{A}\right) \times (\boldsymbol{p} - e\boldsymbol{A})\right]. \tag{3.70}$$

We see that the term that is first order in e is given by

$$\frac{1}{4m^2}[\boldsymbol{\sigma} \times \boldsymbol{\nabla} V_{\mathrm{V}} - \mathrm{i}\omega\, \boldsymbol{\sigma} \times \boldsymbol{p}]\cdot(-e\boldsymbol{A}). \tag{3.71}$$

We can finally transform away the factor $\omega = E_\beta - E_\alpha$ of the second term in the matrix elements. It is clear that

$$(\beta| -\frac{\mathrm{i}\omega}{4m^2}\boldsymbol{\sigma} \times \boldsymbol{p}\,|\alpha) = -\frac{1}{4m^2}\mathrm{i}(\beta|\,[H^{(2)}, \boldsymbol{\sigma} \times \boldsymbol{p}]\,|\alpha), \tag{3.72}$$

where $H^{(2)}$ is the Hamiltonian in the representation where $|\alpha)$ and $|\beta)$ are the Schrödinger wave functions, namely $H^{(2)} = p^2/2m + V_{\mathrm{V}}$, and therefore (3.72) becomes

$$\frac{1}{4m^2}(\beta|\boldsymbol{\sigma}\times\nabla V_V|\alpha) \tag{3.73}$$

which, added to the first term of (3.71), gives finally the current (3.67) by taking out the factor $-e\boldsymbol{A}$.

Had we overlooked the fact that the "electric" field of strong origin is of order $1/e$ relatively to the radiation field, we would have had only the second term in (3.71), which is exactly *one half* of the correct result. This second term is quite often the only one retained in the literature.

Another mistake to be avoided is the following. When one proceeds like Foldy and Wouthuysen but treats H_D+H_I directly instead of H_I, then the $1/m$ expansion no longer coincides with the p/m expansion, as was the case for the diagonalization of H_D. The external field $e\boldsymbol{A}$ is a new independent parameter, and the dimensional argument following (3.33) cannot be used. Then terms $O(mv^3/c^3)$ may come not only from the terms in $1/m^2$, but also from terms in $1/m^3$. Therefore it is incorrect to use the Foldy–Wouthuysen result, which retains only order $1/m^2$. Indeed, the term

$$H=-\frac{(\boldsymbol{p}-e\boldsymbol{A})^4}{8m^3}, \tag{3.74}$$

which would be found by pursuing the Foldy–Wouthuysen procedure to order $1/m^3$, yields a term of order v^3/c^3 in the matrix element for radiative transitions. In fact, it corresponds to the first terms in (3.63), as can immediately be seen by retaining terms proportional to $e\boldsymbol{A}$ in (3.74). The term (3.74) has been consistently retained in the calculation of Friar (1975).

Finally the step of taking into account the v^2/c^2 corrections to the Schrödinger wave function is often overlooked. These corrections contribute through the first term of (3.62), the convective current.

Although terms of order v^3/c^3 may give interesting modifications to the usual naive calculations for radiative transitions (Copley *et al.*, 1969), they are in general less crucial than the v^2/c^2 corrections (3.60) to pionic decays of positive-parity excitations described above. They will contribute to transitions between the ground state and negative-parity excitations, which in general already have contributions of order v/c.

There may, however, be forbidden transitions for which the lowest-order term is zero. The case happens for charmonium decays like $\psi'\to\eta_c\gamma$, $\eta'_c\to J/\psi\gamma$. Terms of order v/c or v^3/c^3 are zero because of parity conservation. In our v/c expansion, it is easy to see that the *spatial* parity of the transition operator is the parity of the power in p/m. Moreover, v^2/c^2 terms coming from the convective term in (3.62) are zero because they do not change the quark spin. The magnetic term contributes at order v^2/c^2, since k/m is of order v^2/c^2, but it gives zero because of the orthogonality of the spatial wave functions.

Finally, one must go to order v^4/c^4. Anyway, it is found that, at orders v^3/c^3 and v^4/c^4 there are new terms not describable by the one-body Dirac equation, because they come from the interplay of the motions of two different quarks. One must then have recourse to the methods described in Section 3.3.

3.2.8 Two-Quark Interactions

In the preceding sections we have discussed the decays that can be described by matrix elements of currents. It is also possible to use the Dirac equation for matrix elements of the weak current–current interaction, relevant to the description of weak nonleptonic decays, and for leptonic annihilation of mesons, both already considered in the framework of the naive quark model in Chapter 2.

The principle of the calculation is exactly the same as before: the quark-field operators of the interaction Hamiltonian are expanded in terms of creation and annihilation operators of quarks in a base of eigenstates of the Dirac Hamiltonian.

Calculations have been done for weak matrix elements, first in the bag model (for a review see Donoghue *et al.*, 1981; Donoghue, Golowich and Holstein, 1986) and also in other potential models (Abe *et al.*, 1980). An interesting qualitative feature of such relativistic calculations is the following. Leaving aside numerical factors and signs, the dependence of the matrix elements of the parity-conserving part of the weak Hamiltonian H_W^{pc} (2.116) on the Dirac wave functions is given by

$$A-B \tag{3.75}$$

in the meson case, and by

$$A+B \tag{3.76}$$

in the baryon case. A and B are integrals over quartic combinations of the ground-state wave functions. When the small components of the wave function are set equal to zero, A tends to the nonrelativistic expression, while B vanishes:

$$B/A=O(v^2/c^2). \tag{3.77}$$

Moreover, A and B are positive. Therefore the ratio of meson to baryon matrix elements is *suppressed* relative to its nonrelativistic expression.

Analogously, it is found that the weak leptonic annihilation of mesons is reduced by a term $O(v^2/c^2)$. The reduction of the meson matrix elements has an intuitive explanation in terms of the so-called helicity suppression (Donoghue *et al.*, 1980). Both the meson-to-meson weak matrix element and the meson-to-leptons annihilation amplitude, deriving from the (V−A)(V

−A) current–current interaction, are controlled by the scattering amplitude

$$\bar{v}_1\gamma^\mu(1-\gamma_5)u_2\bar{u}_3\gamma_\mu(1-\gamma_5)v_4, \tag{3.78}$$

where 1 and 2 are the initial quark and antiquark, 3 and 4 may be the final quarks or leptons. To get this form, it may be necessary to perform a Fierz transformation (Fierz, 1937; Itzykson and Zuber, 1980), which is known to preserve the (V−A)(V−A) structure. The factors $1-\gamma_5$ may be rearranged into right and left projectors $\frac{1}{2}(1\pm\gamma_5)$ acting on the Dirac spinors:

$$4\bar{v}_1\frac{1+\gamma_5}{2}\gamma^\mu\frac{1-\gamma_5}{2}u_2\bar{u}_3\frac{1+\gamma_5}{2}\gamma_\mu\frac{1-\gamma_5}{2}v_4. \tag{3.79}$$

We now take free massless quarks, which correspond to an ultrarelativistic limit, with maximum relativistic effects, as opposed to the calculation of Chapter 2. The projectors $\frac{1}{2}(1\pm\gamma_5)$ will then project on helicity states. $\frac{1}{2}(1-\gamma_5)u$ is a negative-helicity quark, while $\frac{1}{2}\bar{v}(1+\gamma_5)$ is a positive-helicity antiquark. To correctly identify the helicity of antiquarks, we need to take care. In fact, the helicity of a spinor $v(\boldsymbol{p})^*=\mathrm{i}\gamma^2u(\boldsymbol{p})$ is given by

$$-\tfrac{1}{2}\gamma_5, \tag{3.80}$$

which corresponds to $-\frac{1}{2}\boldsymbol{\sigma}^*\cdot\hat{\boldsymbol{p}}$ applied on the antiquark spinor χ^{C*} (1.69), and from $\sigma^2\boldsymbol{\sigma}^*\sigma^2=-\boldsymbol{\sigma}$, to $\frac{1}{2}\boldsymbol{\sigma}\cdot\hat{\boldsymbol{p}}$ on the antiquark spinor in the standard Pauli representation.

On the other hand, a pion considered in its rest frame must have $\boldsymbol{P}=0$ and $\boldsymbol{J}=0$. Therefore the quark and the antiquark should have the same helicity in the pion rest frame, and consequently (3.79) must vanish since it corresponds to opposite helicities of the quark and the antiquark.

3.3 HAMILTONIAN METHODS

Although the Dirac equation in a central static potential is certainly very illustrative of the effects of large internal quark velocities, it misses, as we have seen, important phenomena. Since there are no two-body forces, such important forces as the spin–spin (De Rújula *et al.*, 1975) are lost. Most importantly, there are no momentum eigenstates for the hadrons, so that there is no hope of describing their motion. It is not possible to describe strong decays except in the frame of elementary-emission models.

For a realistic connection to phenomenology, we have to consider frameworks with two-body translational invariant forces, and which describe naturally the strong decay process. We consider two such frameworks: the

Hamiltonian formalism in this section and the Bethe–Salpeter–Mandelstam formalism in Section 3.4. They may be grossly characterized as respectively old-fashioned perturbation theory and Feynman-graph perturbation theory, both adapted to the bound-state problem.

For a long time, the main approach followed by those interested in relativistic quark models has been the Bethe–Salpeter–Mandelstam framework. This is probably because it is manifestly covariant. However, this imposition of covariance has not necessarily led to satisfactory results. One of the pleagues of covariant models is the appearance of many unwanted states in the spectrum.

Once the requirement of manifest covariance is dropped, the Hamiltonian formalism in the context of two-body instantaneous interaction has many advantages. It is in fact the most natural way to keep contact with the nonrelativistic quark model, while going beyond it. As we shall see later, it is also possible to recover the nonrelativistic approximation in the Mandelstam formalism, but the procedure is much more complicated. Indeed, the developments of Chapter 2 on the nonrelativistic quark model were based on the Hamiltonian formalism considered in the lowest order of the v/c expansion (see Section 2.1.1(b)), with the exception of the quark-pair-creation and elementary-emission models of strong decays, which were not derived from a fundamental Hamiltonian. It seems natural to keep to the Hamiltonian framework and to carry on the v/c expansion to higher orders. Close contact is maintained with the naive quark model concepts, because one starts from states containing a minimum number of valence quarks, and additional pairs are introduced perturbatively. This is exactly the spirit of the constituent-quark model as described in Chapter 1.

In the following treatment, we are largely inspired by Sucher, who has renewed the interest in the Hamiltonian method in the context of radiative decays of charmonium, showing that the lowest-order approximation may fail badly (Sucher, 1978; Sucher and Kang, 1978).

Let us make an initial decomposition of the Hamiltonian into

$$H = H_s + H_{ew}, \tag{3.81}$$

where H_s represents the strong-interaction Hamiltonian, including the free part, and H_{ew} represents the additional electroweak interaction responsible for the weak or radiative transitions. For these transitions, we could leave H_s unspecified in Chapter 2, because, in the simplest approximation, the transition-matrix element is simply the matrix element of H_{ew} between the nonrelativistic wave functions, and the strong interaction intervenes only through these wave functions.

For strong decays, as well as for higher-order corrections, it is necessary to specify H_s. Let us assume that H_s is composed of a free part (free Dirac

Hamiltonian in second-quantized version) plus a four-fermion interaction,

$$\begin{aligned} H_s &= \int d\boldsymbol{x}\, q^\dagger(\boldsymbol{x})(-i\boldsymbol{\alpha}\cdot\boldsymbol{\nabla}+m)q(\boldsymbol{x}) \\ &\quad + \frac{1}{2}\sum_i \int d\boldsymbol{x} \int d\boldsymbol{y}\, q^\dagger(\boldsymbol{x})\mathcal{O}_i q(\boldsymbol{x}) V_i(\boldsymbol{x}-\boldsymbol{y}) q^\dagger(\boldsymbol{y})\mathcal{O}_i q(\boldsymbol{y}) \\ &= H_D + H_{Is} \end{aligned} \tag{3.82}$$

where the qs are the quark-field operators and the $\mathcal{O}_i$ are certain matrices in Dirac, color and flavor space. When we write (3.82), we neglect the gluon degrees of freedom. Forces are still described by potentials, a simplifying point of view of the quark-model approach. This may be crude, but it is also a huge simplification. The price to pay is that the potentials $V_i(\boldsymbol{x}-\boldsymbol{y})$ are instantaneous, and we shall not be able to treat retardation effects. In addition, we lack theoretical principles to specify the V_i. We shall only assume that they confine quarks and present color saturation, i.e. there are no long-range strong forces between color singlets (for a discussion about Lorentz-scalar and vector potentials, see Section 1.3.5).

Intuitively, H_s is first responsible for the binding of quarks into hadronic bound states. But it induces also strong-interaction decay, as well as corrections to the wave functions. Considering the full H_s, we shall get unstable hadron states. How can we define precisely the bound states and the decay-matrix element?

We expand the q fields into creation and annihilation operators of free quarks and antiquarks (see (2.3b)), and we separate H_s into two parts:

$$H_s = H_s^{\text{no pair}} + H_s^{\text{pair}}, \tag{3.83}$$

where $H_s^{\text{no pair}}$ conserves the number of quarks, while H_s^{pair} adds or suppresses $q\bar{q}$ pairs. Equation (3.83) expresses the main idea of Sucher's approach.

$H_s^{\text{no pair}}$ will have eigenstates consisting of a fixed number of quarks, in particular color-singlet $q\bar{q}$ and qqq bound states to be identified with the usual *valence quark* approximation of the quark model, but without the limitation of the Schrödinger equation. $H_s^{\text{no pair}}$ will also describe the scattering states of mesons and baryons and possible multiquark bound states. Finally, the scattering states may themselves possess resonances to be identified with unstable multiquark states like diquonia. Let us then treat H_s^{pair} as a perturbation; it will give transition matrix elements for example between a qqq baryon state and a baryon–meson qqq–$q\bar{q}$ state. This is in fact the Cornell model described in Section 2.2.3(d), where additional nonrelativistic approximations were made.

We write the transition-matrix element between a hadron $|i\rangle$ and a two-hadron state $|f\rangle$ as

$$\langle \mathrm{f}|\, H_{\mathrm{s}}^{\mathrm{pair}}\, |\mathrm{i}\rangle. \tag{3.84}$$

As far as the relativistic corrections are concerned, the improvement is that the quark kinetic energy in $H_{\mathrm{s}}^{\mathrm{no\ pair}}$ is the relativistic Dirac one of (3.82). The eigenvalue equation, for example in the two-quark sector,

$$H_{\mathrm{s}}^{\mathrm{no\ pair}}\Psi = E\Psi, \tag{3.85}$$

is a sort of two-body Dirac equation, but still different from the Breit equation. It is interesting to write it in a Dirac-spinor representation. We follow Sucher and introduce the new wave function

$$\psi(\boldsymbol{p}_1,\boldsymbol{p}_2) = \sum_{r_i=1,2} u_{r_1}(\boldsymbol{p}_1) \otimes u_{r_2}(\boldsymbol{p}_2)\langle \boldsymbol{p}_1, r_1; \boldsymbol{p}_2, r_2|\Psi\rangle, \tag{3.86}$$

where the $u_{r_i}(\boldsymbol{p}_i)$ are positive-energy Dirac spinors. We end up with an equation

$$\begin{aligned}\mathscr{H}\psi \equiv \{&(\boldsymbol{\alpha}\cdot\boldsymbol{p}+m)_1 + (\boldsymbol{\alpha}\cdot\boldsymbol{p}+m)_2 \\ &+ \Lambda_+(\boldsymbol{p}_1)\Lambda_+(\boldsymbol{p}_2)\left[\sum_i V_i(1,2)\mathcal{O}_i(1)\mathcal{O}_i(2)\right]\Lambda_+(\boldsymbol{p}_1)\Lambda_+(\boldsymbol{p}_2)\}\psi \\ = E\psi,&\end{aligned} \tag{3.87}$$

where the Λ_+s are the projectors on positive-energy spinors:

$$\Lambda_+(\boldsymbol{p}) = \sum_r u_r(\boldsymbol{p}) \otimes u_r^\dagger(\boldsymbol{p}) \tag{3.88}$$

Equation (3.87) must be supplemented by the condition

$$\Lambda_+(\boldsymbol{p}_1)\Lambda_+(\boldsymbol{p}_2)\psi = \psi. \tag{3.89}$$

Equation (3.87) is similar to the Dirac equation, but it presents the advantage of preserving translational invariance, the one-body static potential being replaced by two-body forces. Therefore one can correctly treat the center-of-mass motion. It differs from the Breit equation, which is also a two-body relativistic equation, by the presence of the Λ_+ projectors. The Breit equation has the same form (3.87), but without the Λ_+ projectors. These projectors have the effect of removing those virtual free-quark pair contributions that appeared in Figure 3.1.

We now proceed to higher orders by treating $H_{\mathrm{s}}^{\mathrm{pair}}$ in the usual perturbation theory. After (3.84), we get the third-order contribution (the second order will not contribute, since it would create two pairs or create and annihilate a pair, giving a loop):

$$\langle \mathrm{f}|\, H_{\mathrm{s}}^{\mathrm{pair}} \frac{1}{E - H_{\mathrm{s}}^{\mathrm{no\ pair}}} H_{\mathrm{s}}^{\mathrm{pair}} \frac{1}{E - H_{\mathrm{s}}^{\mathrm{no\ pair}}} H_{\mathrm{s}}^{\mathrm{pair}}\, |\mathrm{i}\rangle. \tag{3.90}$$

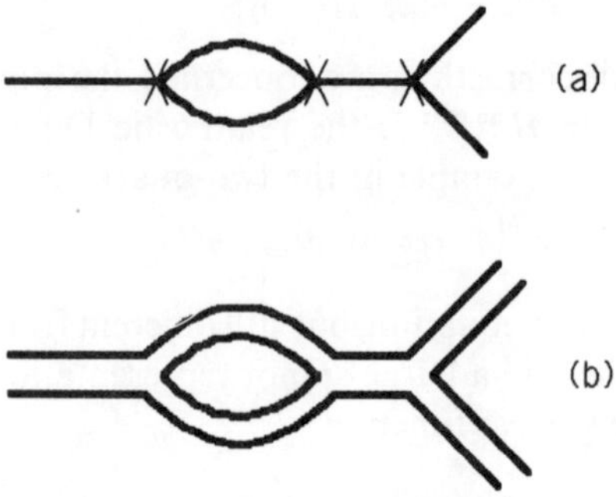

Figure 3.4 Perturbative treatment of the pair-creating part of the Hamiltonian. In (a) the crosses indicate the operation of H_s^{pair}; in (b) the quark structure is detailed.

Intuitively, (3.90) corresponds to the fact that H_s^{pair} introduces a hadronic loop (see Figure 3.4). The crosses in Figure 3.4(a) indicate the operation H_s^{pair}. Figure 3.4(b) shows the quark structure of the hadrons.

These loops can be indefinitely iterated, and the summation leads to a modification of the masses and to finite-width effects of the hadron described in the *mass-matrix* formalism. Extensive calculations have been made by the Cornell group for charmonium, and by Tornqvist *et al.*, for light quarks (see e.g. Eichten *et al.*, 1978; Törnqvist and Zenczykowski, 1984).

It must be noted that the representation of the intermediate states in the loop of Figure 3.4(a) by two free hadron states is an additional approximation, since in principle one has to consider the complete eigenstates of $H_s^{\text{no pair}}$, which introduces an interaction between the free hadrons. Moreover, the additional assumption is always made of choosing a limited number of hadron intermediate states in the loop. Finally, when one considers higher orders than in Figure 3.4 (order five or more), the iteration of loops is only a fraction of the possible diagrams.

Let us now consider the action of H_{ew}. We shall consider only calculations of first order in H_{ew}. Take for instance a current-matrix element for the absorption of a real photon $\alpha+\gamma\rightarrow\beta$. It is given by the same formula as in the Dirac formalism:

$$\langle\psi_\beta|\sum_i \boldsymbol{\alpha}_i\cdot\boldsymbol{\varepsilon}\,e^{i\boldsymbol{k}\cdot\boldsymbol{r}_i}|\psi_\alpha\rangle. \tag{3.91}$$

However, the difference is that the wave function ψ, instead of obeying the Dirac equation, obeys (3.87). Now corrections to (3.91) will be introduced by H_s^{pair} that will give operators depending on the potentials V_i. Once more, the systematic way to proceed is to use perturbation theory. We decompose the total Hamiltonian H into

$$\begin{aligned} H &= H_s^{\text{no pair}} + H_s^{\text{pair}} + H_{\text{ew}} \\ &\equiv H_0 + H_1, \end{aligned} \tag{3.92}$$

where $H_1 = H_s^{\text{pair}} + H_{\text{ew}}$ is the perturbation to consider. We then have the transition matrix in the Dirac-spinor representation:

$$T = \langle \psi_\beta | H_1 + H_1 \frac{1}{E - H_0} H_1 + \dots | \psi_\alpha \rangle, \tag{3.93}$$

where the first order in H_{ew} is to be selected. The second-order terms in (3.93) describe pair effects (Figure 3.5). The dotted line in Figure 3.5 represents the effect of H_s^{pair}. Such an effect is already contained in the Dirac equation, as mentioned in Section 3.2, but the equivalence is not maintained at higher orders.

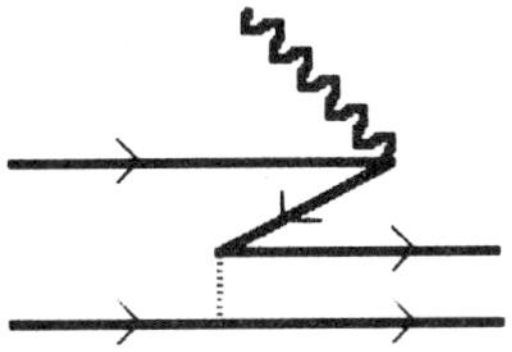

Figure 3.5 Combined perturbation by the electroweak Hamiltonian and the pair-creation part of the strong Hamiltonian.

We finally say some words on the v/c expansion in such a model. Indeed, just as in the case of the Dirac equation, this is the best way to discuss the corrections that it gives to the nonrelativistic approximation. In addition it would be inconsistent to make an expansion in powers in H_s^{pair}, which is an expansion in powers of the strong interactions, without also expanding $H_s^{\text{no pair}}$.

It is clear that the wave equation (3.87) reduces to the two-body Schrödinger equation in the weak-coupling nonrelativistic limit. The naive quark model for transitions consists in taking the lowest nontrivial order in v/c and H_1. In particular, this gives the Cornell model of strong decays (Section 2.2.3(d)) (we recall that the quark-pair-creation model has not been derived from an Hamiltonian and that the elementary-meson-emission model is only an analogy with radiative decay).

Let us now proceed to higher orders. The v/c expansion of (3.87) is in fact easier than that of a Dirac equation, since the Hamiltonian acts in the subspace spanned by positive-energy free-quark spinors. The small components are directly expressible in terms of the large components. The corrections to the two-body Schrödinger equation (Section 1.3.5) are easily found: at order mv^4/c^4 there appear various spin-dependent interactions, including those given by the Dirac equation (3.38) plus the typically two-body spin-dependent forces: spin–spin, tensor and additional spin–orbit forces; one then obtains the generalization of the Breit–Fermi interaction (1.111) to an arbitrary vector potential.

The virtual-pair contributions of the Dirac equation (Figure 3.1) have been discarded in (3.87) and are to be included by perturbation in H_s^{pair}. Since H_s^{pair} adds or substracts quarks, it will not contribute to the energy in first-order perturbation (as noted in Section 1.3.5). Its second-order contribution comes with a denominator $\sim 2m$ (the pair mass).

Let us now show that H_s^{pair} is of order mv^3/c^3 when the potential is vector or scalar. Just as in the Cornell model (2.227 and Figure 2.12), which corresponds to the $\gamma^0 \times \gamma^0$ vector case, in the matrix element between a q and a $qq\bar{q}$, there always contribute as factors three large components and one small component of free Direct spinors; whence the order $O((v/c)V) = O(mv^3/c^3)$. If the potential is scalar (1×1) or $\gamma^0 \times \gamma^0$ then the small component comes from the created pair; in the $\gamma \times \gamma$ case it comes from the preserved quark. This order of H_s^{pair} implies that the change in energy is of order mv^6/c^6 (in particular, the coupled-channel effects are thus in principle very high-order effects). The corrections to the strong decays are of relative order v^6/c^6 (for a further discussion in the charmonium case see Section 4.3.2). The corrections to the space components of the vector-current matrix elements are of order v^3/c^3 (relative order v^2/c^2). Finally, note that this v/c power counting remains valid only so long as all momenta are actually nonrelativistic. In the case of divergent loops, it happens that the momenta become relativistic and the radiative corrections may invalidate the preceding v/c counting estimates (see Section 3.4.8).

3.4 BETHE–SALPETER AMPLITUDES AND THE MANDELSTAM FORMALISM

3.4.1 Introduction

One advantage of the Bethe–Salpeter–Mandelstam formalism (Bethe and Salpeter, 1951; Mandelstam, 1955) is that, since time is treated as being on the same footing as space, covariance can be easily implemented and made manifest, even in simple approximations. However, as we have said, the requirement of manifest covariance is not compelling.

Its main feature is that it relates bound-state matrix elements to Green functions, which in turn may be analysed by Feynman-diagram techniques (covariant or not). These techniques are often more powerful than the time-ordered perturbation theory considered in the previous Section. Instead of insisting on manifest covariance, one may consider semirelativistic schemes; for instance, one can allow for a relativistic quark energy, but still retain an instantaneous interaction, as considered in Section 3.3. Then one would like to compare the two approaches, as we shall do later.

Let us set the general problem. Assuming that we know the Bethe–Salpeter (BS) amplitude for the two-fermion bound state $|a\rangle$,

$$\chi_a(v_1, v_2) = \langle 0| \mathrm{T}\{\psi(v_1)\psi(v_2)\}|a\rangle \tag{3.94}$$

where v_1 and v_2 are space–time points, we want to calculate the matrix element of some dynamical variable. Note that in this section the fermion fields will be denoted by $\psi(x)$. We shall suppose, following Mandelstam (1955), that the dynamical variables are obtained by simple operations from the basic quantities

$$\mathrm{T}\{\ldots\psi(x_i)\ldots\bar{\psi}(y_j)\ldots\varphi(z_k)\ \ldots\}, \tag{3.95}$$

which depend on quark fields $\psi(x)$ and on some other fields $\varphi(z)$. Note that we are now working with Heisenberg fields, while in the preceding section we used the Schrödinger picture. To simplify, we first want to calculate the matrix element of (3.95) between the vacuum and a bound state $|a\rangle$ of two fermions. The use of two-fermion bound states instead of the quark–antiquark ones simplifies the formulae, and is illustrative of the method without changing the main conclusions. The method consists in considering an associated Green function, suitably chosen to present a singularity in momentum space associated with $|a\rangle$, the residue being connected with the matrix element to be evaluated. This Green function is

$$R(x_i, y_j, z_k; v_1, v_2) = \langle 0|\mathrm{T}\{\ldots\psi(x_i)\ldots\bar{\psi}(y_j)\ldots\bar{\psi}(v_1)\bar{\psi}(v_2)\ldots\varphi(z_k)\ldots\}|0\rangle, \tag{3.96}$$

which is obtained by adding to the operators contained in (3.95) two $\bar{\psi}$ operators, corresponding to the two fermion operators in (3.94). The singularity corresponding to $|a\rangle$ is obtained by considering the situation where the times corresponding to v_1 and v_2 are very different from the times of all the x_i, y_j, z_k, the time interval being larger than the time differences within the two groups of coordinates (v_1, v_2) and (x_i, y_j, z_k). In such a situation the T product can be decomposed into two T products corresponding to the two groups of coordinates, and either

$$R \sim \langle 0|\mathrm{T}\{\ldots\psi(x_i)\ \ldots\ \bar{\psi}(y_j)\ \ldots\ \varphi(z_k)\ \ldots\}\mathrm{T}\{\bar{\psi}(v_1)\bar{\psi}(v_2)\}|0\rangle \quad 3.97)$$

or

$$R \sim \langle 0|\mathrm{T}\{\bar{\psi}(v_1)\bar{\psi}(v_2)\}\mathrm{T}\{\ldots\psi(x_i)\ \ldots\ \bar{\psi}(y_j)\ \ldots\ \varphi(z_k)\ \ldots\}|0\rangle. \tag{3.98}$$

In terms of the center-of-mass variable $V=\frac{1}{2}(v_1+v_2)$, these expressions correspond (the x_i, y_j, z_k being fixed) to $|V^0|$ large and $V^0<0$ or $V^0>0$

respectively. We now insert intermediate states between the T products and retain the contribution from $|a\rangle$. For $V^0<0$ we have

$$R_a \sim \sum_{s_a, \boldsymbol{P}_a} \langle 0|\,T\{\ldots\,\psi(x_i)\,\ldots\,\bar{\psi}(y_j)\,\ldots\,\varphi(z_k)\,\ldots\}\,|a\rangle\langle a|\,T\{\bar{\psi}(v_1)\bar{\psi}(v_2)\}\,|0\rangle, \tag{3.99}$$

the sum being taken on the spin and momentum of $|a\rangle$. In the first factor we have the matrix element that is sought, and, in the second factor, the *conjugate* of the BS amplitude (3.94),

$$\bar{\chi}_a(v_1, v_2) \equiv \langle a|\,T\{\bar{\psi}(v_1)\bar{\psi}(v_2)\}\,|0\rangle \tag{3.100}$$

In (3.100) translation invariance allows extraction of the center-of-mass dependence, P_a being the four momentum of the bound state $|a\rangle$,

$$\bar{\chi}_a(v_1, v_2) = e^{-iP_a\cdot V}\bar{\chi}_a(P_a, v_1 - v_2), \tag{3.101}$$

whence, for $V^0<0$ and large,

$$R_a \sim \sum e^{-iP_a\cdot V}\langle 0|\,T\{\ldots\,\psi(x_i)\,\ldots\,\bar{\psi}(y_j)\,\ldots\,\varphi(z_k)\,\ldots\}\,|a\rangle\,\bar{\chi}_a(P_a, v_1 - v_2). \tag{3.102}$$

On the other hand, for $V^0>0$,

$$R_a \sim \langle 0|\,T\{\bar{\psi}(v_1)\bar{\psi}(v_2)\}\,|a\rangle\,\ldots \quad = 0, \tag{3.103}$$

because $|a\rangle$ is a two-fermion bound state and ψ does not annihiliate fermions.

This simple behavior in V^0 given by (3.102) and (3.103) corresponds to a pole in the Fourier transform of R, relative to the coordinates V. With $P=p_1+p_2$ being the conjugate coordinates, it corresponds to a pole at $P=P_a$. More precisely, let us normalize the states as before,

$$\langle a(\boldsymbol{P})|a(\boldsymbol{P}')\rangle = \delta(\boldsymbol{P}-\boldsymbol{P}'), \tag{3.104}$$

and define the Fourier transform

$$\tilde{R}(P) = \frac{1}{(2\pi)^2}\int d^4V R(V) e^{iP\cdot V} \tag{3.105}$$

We then get

$$\tilde{R}(P) = (2\pi)^2 \frac{i}{2\pi} \frac{\langle 0|\,T\{\ldots\,\psi(x_i)\ldots\,\bar{\psi}(y_j)\,\ldots\,\varphi(z_k)\,\ldots\}\,|a\rangle\,\bar{\chi}_a(P, v_1 - v_2)}{P_0 - (\boldsymbol{P}^2 + m_a^2)^{1/2}} + \text{Regular terms} \tag{3.106}$$

This analysis is useful in estimating the matrix element if we have an independent way to estimate the residue. There is indeed a possibility to achieve this by perturbative methods based on Feynman diagrams. In fact it is possible to exhibit the singularities of R using an expression for R that derives from diagram analysis:

$$R(x_i,y_j,z_k;v_1,v_2) = \int d^4v_3 d^4v_4\ G(x_i,y_j,z_k;v_3,v_4)K(v_3,v_4;v_1,v_2), \quad (3.107)$$

where $K(v_3,v_4;v_1,v_2)$ is the usual two-body propagator, i.e. the Green function,

$$K(v_3,v_4;v_1,v_2) = \langle 0|\, \mathrm{T}\{\psi(v_3)\psi(v_4)\bar{\psi}(v_1)\bar{\psi}(v_2)\}\, |0\rangle, \qquad (3.108)$$

and $G(x_i,y_j,z_k;v_3,v_4)$ is the sum of all diagrams in the diagrammatic expansion of R that are irreducible. These are diagrams that cannot be divided into two parts, one being a diagram for K, and the two parts being connected by two fermion lines. Note that this particular notion of irreducibility is relative to a given two-fermion channel, here $\psi(v_1)\ \psi(v_2)$. Following Mandelstam (1955), we call G the irreducible kernels and K the Green functions throughout this section. The decomposition (3.107) can be understood by decomposing a general diagram for R as in Figure 3.6.

Grossly speaking, the upper blob corresponds to G, and the lower square to K. However, some care must be taken in the definition of these blobs and squares. The graphs we are considering are graphs for Green functions where the external lines are associated with propagators. This is indicated by the dots on which the external lines finish. Similarly, graphs for K include not only propagators corresponding to the external lines v_1 and v_2, but also *those corresponding to the two other fermion lines, which connect them to the upper blob.* Therefore, on factorizing K, it must be understood that we include the

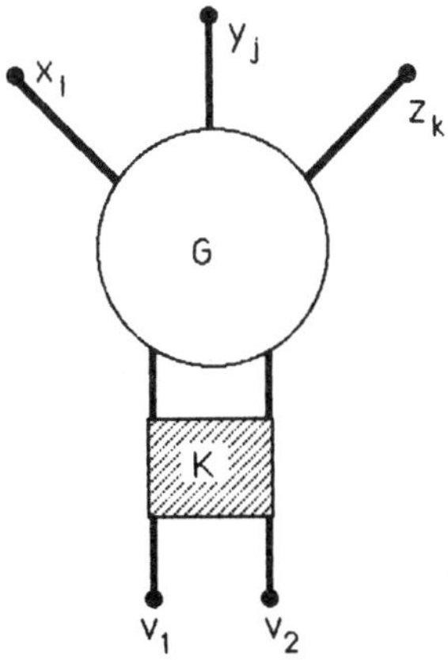

Figure 3.6 Decomposition of a general Green function into a part G irreducible with respect to a two-quark channel, and the two-quark propagator K.

propagators corresponding to these latter lines, and G is defined accordingly. We shall give concrete applications of these prescriptions in Sections 3.4.3 and 3.4.5. We can see that, in the expression (3.107) or rather in its Fourier transform $\tilde{R}(P)$ with respect to $V=\frac{1}{2}(v_1+v_2)$, the singularity in P due to the bound state $|a\rangle$ can result only from the corresponding singularity of the two-body propagator or rather from that of its Fourier transform with respect to V:

$$\tilde{K}(P)=(2\pi)^2\frac{i}{2\pi}\frac{\chi_a(v_3,v_4)\,\bar{\chi}_a(P,v_1-v_2)}{P_0-(\boldsymbol{P}^2+m_a^2)^{1/2}}. \tag{3.109}$$

On the other hand, since G is nonsingular, it can safely be expanded to a given order in the coupling constant.

From (3.106), (3.107) and (3.109), we obtain the final result

$$\begin{aligned}&\langle 0|\,\mathrm{T}\{\ldots\,\psi(x_i)\,\ldots\,\bar{\psi}(y_j)\,\ldots\,\varphi(z_k)\,\ldots\}\,|a\rangle\\&\quad=\int d^4v_3\,d^4v_4\,G(x_i,y_j,z_k;v_3,v_4)\,\chi_a(v_3,v_4).\end{aligned} \tag{3.110}$$

This method can be immediately generalized to a transition-matrix element between bound states $|a\rangle$ and $|b\rangle$,

$$\langle b|\,\mathrm{T}\{\ldots\,\psi(x_i)\ldots\bar{\psi}(y_j)\ldots\,\varphi(z_k)\,\ldots\}\,|a\rangle, \tag{3.111}$$

by considering the Green function

$$\begin{aligned}&R'(x_i,y_j,z_k;u_1,u_2,v_1,v_2)\\&=\langle 0|\,\mathrm{T}\{\psi(u_1)\psi(u_2)\,\ldots\,\psi(x_i)\,\ldots\,\bar{\psi}(y_j)\,\ldots\,\varphi(z_k)\,\ldots\,\bar{\psi}(v_1)\bar{\psi}(v_2)\}\,|0\rangle\end{aligned} \tag{3.112}$$

and looking for its simultaneous singularities due to the bound states $|a\rangle$ and $|b\rangle$ in the momenta respectively conjugate to $U=\frac{1}{2}(u_1+u_2)$ and $V=\frac{1}{2}(v_1+v_2)$. The final answer is

$$\begin{aligned}&\langle b|\,\mathrm{T}\{\ldots\,\psi(x_i)\,\ldots\,\bar{\psi}(y_j)\,\ldots\,\varphi(z_k)\,\ldots\}\,|a\rangle\\&=\int d^4u_3\,d^4u_4\,d^4v_3 d^4v_4\,\bar{\chi}_b(u_3,u_4)\;G'(x_i,y_j,z_k;u_3,u_4,v_3,v_4)\,\chi_a(v_3,v_4),\end{aligned} \tag{3.113}$$

where G' is now the sum of all the graphs for the Green function R' that are irreducible with respect to the two channels u_1, u_2 and v_1, v_2.

Using this approach the result is automatically covariant if we follow the following two prescriptions.

(i) use covariant one-particle propagators and vertices in the Feynman rules giving the expansion of G in powers of the coupling constant; and
(ii) use a covariant integral equation for the BS amplitudes, which is realized if the kernel is covariant as well as the one-particle propagators.

It is also possible to directly implement covariance of the BS amplitudes by expanding them, as usual, into a certain number of invariant quantities, or by applying a Lorentz boost to the amplitudes defined in the rest frame. The latter method seems natural, since one may have a better intuition of what happens in the rest frame, and since the effect of a Lorentz boost on BS amplitudes is straightforward. If Λ is the Lorentz transformation of coordinates that brings the hadron at rest into the moving hadron, we have

$$\chi_{\boldsymbol{P}}^{\alpha\beta}(x_1,x_2) = S^{\alpha\alpha'}(\Lambda)S^{\beta\beta'}(\Lambda)\chi_{\boldsymbol{P}=0}^{\alpha'\beta'}(\Lambda^{-1}x_1,\Lambda^{-1}x_2) \tag{3.114}$$

where α, β, α' and β' are Dirac indices and $S(\Lambda)$ is the matrix of the Lorentz transformation in the Dirac space. The simplicity of the law (3.114) is a great advantage of the BS formalism over the usual wave functions, but the price to pay is that $\chi_{\boldsymbol{P}=0}$ must be known for nonzero relative time.

3.4.2 Meson Annihilation into Leptons

Leptonic annihilation is a very special case in the Mandelstam formalism, because the kernel G is known exactly. In fact, the Green function R is the two-body propagator, and its singularities are completely given by the BS amplitude. The irreducible kernel G is the identity. Indeed, the quantity under consideration is

$$\langle 0|j_\mu(x)\,|\mathrm{a}\rangle, \tag{3.115}$$

where j_μ is some bilinear current like

$$j_\mu = \bar{\psi}\gamma_\mu\psi \quad \text{or} \quad \bar{\psi}\gamma_\mu\gamma_5\psi, \tag{3.116}$$

and $|\mathrm{a}\rangle$ is a meson. We could more generally consider $j_\mu = \bar{\psi}\mathcal{O}_\mu\psi$, where $\mathcal{O}_\mu$ is any Dirac operator that governs physically interesting meson–vacuum matrix elements. Following the general derivation of Section 3.4.1., we start from the basic form

$$\langle \mathrm{o}|\mathrm{T}\{\psi(x)\bar{\psi}(y)\}\,|\mathrm{a}\rangle, \tag{3.117}$$

and we have to consider

$$R(x,y;u_1,v_2) = \langle 0|\,\mathrm{T}\{\psi(u_1)\psi(x)\bar{\psi}(y)\bar{\psi}(v_2)\}\,|0\rangle, \tag{3.118}$$

where we have added the operators ψ and $\bar{\psi}$ because we want to consider a meson $\mathrm{q}\bar{\mathrm{q}}$ intermediate state. Now R is immediately recognized to be the two-body propagator, and then, according to (3.107), G is exactly the identity, with

$$G(x,y;u_3,v_4) = \delta(x-u_3)\delta(y-v_4), \tag{3.119}$$

$$R(x,y;u_1,v_2) = \int \mathrm{d}^4u_3\,\mathrm{d}^4v_4\,G(x,y;u_3,v_4)\,K(u_1,u_3;v_4,v_2), \tag{3.120}$$

where the variables inside K are written in the same order as in (3.108), although this time we have to deal with fermion–antifermion propagation instead of two-fermion propagation. The fact that G is the identity means that, according to (3.110), (3.117) is directly related to the BS amplitude. Of course, this could have been seen immediately, since (3.117) is recognized to be the BS amplitude itself. We just wanted to make the connection with the general formalism for this particular application. Now if $j_\mu = \bar{\psi}\mathcal{O}_\mu\psi$, and with

$$\chi_{\rm a}(x, y) = \langle 0|\,\mathrm{T}\{\psi(x)\bar{\psi}(y)\}\,|{\rm a}\rangle, \tag{3.121}$$

we obtain

$$\langle 0|\,j_\mu(x)\,|{\rm a}\rangle = \mathrm{Tr}\,[\mathcal{O}_\mu\chi_{\rm a}(x=y)]. \tag{3.122}$$

This remarkably simple and exact result suggest that leptonic annihilation is the simplest thing to calculate in a relativistic model. However, (3.122) is useful only if one can start from the fundamental interaction, and derive from it the BS amplitude $\chi_{\rm a}$. The fact that it is exact is important only if we can evaluate this amplitude exactly, or if we have a well-defined approximation procedure. In practice this is not the case; one does not know the interaction kernel from first principles, and tries to choose it so as to reproduce the largest set of data, including the spectrum, the radiative transitions, etc., as well as the leptonic annihilation. Moreover, the BS equation itself cannot be solved exactly. It then appears that leptonic annihilation, as well as other hadronic properties, are affected by very drastic assumptions and approximations that are necessary on technical grounds, but difficult to justify theoretically.

3.4.3 Transition-Matrix Elements of Currents

With hadron–hadron matrix elements of currents, we enter a category of quantities for which the irreducible kernel G central to the Mandelstam formalism is known only in the lowest order of perturbation theory in powers of the coupling constant. It is clear that such an approximation is not justified in the domain of strong interactions, and the situation is exactly the same as for the Hamiltonian perturbation theory of Section 3.3. The calculations can only be justified at the moment by their possible phenomenological success. An important point, however, is that they offer another type of perturbative approximation, different from the naive quark model, and different from the Hamiltonian perturbation theory. Of course, one may object that if everything was consistently expanded in powers of the interaction, the results should be the same in the various perturbative schemes, and that, in particular, the lowest-order approximation should be identical with the naive quark model. We shall indeed demonstrate this latter assertion in Section 3.4.6. But here we consider the BS amplitudes as given quantities, and we do not look for their own perturbative expansion. In this respect, our attitude is exactly the same as that taken in the Hamiltonian formalism, when we refrained from expanding perturbatively the solution of the wave equation (3.85). The lowest-order approximations of the two formalisms are then different from each other, and different from the naive quark model.

Extensive calculations have been made by the Hamburg group (see e.g. Böhm, Joos and Krammer, 1972, 1973a,b; 1974), using the lowest-order approximation for the irreducible kernel G, but nevertheless deriving the BS amplitudes in a strong-binding limit. These calculations are the most systematic attempt made within the Mandelstam formalism (see also Llewellyn-Smith, 1969a; Mitra, 1981; Mitra and Santhanam, 1981). They have been extended to weak matrix elements (Flamm and Schöberl, 1978). The main advantage of working in this lowest-order approximation is that the kernel is very simple, and all the dependence on the interaction is included in the BS amplitudes.

Let us now turn to actual calculations. The quantities we want to calculate can be straightforwardly derived from the matrix element,

$$\langle \mathrm{b} | \, \mathrm{T}\{\psi(x)\bar{\psi}(y)\} \, | \mathrm{a} \rangle, \tag{3.123}$$

to which we apply the methods of Section 3.4.1. The lowest approximation for the corresponding Green function R' (3.112) can be read from the graph of Figure 3.7.

We identify the kernel G' as understood above. The hatched rectangles denote the two-body propagators. But since both contain the fermion

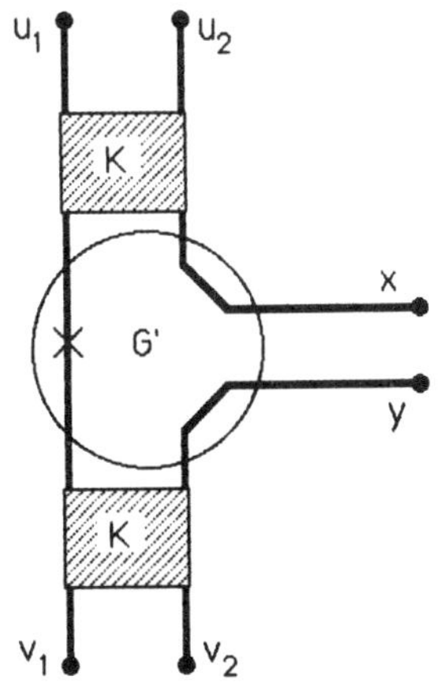

Figure 3.7 The simplest approximation to a Green function G' irreducible with respect to two-meson channels; the cross indicates an inverse quark propagator.

propagator to be associated with the line connecting them, the kernel G' must contain a factor $(\mathrm{i}S)^{-1}$ (indicated by a cross), according to,

$$\mathrm{i}S = \mathrm{i}S(\mathrm{i}S)^{-1}\mathrm{i}S. \tag{3.124}$$

The products must be understood in the sense of convolutions for spatial variables and matrix products for spin indices:

$$f \cdot g(u-w) = \int \mathrm{d}v\, f(u-v)\, g(v-w). \tag{3.125}$$

On the other hand, for the lines ending on x and y, we just have to write δ-functions identifying the variables relative to K, u_4 and v_4, with x and y (including spin variables). Finally,

$$G'(x,y;u_3,u_4,v_3,v_4) = -\mathrm{i}S^{-1}(v_3-u_3)\,\delta^4(x-u_4)\,\delta^4(y-v_4). \tag{3.126}$$

we still have to evaluate $S^{-1}(v_3-u_3)$. In fact, from the definition,

$$\delta^4(x-z) = \int \mathrm{d}^4y\, S(x-y)S^{-1}(y-z), \tag{3.127}$$

and by applying to both sides the Dirac operator $\mathrm{i}\gamma_\mu\partial^\mu - m$, and recalling the equation for the propagator,

$$(\mathrm{i}\partial\!\!\!/_x - m)S(x-y) = \delta^4(x-y), \tag{3.128}$$

we get

$$\begin{aligned}(\mathrm{i}\partial\!\!\!/_x - m)\delta^4(x-z) &= \int \mathrm{d}^4y\, \delta^4(x-y)S^{-1}(y-z) = S^{-1}(x-z),\\ S^{-1}(x-z) &= (\mathrm{i}\partial\!\!\!/_x - m)\,\delta^4(x-z).\end{aligned} \tag{3.129}$$

Inserting (3.129) into (3.126) and G' into (3.113), we obtain the answer for the current-matrix element (we have made now explicit the spinor indices for notational clarity):

$$\langle \mathrm{b}|\mathrm{T}\{\psi_\alpha(x)\bar{\psi}_\beta(y)\}\,|\mathrm{a}\rangle = -\mathrm{i}\int \mathrm{d}^4u_3\, \bar{\chi}_\mathrm{b}^{\gamma'\beta}(u_3,y)(\mathrm{i}\partial\!\!\!/_{u_3} - m)^{\gamma'\gamma}\chi_\mathrm{a}^{\gamma\alpha}(u_3,x). \tag{3.130}$$

Of course, in all the preceding expressions, we have ignored the fact that a second graph is possible, by opening the line u_1–v_1 instead of u_2–v_2. The final correct expression is therefore

$$\begin{aligned}\langle \mathrm{b}|\mathrm{T}\{\psi_\alpha(x)\bar{\psi}_\beta(y)\}\,|\mathrm{a}\rangle = &-\mathrm{i}\int \mathrm{d}^4u_3\, \bar{\chi}_\mathrm{b}^{\gamma'\beta}(u_3,y)(\mathrm{i}\partial\!\!\!/_{u_3} - m)^{\gamma'\gamma}\chi_\mathrm{a}^{\gamma\alpha}(u_3,x)\\ &-\mathrm{i}\int \mathrm{d}^4u_4\, \bar{\chi}_\mathrm{b}^{\beta\gamma'}(y,u_4)(\mathrm{i}\partial\!\!\!/_{u_4} - m)^{\gamma'\gamma}\chi_\mathrm{a}^{\alpha\gamma}(x,u_4).\end{aligned} \tag{3.131}$$

Suppose for the sake of definiteness that we have to calculate the electromagnetic current of a unit-charge bound state formed by a neutral fermion plus a charged one. We get

$$\langle \mathrm{b}|j_\mu(x)|\mathrm{a}\rangle = -\mathrm{i}\int \mathrm{d}^4u_3\,\bar{\chi}_{\mathrm{b}}^{\gamma'\alpha'}(u_3,x)(\mathrm{i}\not{\partial}_{u_3}-m)^{\gamma'\gamma}\gamma_\mu^{\alpha'\alpha}\chi_{\mathrm{a}}^{\gamma\alpha}(u_3,x). \quad (3.132)$$

This expression gives at the same time a condition of normalization for the BS amplitudes. Suppose that $|\mathrm{a}\rangle$ and $|\mathrm{b}\rangle$ are the same state, with possibly different momenta. Since the charge of the bound states is $+1$, $Q|\mathrm{a}\rangle = |\mathrm{a}\rangle$,

$$\langle \mathrm{b}|\,Q\,|\mathrm{a}\rangle = \delta(\boldsymbol{P}-\boldsymbol{P}'), \quad (3.133)$$

but from the definition of the charge operator, $Q=\int \mathrm{d}\boldsymbol{x}\,j_0(\boldsymbol{x})$ and from translational invariance, we also have

$$\langle \mathrm{b}|\,Q\,|\mathrm{a}\rangle = \delta(\boldsymbol{P}-\boldsymbol{P}')(2\pi)^3\langle \mathrm{b}|j_0(0)\,|\mathrm{a}\rangle. \quad (3.134)$$

Therefore,

$$\langle \mathrm{b}|j_0(0)\,|\mathrm{a}\rangle = \frac{1}{(2\pi)^3}. \quad (3.135)$$

This makes sense only if Q is actually a constant of motion, and therefore if j_μ is conserved. Conservation can be verified directly from (3.132) using the BS equation in the *ladder* approximation.

Llewellyn-Smith (1969b) describes how to find the normalization of the Bethe–Salpeter amplitudes directly from the normalization of the states, without recourse to a conserved charge.

3.4.4 The $\pi^0 \to \gamma\gamma$ decay

The $\pi^0 \to \gamma\gamma$ decay is a special case of meson annihilation, although more complicated than the meson annihilation into a lepton pair considered in Sections 2.1.3 and 3.4.2. The quasi-Goldstone character of the pion (see Section 1.2.4) and the fact that this decay is related to a special concept in quantum field theory, the *chiral anomaly* (or axial anomaly), makes it crucial to study this decay with a relativistic framework; this is the reason why it is treated in this Chapter. Indeed the nonrelativistic calculation according to the lines of Section 2.1.3 gives a result that is much too large. A calculation (Van Royen and Weisskopf, 1967), which uses vector-meson dominance and the $\mathrm{V}\pi\gamma$ coupling as well as the $\mathrm{V}\gamma$ coupling computed in the quark model (V representing the ω and ρ vector mesons), gives rather good results, but this calculation is only partly nonrelativistic.

The axial anomaly was discovered by Adler, Bell and Jackiw (Adler, 1969; Bell and Jackiw, 1969), and we shall give an introduction to it in the course of the following discussion.

We must here consider the quantity

$$\langle 0|\,\mathrm{T}(j_\mu^{\mathrm{em}}(x)\,j_\nu^{\mathrm{em}}(y))\,|\mathrm{a}\rangle, \quad (3.136)$$

which is not so simple as (3.115). From (3.96), its evaluation involves a Green function that is of the sixth power in the quark field, close to what is needed to evaluate (3.123):

$$R(x,y,v_1,v_2) = \langle 0|\,\mathrm{T}(j_\mu^{\mathrm{em}}(x)\,j_\nu^{\mathrm{em}}(y)\,\bar{q}(v_1)q(v_2))\,|0\rangle. \tag{3.137}$$

To compute $\pi^0 \to \gamma\gamma$ we shall have to consider the linear combinations of the $q_\alpha(v_1)\,q_\beta(v_2)$ matrix elements (α, $\beta = 1, \ldots, 4$ are indices in Dirac space) that have the necessary quantum numbers for the π^0 intermediate state to contribute to the Green function (3.137). In this way we shall be able to compute the residue and hence the $\pi^0 \to \gamma\gamma$ amplitude. There are two possible choices that give a known expression similar to (3.100): the axial current and the pseudoscalar density, since

$$\langle \pi^0 | \frac{1}{2} \sum_i \{\bar{u}_i \gamma_\mu \gamma_5 u_i - \bar{d}_i \gamma_\mu \gamma_5 d_i\} |0\rangle = \frac{-\mathrm{i} q_\mu F_\pi}{(2E_\pi)^{1/2}(2\pi)^{3/2}}, \tag{3.138}$$

$$\langle \pi^0 | \mathrm{i} \sum_i \{m_\mathrm{u} \bar{u}_i \gamma_5 u_i - m_\mathrm{d} \bar{d}_i \gamma_5 d_i\} |0\rangle = \frac{m_\pi^2 F_\pi}{(2E_\pi)^{1/2}(2\pi)^{3/2}}, \tag{3.139}$$

where the index $i = 1, \ldots, 3$ runs over colors, m_u and m_d are the current masses, E_π is the pion energy, the factor $(2E_\pi)^{-1/2}(2\pi)^{-3/2}$ is due to the normalization (3.104) of the pion state, and $F_\pi = 93$ MeV. Equation (3.139) is deduced from (3.138) through the *normal* Ward identity, which relates the divergence of the axial current and the pseudoscalar density:

$$\frac{\mathrm{d}}{\mathrm{d}v^\mu}(\bar{q}(v)\gamma^\mu\gamma_5 q(v)) = 2\mathrm{i}m(\bar{q}(v)\gamma_5 q(v)). \tag{3.140}$$

As we shall see, the axial anomaly consists in the fact that (3.140) is not exact: we must include a correction of order $\alpha = e^2/4\pi$ that leaves (3.138) and (3.139) unchanged.

We must now compute (3.137) with the two linear combinations of the $\bar{q}_\alpha(v_1)q_\beta(v_2)$ to be considered from (3.138) and (3.139): the pseudoscalar density $\bar{q}(v)\gamma_5 q(v)$ and the axial density $\bar{q}(v)\gamma_\mu\gamma_5 q(v)$. Comparison of these two calculations leads to a contradiction with (3.140), and reveals the anomaly. We shall now proceed with the calculation of these two Green functions, exhibit and discuss the anomaly, and then we shall take advantage of the anomaly to compute $\pi^0 \to \gamma\gamma$.

Let us start with the calculation involving the pseudoscalar density. We define

$$R_{\mu\nu}^{\mathrm{P}}(x,y,v) = \langle 0|\,\mathrm{T}(j_\mu^{\mathrm{em}}(x)\,j_\nu^{\mathrm{em}}(y)\,\bar{q}_i(v)\gamma_5 q_i(v))\,|0\rangle. \tag{3.141}$$

where the index i labels the color of the quark q_i. To zeroth order in the strong

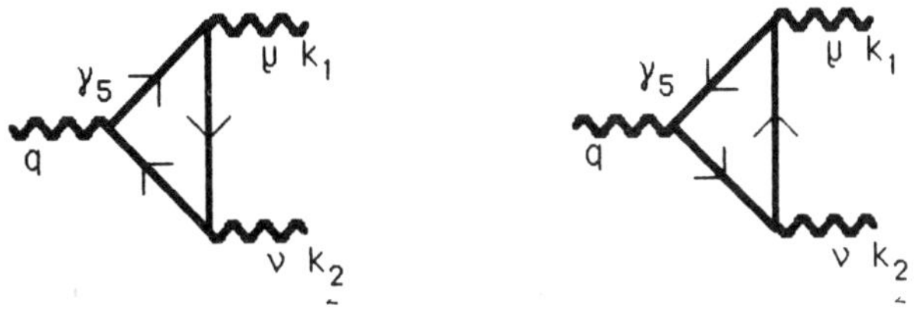

Figure 3.8

coupling constant, (3.141) is estimated from the Feynman diagrams in Figure 3.8. We denote

$$\tilde{R}^{\mathrm{P}}_{\mu\nu}(k_1,k_2)=\int \mathrm{d}^4x\,\mathrm{d}^4y\,\mathrm{e}^{\mathrm{i}(k_1\cdot x+k_2\cdot y)}R^{\mathrm{P}}_{\mu\nu}(x,y,0), \tag{3.142}$$

$$\tilde{R}^{\mathrm{P}(0)}_{\mu\nu}(k_1,k_2)=\tilde{R}^{\mathrm{P}_1(0)}_{\mu\nu}(k_1,k_2)+\tilde{R}^{\mathrm{P}_2(0)}_{\mu\nu}(k_1,k_2), \tag{3.143}$$

where $\tilde{R}^{\mathrm{P}(0)}_{\mu\nu}(k_1,k_2)$ is $\tilde{R}^{\mathrm{P}}_{\mu\nu}(k_1,k_2)$ to lowest order in α_s. $R^{\mathrm{P}_1(0)}$ and $R^{\mathrm{P}_2(0)}$ correspond to the two diagrams in Figure 3.8, and $k_1^2=k_2^2=0$ for real photons. We have

$$\tilde{R}^{\mathrm{P}_1(0)}_{\mu\nu}(k_1,k_2)$$
$$=e^2Q^2\int\frac{\mathrm{d}^4p}{(2\pi)^4}\,\mathrm{Tr}\left\{\frac{(\not{q}+\not{p}-\not{k}_1+m)\gamma_\mu(\not{q}+\not{p}+m)\gamma_5(\not{p}+m)\gamma_\nu}{[(q+p-k_1)^2-m^2][(q+p)^2-m^2][p^2-m^2]}\right\}, \tag{3.144}$$

where $q=k_1+k_2$, and Q is the quark charge in e units. Changing p into $-p$ in the second diagram leads to

$$\tilde{R}^{\mathrm{P}_1(0)}_{\mu\nu}(k_1,k_2)=\tilde{R}^{\mathrm{P}_2(0)}_{\mu\nu}(k_1,k_2) \tag{3.145}$$

The trace numerator of (3.144) is equal to

$$4\mathrm{i}\varepsilon_{\mu\nu\rho\sigma}k_1^\rho k_2^\sigma, \tag{3.146}$$

where $\varepsilon_{0123}=-1$. Then

$$\tilde{R}^{\mathrm{P}_1(0)}_{\mu\nu}(k_1,k_2)$$
$$=4\mathrm{i}me^2Q^2\varepsilon_{\mu\nu\rho\sigma}k_1^\rho k_2^\sigma\int\frac{\mathrm{d}^4p}{(2\pi)^4}\frac{1}{[(q+p-k_1)^2-m^2][(q+p)^2-m^2][p^2-m^2]} \tag{3.147}$$

$$=4\mathrm{i}me^2Q^2\varepsilon_{\mu\nu\rho\sigma}k_1^\rho k_2^\sigma\int_0^1\mathrm{d}x\int_0^x\mathrm{d}y\,\frac{1}{m^2-xyq^2}.$$

In the vicinity of $q^2=0$ we get, using (3.145),

$$\tilde{R}^{P(0)}_{\mu\nu}(k_1,k_2) = \frac{e^2 Q^2}{4\pi^2 m}\varepsilon_{\mu\nu\rho\sigma}k_1^\rho k_2^\sigma\left(1+O\left(\frac{q^2}{m^2}\right)\right). \tag{3.149}$$

Let us now perform the calculation of (3.137) with the axial-vector current. We define

$$R^A_{\mu\nu\rho}(x,y,z) = \langle 0|\, T(j^{em}_\mu(x)\, j^{em}_\nu(y)\, \bar{q}_i(z)\gamma_\rho\gamma_5 q_i(z))\,|0\rangle, \tag{3.150}$$

$$\tilde{R}^A_{\mu\nu\rho}(k_1,k_2) = \int d^4x\, d^4y\, e^{i(k_1\cdot x + k_2\cdot y)} R^A_{\mu\nu\rho}(x,y,0), \tag{3.151}$$

To lowest order in α_s these are given by the diagrams in Figure 3.9.

As the pion field has the quantum numbers of the divergence of the axial current, we are interested in the expression

$$q^\rho \tilde{R}^{A(0)}_{\mu\nu\rho}(k_1,k_2) = q^\rho \tilde{R}^{A_1(0)}_{\mu\nu\rho}(k_1,k_2) + q^\rho \tilde{R}^{A_2(0)}_{\mu\nu\rho}(k_1,k_2) \tag{3.152}$$

where $q = k_1 + k_2$, $\tilde{R}^{A(0)}_{\mu\nu\rho}$ represents $\tilde{R}^A_{\mu\nu\rho}$ to lowest order in α_s, and the two terms on the right-hand side correspond to the two diagrams in Figure 3.9, which are equal.

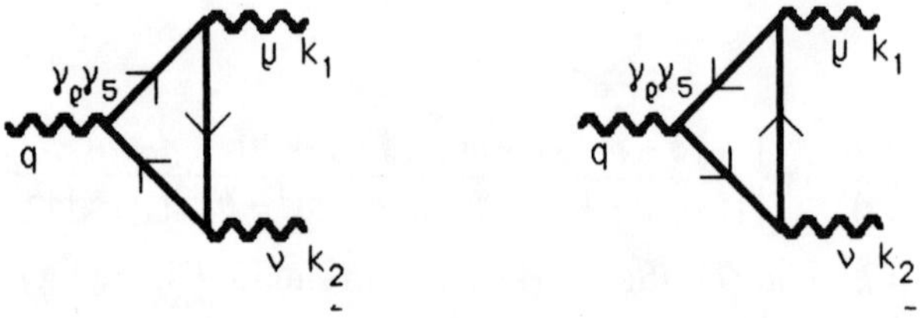

Figure 3.9

Before going into the calculation, let us note what would be expected if the identity (3.140) were exact. We replace the pseudoscalar density in (3.141) by the divergence of the axial current. Next we note that the currents in (3.150) commute with each other such that the divergence can be taken out of the T product, and we end with the normal Ward identity:

$$q^\rho \tilde{R}^A_{\mu\nu\rho}(k_1,k_2) = 2m\tilde{R}^P_{\mu\nu}(k_1,k_2). \tag{3.153}$$

We shall now see that this Ward identity has to be corrected. We have

$$q^\rho \tilde{R}^{A_1(0)}_{\mu\nu\rho}(k_1,k_2)$$
$$= e^2Q^2 \int \frac{d^4p}{(2\pi)^4} \mathrm{Tr}\left\{\frac{(\not{p}+m)\gamma_\mu(\not{p}+\not{k}_1+m)(\not{k}_1+\not{k}_2)\gamma_5(\not{p}-\not{k}_1+m)\gamma_\nu}{[p^2-m^2][(p+k_1)^2-m^2][(p-k_1)^2-m^2]}\right\}, \tag{3.154}$$

and from

$$
\begin{aligned}
&(\not p+\not k_1+m)(\not k_1+\not k_2)\gamma_5(\not p-\not k_2+m)\\
&=(\not p+\not k_1+m)(\not p+\not k_1-m-\not p+\not k_2+m)\gamma_5(\not p-\not k_2+m)\\
&=2m(\not p+\not k_1+m)\gamma_5(\not p-\not k_2+m)\\
&\quad+((p+k_1)^2-m^2)\gamma_5(\not p-\not k_2+m)+(\not p+\not k_1+m)\gamma_5((p-k_2)^2-m^2)
\end{aligned}
\tag{3.155}
$$

we obtain

$$
\begin{aligned}
&q^\rho \tilde{R}^{\mathrm{A}_1(0)}_{\mu\nu\rho}(k_1,k_2)=2m\tilde{R}^{\mathrm{P}_1(0)}_{\mu\nu}(k_1,k_2)\\
&+e^2Q^2\int\frac{\mathrm{d}^4p}{(2\pi)^4}\,\mathrm{Tr}\left\{\frac{(\not p+m)\gamma_\mu\gamma_5(\not p-\not k_2+m)\gamma_\nu}{[p^2-m^2][(p-k_2)^2-m^2]}-\frac{(\not p+\not k_1+m)\gamma_\nu\gamma_5(\not p+m)\gamma_\mu}{[(p+k_1)^2-m^2][p^2-m^2]}\right\}.
\end{aligned}
\tag{3.156}
$$

The question is thus whether the second term in (3.156) vanishes or cancels the analogous term in $q^\rho \tilde{R}^{\mathrm{A}_2(0)}_{\mu\nu\rho}$. This seems superficially to be the case, since a change of variables $p\to p+k_1$ for the first term and $p\to p+k_2$ for the second term in the braces in (3.156) transforms the terms in $q^\rho \tilde{R}^{\mathrm{A}_1(0)}_{\mu\nu\rho}$ into the opposite of the terms in $q^\rho \tilde{R}^{\mathrm{A}_2(0)}_{\mu\nu\rho}$. But *this argument is incorrect* owing to the linear divergences in (3.156). The terms that seem to cancel each other are not well defined—they are infinite. We must first regularize the divergent integrals. We use the gauge-invariant Pauli–Villars (1949) regularization procedure, which consists in subtracting from the integrand in (3.154) the same integrand with the quark mass m replaced by a heavy regulator mass M. Now the integral converges, we can use (3.155), change the variables $p\to p+k_1$ and $p\to p+k_2$, and the cancellation between the terms in $q^\rho \tilde{R}^{\mathrm{A}_1(0)}_{\mu\nu\rho}$ and $q^\rho \tilde{R}^{\mathrm{A}_2(0)}_{\mu\nu\rho}$ does indeed occur. The result is

$$
q^\rho(\tilde{R}^{\mathrm{A}(0)}_{\mu\nu\rho}(m)-\tilde{R}^{\mathrm{A}(0)}_{\mu\nu\rho}(M))=2m\tilde{R}^{\mathrm{P}(0)}_{\mu\nu}(m))-2M\tilde{R}^{\mathrm{P}(0)}_{\mu\nu}(M),
\tag{3.157}
$$

where we have made explicit the masses in the diagrams. From (3.149) we have

$$
\lim_{M\to\infty} 2M\tilde{R}^{\mathrm{P}(0)}_{\mu\nu}(M)=\frac{e^2Q^2}{2\pi^2}\,\varepsilon_{\mu\nu\rho\sigma}k_1^\rho k_2^\sigma,
\tag{3.158}
$$

and, denoting the regularized Green function (3.154) by $\tilde{R}^{\mathrm{A(R)}(0)}_{\mu\nu\rho}(k_1,k_2)$, we now have

$$
q^\rho \tilde{R}^{\mathrm{A(R)}(0)}_{\mu\nu\rho}(k_1,k_2)=2m\tilde{R}^{\mathrm{P}(0)}_{\mu\nu}(k_1,k_2)-\frac{2\alpha Q^2}{\pi}\,\varepsilon_{\mu\nu\rho\sigma}k_1^\rho k_2^\sigma
\tag{3.159}
$$

Comparing (3.159 with the normal Ward identity (3.153), we see that a new term arises in (3.159)—an anomalous term. Identities such as (3.159) are called *anomalous axial Ward identities.* Note that from (3.149) $q^\rho R^{\mathrm{A(R)}(0)}_{\mu\nu\rho}(k_1,k_2)$ vanishes in the limit $q^2\to 0$ since

$$
2m\tilde{R}^{\mathrm{P}(0)}_{\mu\nu}(m)-2M\tilde{R}^{\mathrm{P}(0)}_{\mu\nu}(M)=O(q^2/m^2)+O(q^2/M^2).
\tag{3.160}
$$

In terms of operators, we can incorporate the result (3.159) by writing instead of (3.140) the *anomalous Ward identity*

$$\partial_\mu(\bar{q}_i\gamma^\mu\gamma_5 q_i) = 2im\bar{q}_i\gamma_5 q_i - \frac{\alpha}{2\pi}Q^2\,\varepsilon_{\mu\nu\rho\sigma}F^{\mu\nu}F^{\rho\sigma}, \tag{3.161}$$

where $F^{\mu\nu}$ is the electromagnetic-field tensor

$$F^{\mu\nu} = \partial^\mu A^\nu - \partial^\nu A^\mu. \tag{3.162}$$

Note that (3.159) concerns only the Green functions to lowest order in the strong coupling constant. But a crucial result—the nonrenormalizability of the anomaly (Adler, 1970)—implies that the relation (3.159) for the exact Green function and the Ward identity (3.161) are *true to all orders* in the coupling strength.

The anomalous term in (3.161) does not change matrix elements of the axial current other than the coupling to two photons (except for changes at higher orders in e^2). Thus the successes of the normal Ward identity are preserved and, in particular, the relation between (3.138) and (3.139) is still valid.

We must now compute the $\pi^0\to\gamma\gamma$ decay amplitude. We might start from the Green function (3.141) or (3.150) and use (3.106). However, we know (3.141) and (3.150) only to lowest order in α_s. It is thus better to use their linear combination that gives the anomaly, since the latter is known to all orders, thanks to the nonrenormalizability theorem. From (3.161), we have

$$\begin{aligned}&\int d^4x\,d^4y\,e^{i(k_1\cdot x+k_2\cdot y)}\langle 0|\,T\{j_\mu^{em}(x)j_\nu^{em}(y)[2im_u\bar{u}_i\gamma_5 u_i - 2im_d\bar{d}_i\gamma_5 d_i\\ &-\partial^\rho(\bar{u}_i\gamma_\rho\gamma_5 u_i - \bar{d}_i\gamma_\rho\gamma_5 d_i)](0)\}\,|0\rangle\\ &= \frac{\alpha}{2\pi}(Q_u^2 - Q_d^2)\int d^4x\,d^4y\,e^{i(k_1\cdot x+k_2\cdot y)}\langle 0|\,T\{j_\mu^{em}(x)j_\nu^{em}(y)\,\varepsilon_{\alpha\beta\gamma\delta}\,F^{\alpha\beta}(0)F^{\gamma\delta}(0)\}|0\rangle.\end{aligned} \tag{3.163}$$

We now consider the *pion pole contribution* to the left-hand side of this equation. Using the normalization (2.3a), the Mandelstam formula (3.106) and the amplitudes (3.138) and (3.139), we obtain for the *left-hand side* of (3.163),

$$\begin{aligned}&2im_u\tilde{R}^{Pu}_{\mu\nu}(k_1,k_2) - 2im_d\tilde{R}^{Pd}_{\mu\nu}(k_1,k_2) - iq^\rho\tilde{R}^{Au}_{\mu\nu\rho}(k_1,k_2) + iq^\rho\tilde{R}^{Ad}_{\mu\nu\rho}(k_1,k_2)\\ &= -\frac{i}{3}F_\pi\left(\frac{m_\pi^2-q^2}{m_\pi^2-q^2}\right)\int d^4x\,e^{ik_1\cdot x}\,\langle 0|\,T(j_\mu^{em}(x)j_\nu^{em}(0))\,|\pi^0\rangle(2E_\pi)^{1/2}(2\pi)^{3/2},\end{aligned} \tag{3.164}$$

where $q = k_1 + k_2$ and we have made explicit the quark fields u and d in $\tilde{R}^P$ and $\tilde{R}^A$. To estimate the non-pion pole terms we have used the important hypothesis of *partial conservation of axial current* (PCAC). We consider (3.164) at $q^\rho = 0$. The only contribution comes from the pseudoscalar terms, since the axial-vector ones contain a factor q^ρ. The pseudoscalar densities are multiplied by the very small current masses. The PCAC hypothesis can be expressed by saying that *the only contribution that does not vanish* when the current masses go to zero is the *Goldstone-boson pole* term, i.e. the pion pole term. In other terms the non-pion pole contribution vanishes at $q^\rho = 0$ in the exact chiral limit m_u, $m_d \to 0$. Furthermore, the non-pion pole terms can reasonably be assumed to vary slowly in function of q^2 in the region where q^2 is of the order of m_π^2. We therefore neglect them. We thus have from (3.163) and (3.164),

$$F_\pi \int d^4x\, e^{ik_1 \cdot x} \langle 0|\, T(j_\mu^{em}(x) j_\nu^{em}(0))\, |\pi^0\rangle = \frac{3i}{(2E_\pi)^{1/2}(2\pi)^{3/2}}$$

$$\times \frac{\alpha}{2\pi}(Q_u^2 - Q_d^2) \int d^4x\, d^4y\, e^{i(k_1 \cdot x + k_2 \cdot y)} \langle 0|\, T\{j_\mu^{em}(x) j_\nu^{em}(y)\, \varepsilon_{\alpha\beta\gamma\delta} F^{\alpha\beta}(0) F^{\gamma\delta}(0)\}\, |0\rangle \tag{3.165}$$

We now apply the photon reduction formula (Bjorken and Drell, 1964) to the left-hand side of (3.165) and the field equation $\Box A = ej^{tr}$.

$$M(\pi^0 \to \gamma\gamma) = \frac{e^2}{2E_\gamma} \int d^4x\, e^{ik_1 \cdot x}\, \varepsilon_1^\mu \varepsilon_2^\nu \langle 0| T(j_\mu^{em}(x) j_\nu^{em}(0)) |\pi^0\rangle, \tag{3.166}$$

where $M(\pi^0 \to \gamma\gamma)$ is the $\pi^0 \to \gamma\gamma$ decay amplitude according to the notation (2.27). Analogously,

$$\frac{e^2 \varepsilon_1^\mu \varepsilon_2^\nu}{(2\pi)^3 2E_\gamma} \int d^4x\, d^4y\, e^{i(k_1 \cdot x + k_2 \cdot y)} \langle 0|\, T\{j_\mu^{em}(x) j_\nu^{em}(y)) \varepsilon_{\alpha\beta\gamma\delta} F^{\alpha\beta}(0) F^{\gamma\delta}(0)\}\, |0\rangle$$

$$= \langle \gamma_1 \gamma_2 |\, \varepsilon_{\mu\nu\rho\sigma} F^{\mu\nu}(0) F^{\rho\sigma}(0)\, |0\rangle, \tag{3.167}$$

where we have again used the photon reduction formula.

Now using the expression (2.14) for the photon-field matrix element and (3.162) to estimate the right-hand side of (3.167), we get the pion-to-two-photon amplitude:

$$M(k_1, k_2, \varepsilon_1, \varepsilon_2) = \frac{3i\alpha}{\pi F_\pi} (Q_u^2 - Q_d^2) \frac{\varepsilon_{\mu\nu\rho\sigma} \varepsilon_1^\mu \varepsilon_2^\nu k_1^\rho k_2^\sigma}{2E_\gamma (2m_\pi)^{1/2} (2\pi)^{3/2}}, \tag{3.168}$$

where $\alpha = e^2/4\pi$, and the factor 3 comes from the color. To compute the width, we use (2.30). The two photons being identical, we decide to label 1 the photon in one hemisphere and replace the factor 4 in (2.31a) by 2. Assuming that the photons go in the z direction, we must add the amplitudes $k_1^0 k_2^3$ and $k_1^3 k_2^0$, and we must also add two polarizations. The result is finally

$$\Gamma(\pi^0 \to \gamma\gamma) = \frac{[3(Q_u^2 - Q_d^2)\alpha]^2 m_\pi^3}{64\pi^3 F_\pi^2} \tag{3.169}$$

From

$$3(Q_u^2 - Q_d^2) = 3(\tfrac{4}{9} - \tfrac{1}{9}) = 1, \tag{3.170}$$

we get

$$\Gamma(\pi^0 \to \gamma\gamma) = 7.6\,\text{eV}, \tag{3.171}$$

in close agreement with experiment:

$$\Gamma(\pi^0 \to \gamma\gamma) = 7.64\,\text{eV}. \tag{3.172}$$

If we did not have three colors, we would miss the factor 3 in (3.170) and the agreement with experiment. This was historically one of the *decisive proofs of the existence of color.*

The same calculation can be done for $\eta \to \gamma\gamma$, also with good success. It is worth stressing once again that the very good agreement obtained with experiment is due to three very special features: the fact that the anomalous Ward identity (3.161) is valid to all orders in the strong interactions, the good agreement of PCAC with experiment, and the existence of the color quantum number.

3.4.5 Strong-Interaction Decays

For strong-interaction decays, we encounter slightly different problems, because the scattering amplitude refers to "in" or "out" states made of several bound states. Nevertheless, the method of Mandelstam can be extended to treat such scattering states (see Mandelstam, 1955—the final part of the paper). But we think it useful to use a general formulation, which is able to handle any type of physical matrix element rather directly. This is the generalized *reduction formula*, a generalization of the usual reduction formula to bound states, which was proposed by Nishijima (1958) and Zimmermann (1958). The usual reduction formula (see e.g. Gasiorowicz, 1966) allows one to substitute for each particle state an interpolating field to be included under the T product. Analogously, one can substitute each bound state by an interpolating field. If we consider a two-fermion bound state, this field will be the product of two fermion fields, with a normalization factor related to the BS amplitude of the bound state at the origin. This formalism has been applied to the decay of one meson into two mesons by Kitazoe and Teshima (1968).

Let us assume, for the sake of simplicity, that we have to deal with a decay $a \to b + c$, where a, b and c are neutral scalar mesons. An interpolating field for these mesons can be constructed from the quark fields:

$$\varphi_a(U) = \lim_{\substack{\varepsilon \to 0 \\ \varepsilon^2 < 0}} \frac{\psi_\alpha(U+\varepsilon)\bar{\psi}_\alpha(U-\varepsilon)}{(2E_a)^{1/2}(2\pi)^{3/2}\langle 0|\,\psi_\alpha(\varepsilon)\bar{\psi}_\alpha(-\varepsilon)\,|a\rangle}, \tag{3.173}$$

with contraction of the spinor indices. This composite field is a scalar Hermitian field with a proper choice of the phase of $|a\rangle$ making real the denominator expression

$$\langle 0|\,\psi_\alpha(\varepsilon)\bar{\psi}_\alpha(-\varepsilon)\,|a\rangle \to \mathrm{Tr}\,[\chi_a(0,0)]. \tag{3.174}$$

This field is also properly normalized to

$$\langle 0|\,\varphi_a(U)\,|a\rangle = \frac{1}{(2E)^{1/2}}\frac{1}{(2\pi)^{3/2}}\,e^{-iP_a\cdot U}. \tag{3.175}$$

We can now express the S matrix, through the reduction formula, in the following form:

$$\begin{aligned}\langle b,c|S\,|a\rangle = i^3 &\frac{1}{F_a(0)F_b(0)F_c(0)}\int d^4U\,d^4V d^4W f_b^*(V) f_c^*(W)\\ &\times (\overrightarrow{\Box_V+m_b^2})(\overrightarrow{\Box_W+m_c^2})\,\langle 0|\,T\{\psi(V)\bar{\psi}(V)\psi(W)\bar{\psi}(W)\\ &\times \psi(U)\bar{\psi}(U)\,|0\rangle\,(\overleftarrow{\Box_U+m_a^2}) f_a(U),\end{aligned} \tag{3.176}$$

where $F_i(0) = (2\pi)^{3/2}(2E_i)^{1/2}\,\mathrm{Tr}[\chi_i(0,0)]$ (invariant) and $\Box_U$ is the d'Alembertian relative to the coordinates U. The f_i are the wave functions describing the center-of-mass motion of the mesons, normalized in the standard way for momentum eigenstates:

$$f_a(U) = \frac{1}{(2E_a)^{1/2}}\frac{1}{(2\pi)^{3/2}}\,e^{-iP_a\cdot U}, \tag{3.177}$$

in accordance with (3.175).

In the Green function

$$R = \langle 0|\,T\{\psi(V)\bar{\psi}(V)\psi(W)\bar{\psi}(W)\psi(U)\bar{\psi}(U)\}\,|0\rangle \tag{3.178}$$

a contraction is understood for the spinor indices of the $\psi\bar{\psi}$ couples.

It can easily be seen that the operations on R indicated in (3.176), starting from the integral, amount to picking out, in the simultaneous Fourier transform of R with respect to U, V and W, the singularities located at $P_1^2 = P_a^2 = m_a^2$, etc. We can exhibit these singularities by the same methods already used in Section 3.4.1. A slight difference is found owing to the fact that we have to deal with mesons: there are two singularities that are relevant for the present calculation. For $U^0 > 0$ and large, we can write

$$R \approx \sum_{P_a,s_a} \langle 0|\,T\{\psi(U)\bar{\psi}(U)\}|a\rangle\langle a|\,T\{\psi(V)\bar{\psi}(V)\psi(W)\bar{\psi}(W)\}\,|0\rangle, \tag{3.179}$$

while for $U^0 < 0$ and large,

$$R \approx \sum_{\boldsymbol{P}_a, s_a} \langle 0|\mathrm{T}\{\psi(V)\bar{\psi}(V)\psi(W)\bar{\psi}(W)\}|a\rangle\langle a|\mathrm{T}\{\psi(U)\bar{\psi}(U)\}|0\rangle. \quad (3.180)$$

These intermediate states will lead to poles in the Fourier transform, which is given by

$$\tilde{R}(P_1, V, W) = \frac{1}{(2\pi)^2}\int \mathrm{d}^4U\, \mathrm{e}^{-\mathrm{i}P_1\cdot U} R(U,V,W). \quad (3.181)$$

Defining

$$\bar{\chi}_a(x,y) = \langle a|\mathrm{T}\{\bar{\psi}(y)\psi(x)\}|0\rangle, \quad (3.182)$$

we find the poles corresponding to the state $|a\rangle$:

$$(2\pi)^2\frac{-1}{2\pi\mathrm{i}}[-\mathrm{Tr}\,\chi_{\boldsymbol{P}_a}(0)]\frac{1}{P_1^0-(\boldsymbol{P}_1^2+m_a^2)^{1/2}+\mathrm{i}\varepsilon} \times \langle 0|\mathrm{T}\{\psi(V)\bar{\psi}(V)\psi(W)\bar{\psi}(W)\}|a\rangle \quad (3.183)$$

and

$$(2\pi)^2\frac{1}{2\pi\mathrm{i}}\mathrm{Tr}\,\chi_{\boldsymbol{P}_a}(0)\frac{1}{P_1^0+(\boldsymbol{P}_1^2+m_a^2)^{1/2}-\mathrm{i}\varepsilon} \times \langle a|\mathrm{T}\{\psi(V)\bar{\psi}(V)\psi(W)\bar{\psi}(W)\}|0\rangle. \quad (3.184)$$

An identical analysis can be performed for singularities due to b and c. It can be seen immediately that the respective integrations over $\int \mathrm{d}U\,\mathrm{e}^{-\mathrm{i}P_a\cdot U}$, $\int \mathrm{d}V\mathrm{e}^{\mathrm{i}P_b\cdot V}$ and $\int \mathrm{d}W\mathrm{e}^{\mathrm{i}P_c\cdot W}$ pick up respectively the poles at $P_1 = P_a$, $P_2 = -P_b$ and $P_3 = -P_c$. Exhibiting these poles according to

$$R = \int \mathrm{d}^4P_i\,\mathrm{e}^{\mathrm{i}(P_1\cdot U - P_2\cdot V - P_3\cdot W)} \times \frac{-1}{2\pi\mathrm{i}}\frac{1}{P_1^0-E_a}\frac{1}{2\pi\mathrm{i}}\frac{1}{P_2^0-E_b}\frac{1}{2\pi\mathrm{i}}\frac{1}{P_3^0-E_c} \times[-\mathrm{Tr}\,\bar{\chi}_{\boldsymbol{P}_a}(0)\,\mathrm{Tr}\,\chi_{\boldsymbol{P}_b}(0)\,\mathrm{Tr}\,\chi_{\boldsymbol{P}_c}(0)]\,K_{abc}, \quad (3.185)$$

we find

$$K_{abc} = -\frac{\mathrm{Tr}\,\bar{\chi}_{\boldsymbol{P}_a}(0)}{\mathrm{Tr}\,\chi_a(0,0)}\langle b,c|S|a\rangle. \quad (3.186)$$

But, since χ is real, we have

$$\mathrm{Tr}\,\bar{\chi}_{\boldsymbol{P}_a}(0) = -\mathrm{Tr}\,\chi_a(0,0). \quad (3.187)$$

Then

$$K_{abc} = \langle b,c|S|a\rangle. \quad (3.188)$$

It remains for us to exhibit the singularities described by (3.185) in the diagrammatic expansion of R. We can write, in a shorthand notation,

$$R = \int \mathrm{d}\tau\, K_a K_b K_c G'', \tag{3.189}$$

where K_a, K_b and K_c are the two-body propagators corresponding to the graph of Figure 3.10, with the restriction that $u_1 = u_2$, etc.

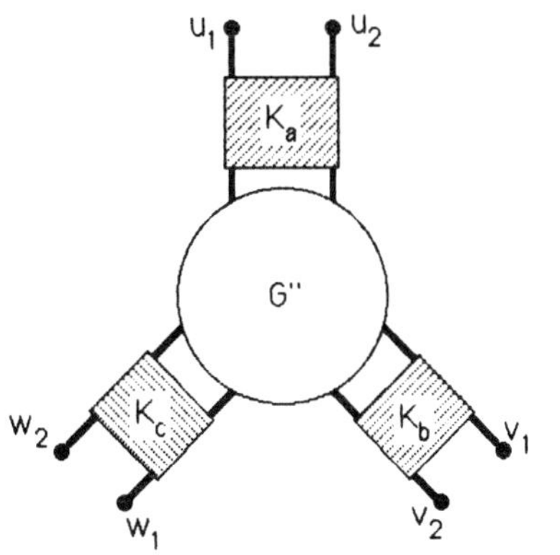

Figure 3.10 The problem of a three-meson vertex: decomposition of the Green function into a two-particle-irreducible part G'' and two-quark propagators K.

Equation (3.189) can be put into the form (3.185) by extraction of the singularities of the Fourier transforms of K_a, K_b and K_c. This will be done in a parallel way to the extraction of the singularities in R. For definiteness, we set

$$K_a(u_1,u_2,u_3,u_4) = \langle 0|\, \mathrm{T}\{\psi(u_1)\bar{\psi}(u_2)\bar{\psi}(u_3)\psi(u_4)\}\, |0\rangle, \tag{3.190}$$

with analogous expressions for K_b and K_c, and we define the Fourier transforms with respect to $\frac{1}{2}(u_1+u_2)$, $\frac{1}{2}(v_1+v_2)$ and $\frac{1}{2}(w_1+w_2)$, with variables denoted by P_1, P_2 and P_3. We find that, for example, $\tilde{K}_a(P_1)$ has two poles at $P_1^0 = E_a - \mathrm{i}\varepsilon$,

$$\tilde{K}_a(P_1) \approx (2\pi)^2 \frac{-1}{2\pi \mathrm{i}} \frac{\chi_a(u_4,u_3)(-1)\bar{\chi}_{\boldsymbol{P}_a}(u_1-u_2)}{P_1^0 - E_a + \mathrm{i}\varepsilon}, \tag{3.191}$$

and at $P_1^0 = -E_a + \mathrm{i}\varepsilon$,

$$\tilde{K}_a(P_1) \approx (2\pi)^2 \frac{1}{2\pi \mathrm{i}} \frac{\bar{\chi}_{\boldsymbol{P}_a}(u_1-u_2)\chi_a(u_4,u_3)}{P_1^0 + E_a - \mathrm{i}\varepsilon}, \tag{3.192}$$

Starting from (3.189), a straightforward comparison with (3.185) shows that

$$K_{abc} = \int d^4u_3\, d^4u_4\, d^4v_3\, d^4v_4\, d^4w_3\, d^4w_4\; \chi_a(u_4,u_3)\bar{\chi}_b(v_4,v_3) \qquad (3.193)$$
$$\bar{\chi}_c(w_4,w_3)\, G''(u_3,u_4,v_3,v_4,w_3,w_4).$$

The lowest approximation for G'' corresponds to taking only the so-called *triangle graphs* of Figure 3.11.

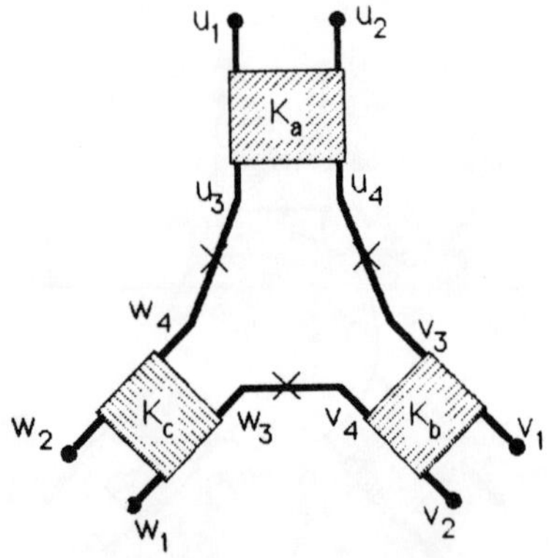

Figure 3.11 The triangle graph. It is the simplest approximation to the irreducible Green function G''. The crosses indicate inverse quark propagators.

To get G'' from Figure 3.11, we have to consider, as for Figure 3.7, that the K_i include the fermion propagators in all their legs, and that therefore the propagators of the three internal quark lines would be counted twice were it not for the inclusion of a factor $-iS^{-1}$ in G'', which we have denoted by the crosses:

$$G'' = (-i)^3 S^{-1}(v_3-u_4) \otimes S^{-1}(w_3-v_4) \otimes S^{-1}(u_3-w_4). \qquad (3.194)$$

Using the expression for S^{-1} (3.129), we finally end up with

$$\langle b,c|S|a\rangle = K_{abc}$$
$$= (-i)^3 \int d^4u_3\, d^4v_3\, d^4w_3 \qquad (3.195)$$
$$\times\, \mathrm{Tr}\,[(i\not{\partial}_{u_3}-m)\bar{\chi}_c(u_3,w_3)(i\not{\partial}_{w_3}-m)\bar{\chi}_b(w_3,v_3)\,(i\not{\partial}_{v_3}-m)\chi_a(v_3,u_3)].$$

The notable feature of this lowest-order formula is that it gives an equivalent of the quark-pair creation (see Figure 3.11) in terms of the sole Bethe–Salpeter amplitudes that contain all the effect of the interaction.

3.4.6 Recovering the Naive Quark Model in the Mandelstam Formalism

As already emphasized in Section 3.4.3, all the formalisms giving a scheme of successive approximations must give the naive quark model in the lowest-order approximation. However, we must make this notion of lowest-order approximation more precise. It is not sufficient to take the first-order perturbation theory in H_s^{pair} in the Hamiltonian formalism, or to take the lowest-order approximation to the kernel G in the Mandelstam formalism, because the BS amplitude or the wave function (3.86) can be still expanded in powers of the strong potential. It is only after this additional expansion has been performed that we can expect to recover the naive quark model.

This additional expansion was rather easy in the case of the Hamiltonian formalism of Section 3.3 (see end of the section). In the case of the Mandelstam formalism, things are more involved, because we have first to connect the four-dimensional BS amplitudes with the usual three-dimensional wave functions, and we have to reduce the four-dimensional integrals to three-dimensional ones, as encountered in the naive quark model. It is possible to do this in an exact and reasonably simple manner in the case of an instantaneous interaction, as described by (3.82). Since our aim is not to discuss a fully relativistic theory, but rather to compare various formalisms, it is quite natural to apply them to the same interaction (3.82).

The first point to realize is that the BS amplitude for *equal times* reduces to the ordinary Schrödinger wave function in the nonrelativistic limit apart from statistical factors of normalization. One gets this relation by expanding the state $|a\rangle$ in the Fock space.

The nonrelativistic limit of the leptonic-annihilation matrix element (3.122) is immediately found from the above remaks, since it directly implies the BS amplitude for equal times, $x^0 = y^0$. The nonrelativistic limit will replace $\chi_a(x=y)$ by the wave function at the origin $\psi(\boldsymbol{x}=0, \boldsymbol{y}=0)$, as found in Section 2.1.3.

The nonrelativistic limit of a current hadron-to-hadron matrix element is more involved, because we have nontrivial integrations over time variables. The derivation has been given by Brodsky and Primack (1969). It is most easily done by going to momentum space. After momentum conservation has been taken into account there remains one integration over the time component of a four-momentum, which is straightforwardly performed by using the Cauchy residue theorem. The desired three-dimensional expression is thereby obtained. The final step is to write this expression in terms of the nonrelativistic wave functions. The same general line of calculation holds for the strong decay in the triangle approximation, although with technical complications.

Let us consider the matrix element (3.132) for $x=0$ (the x dependence is trivial) and pass to the Fourier transforms

$$\chi(x_1,x_2)=\frac{1}{(2\pi)^4}\int d^4p_1\, d^4p_2\, e^{-i(p_1\cdot x_1+p_2\cdot x_2)}\tilde{\chi}(p_1,p_2), \tag{3.196}$$

$$\bar{\chi}(x_1,x_2)=\frac{1}{(2\pi)^4}\int d^4p_1'\, d^4p_2'\, e^{i(p_1'\cdot x_1+p_2'\cdot x_2)}\tilde{\bar{\chi}}(p_1',p_2'). \tag{3.197}$$

From translational invariance, we can write

$$\tilde{\chi}(p_1,p_2)=\delta^4(p_1+p_2-p)\,\tilde{\chi}_P^R(p_1,p_2), \tag{3.198}$$

$$\tilde{\bar{\chi}}(p_1',p_2')=\delta^4(p_1'+p_2'-P')\,\tilde{\bar{\chi}}_{P'}^R(p_1',p_2'), \tag{3.199}$$

where $\tilde{\chi}_P^R$ and $\tilde{\bar{\chi}}_{P'}^R$ denote the relative wave functions, with the constraints $p_1+p_2=P$ and $p_1'+p_2'=P'$. It is then straightforward to find

$$\langle b|\,j_\mu(0)\,|a\rangle=$$

$$-i\frac{1}{(2\pi)^4}\int d^4p_1\,\tilde{\bar{\chi}}^R(p_1,P'-p_1)(\not{p}_1-m)\otimes\gamma_\mu^{(2)}\tilde{\chi}^R(p_1,P-p_1). \tag{3.200}$$

where we have taken the fermion 2 as the charged one. The important point is that we can now write

$$\tilde{\chi}_P^R(p_1,p_2)=\frac{1}{\not{p}_1-m}\otimes\frac{1}{\not{p}_2-m}\Gamma_P(p_1,p_2), \tag{3.201}$$

where $\Gamma_P(p_1,p_2)=\Gamma_P(p_1,P-p_1)$ does not depend on the time component of the relative momentum, say p_1^0. Since, on the other hand, the propagators are simple pole expressions in the variable p_1^0, the p_1^0 integration is very easily carried out in (3.200).

The expression (3.201) for the χs is a direct consequence of the Bethe–Salpeter equation for an instantaneous potential:

$$\begin{aligned}\tilde{\chi}_P^R(p_1,p_2)&=\frac{1}{\not{p}_1-m}\otimes\frac{1}{\not{p}_2-m}\frac{1}{2\pi i}\\&\quad\times\int d^4q\,\tilde{V}(q)\tilde{\chi}_P^R(p_1+q,p_2-q)\\&=\frac{1}{\not{p}_1-m}\otimes\frac{1}{\not{p}_2-m}\frac{1}{2\pi i}\\&\quad\times\int d^4q\,\tilde{V}(q)\int dq^0\,\tilde{\chi}_P^R(p_1+q,p_2-q).\end{aligned} \tag{3.202}$$

In (3.202), $\tilde{V}(q)$ is the Fourier transform of the four-dimensional potential $V(x-y)$, which is related to the usual potential $V(\boldsymbol{x}-y)$ by

$$V(x-y) = V(\boldsymbol{x}-\boldsymbol{y})\,\delta(x^0-y^0). \tag{3.203}$$

In (3.203) we ignore numerical factors, since our aim is simply to sketch the derivation of the nonrelativistic model for an unspecified potential.

It is clear that

$$\Gamma \equiv \int \mathrm{d}\boldsymbol{q}\, \tilde{V}(\boldsymbol{q}) \int \mathrm{d}q^0\, \tilde{\chi}_P^{\mathrm{R}}(p_1+q, p_2-q) \tag{3.204}$$

does not depend on the time component of the relative momentum, $p_1^0-p_2^0$, since $\tilde{\chi}_P^{\mathrm{R}}(p_1+q,\, p_2-q)$ depends only on

$$p_1^0-p_2^0+2q^0, \tag{3.205}$$

and this dependence on $p_1^0-p_2^0$ is eliminated, after integration over q^0, by a translation of the integration variable.

After having brought the expression (3.201) (and similarly that for χ) into (3.200), there remain, as regards the integration over p_1^0, simple pole terms due to the propagators, since Γ_P can be factored out of this integration. A necessary intermediate step is to display these poles in p_1^0 in the propagator denominators, to allow the use of the method of residues. This introduces the projectors on positive- and negative-energy spinors, according to

$$\begin{aligned} S(p) &= \frac{1}{\not{p}-m+\mathrm{i}\varepsilon} = \frac{\not{p}+m}{p^2-m^2+\mathrm{i}\varepsilon} \\ &= \frac{\Lambda_+(\boldsymbol{p})\,\gamma_0}{p_0-(\boldsymbol{p}^2+m^2)^{1/2}+\mathrm{i}\varepsilon} + \frac{\Lambda_-(\boldsymbol{p})\,\gamma_0}{p_0+(\boldsymbol{p}^2+m^2)^{1/2}-\mathrm{i}\varepsilon}. \end{aligned} \tag{3.206}$$

After integration over the poles, there remains a product of the Γs and the electromagnetic vertex γ_μ, with insertion of various projectors $\Lambda_\pm$ between them, the whole being integrated over $\mathrm{d}\boldsymbol{p}_1$. Finally the Γs must be related to the usual nonrelativistic wave functions.

As stated before, the nonrelativistic wave function is the limit of the equal-time BS amplitude in the nonrelativistic approximation. This, as well as the desired expressions for the Γs, may be obtained by returning to the BS equation (3.202). In momentum space, the equal-time BS amplitude corresponds, after elimination of the center-of-mass wave function, to

$$\varphi_P(\boldsymbol{p}_1,\boldsymbol{p}_2) = \int \mathrm{d}p_1^0\, \chi_P^{\mathrm{R}}(p_1, P-p_1). \tag{3.207}$$

This expression is found to be proportional, in the nonrelativistic limit, to the usual nonrelativistic *internal* wave function. On the other hand, by a trivial change of integration variable

$$q^0 \to p_1^0 + q^0, \tag{3.208}$$

we have

$$\int \mathrm{d}q^0\, \chi_P^{\mathrm{R}}(p_1+q, p_2-q) = \varphi_P(\boldsymbol{p}_1+\boldsymbol{q}, \boldsymbol{p}_2-\boldsymbol{q}). \tag{3.209}$$

Therefore

$$\Gamma_P(\boldsymbol{p}_1, \boldsymbol{p}_2) = \int \mathrm{d}\boldsymbol{q}\tilde{V}(\boldsymbol{q})\, \varphi_P(\boldsymbol{p}_1+\boldsymbol{q}, \boldsymbol{p}_2-\boldsymbol{q}) \tag{3.210}$$

which yields an expression for Γ_P in terms of φ_P. Moreover, we get an integral equation for $\varphi_P(\boldsymbol{p}_1,\boldsymbol{p}_2)$ by integrating (3.202) over p_1^0:

$$\begin{aligned}\varphi_P(\boldsymbol{p}_1, \boldsymbol{p}_2) = &\int \frac{\mathrm{d}p_1^0}{2\pi \mathrm{i}} \frac{1}{\not{p}_1 - m} \otimes \frac{1}{\not{p}_2 - m} \\ &\times \int \mathrm{d}\boldsymbol{q}\tilde{V}(\boldsymbol{q})\, \varphi_P(\boldsymbol{p}_1+\boldsymbol{q}, \boldsymbol{p}_2-\boldsymbol{q}).\end{aligned} \tag{3.211}$$

It is clear that the integration on the right-hand side applies only to the propagators, and it can be performed by similar methods as used to integrate over p_1^0 in the current-matrix element, namely by exhibiting the propagator poles according to (3.206). This gives

$$\begin{aligned}&\varphi_P(\boldsymbol{p}_1, \boldsymbol{p}_2) \\ &= \left\{ \frac{\Lambda_+(\boldsymbol{p}_1)\gamma^0 \otimes \Lambda_+(\boldsymbol{p}_2)\gamma^0}{P^0 - (\boldsymbol{p}_1^2+m^2)^{1/2} - (\boldsymbol{p}_2^2+m^2)^{1/2}} \right. \\ &\quad \left. - \frac{\Lambda_-(\boldsymbol{p}_1)\gamma^0 \otimes \Lambda_-(\boldsymbol{p}_2)\gamma^0}{P^0 + (\boldsymbol{p}_1^2+m^2)^{1/2} + (\boldsymbol{p}_2^2+m^2)^{1/2}} \right\} \\ &\quad \times \int \mathrm{d}\boldsymbol{q}\tilde{V}(\boldsymbol{q})\, \varphi_P(\boldsymbol{p}_1+\boldsymbol{q}, \boldsymbol{p}_2-\boldsymbol{q})\end{aligned} \tag{3.212.}$$

This equation displays the fact that

$$\Lambda_\pm(\boldsymbol{p}_1) \otimes \Lambda_\mp(\boldsymbol{p}_2)\, \varphi \equiv \varphi_{\pm\mp} = 0; \tag{3.213}$$

and for the remaining projections,

$$\Lambda_\pm(\boldsymbol{p}_1) \otimes \Lambda_\pm(\boldsymbol{p}_2)\, \varphi \equiv \varphi_{\pm\pm}, \tag{3.214}$$

we find

$$\begin{aligned}&[P^0 \mp (\boldsymbol{p}_1^2+m^2)^{1/2} \mp (\boldsymbol{p}_2^2+m^2)^{1/2}]\varphi_{\pm\pm} \\ &= \Lambda_\pm(\boldsymbol{p}_1)\gamma^0 \otimes \Lambda_\pm(\boldsymbol{p}_2)\gamma^0 \int \mathrm{d}\boldsymbol{q}\tilde{V}(\boldsymbol{q})\varphi_P(\boldsymbol{p}_1+\boldsymbol{q}, \boldsymbol{p}_2-\boldsymbol{q}),\end{aligned} \tag{3.215}$$

which is the Salpeter equation. If V is taken to be the time component of a vector (as in Coulomb exchange), it is easily seen that φ_{++} obeys in the nonrelativistic limit the Schrödinger equation

$$\left(E - \frac{\boldsymbol{p}_1^2}{2m} - \frac{\boldsymbol{p}_2^2}{2m}\right)\varphi_{++} = \int \mathrm{d}\boldsymbol{q}\ \tilde{V}'(\boldsymbol{q})\, \varphi_{++}(\boldsymbol{p}_1+\boldsymbol{q}, \boldsymbol{p}_2-\boldsymbol{q}) \tag{3.216}$$

where V' is related to V by $V = V'\gamma^0 \otimes \gamma^0$. This establishes the desired connection between the equal-time BS amplitude and the nonrelativistic internal wave function. At the same time, we find

$$\varphi_{--}/\varphi_{++} = O(v^4/c^4). \tag{3.217}$$

The Salpeter equation yields also for the $\pm\pm$ projections of Γ a new expression, more tractable than (3.210):

$$\Lambda_{\pm}(\boldsymbol{p}_1)\gamma^0 \otimes \Lambda_{\pm}(\boldsymbol{p}_2)\gamma^0 \Gamma = [P^0 \mp (\boldsymbol{p}_1^2 + m^2)^{1/2} \mp (\boldsymbol{p}_2^2 + m^2)^{1/2}]\,\varphi_{\pm\pm}. \tag{3.218}$$

This last expression allows $(\gamma^0 \otimes \gamma^0 \Gamma)_{\pm\pm}$ to be expressed in terms of the φs without explicit need for the potential, in contrast with (3.210), which must still be used for the other projections $(\gamma^0 \otimes \gamma^0 \Gamma)_{+-}$ and $(\gamma^0 \otimes \gamma^0 \Gamma)_{-+}$. With the help of (3.210) or (3.218), we finally get

$$\begin{aligned}\langle \mathrm{b}|\, j_\mu(0)\, |\mathrm{a}\rangle \sim & \int \mathrm{d}\boldsymbol{p}_1\, \varphi^{\mathrm{b}}_{++}(\boldsymbol{p}_1, \boldsymbol{P}' - \boldsymbol{p}_1)^\dagger \gamma_0^{(2)} \gamma_\mu^{(2)} \varphi^{\mathrm{a}}_{++}(\boldsymbol{p}_1, \boldsymbol{P} - \boldsymbol{p}_1) \\ & - \int \mathrm{d}\boldsymbol{p}_1\, \varphi^{\mathrm{b}}_{--}(\boldsymbol{p}_1, \boldsymbol{P}' - \boldsymbol{p}_1)^\dagger \gamma_0^{(2)} \gamma_\mu^{(2)} \varphi^{\mathrm{a}}_{--}(\boldsymbol{p}_1, \boldsymbol{P} - \boldsymbol{p}_1) \\ & + \text{potential-dependent terms.}\end{aligned} \tag{3.219}$$

In the nonrelativistic limit, there remains only the first term, which furthermore reduces to the usual nonrelativistic expression. The third term, dependent on the potential, corresponds to the Z-graph corrections described in (3.93) and in Figure 3.5, and comes through $(\gamma^0 \otimes \gamma^0 \Gamma)_{+-}$ and $(\gamma^0 \otimes \gamma^0 \Gamma)_{-+}$.

The derivation of the nonrelativistic expression for strong decays from (3.195) is quite analogous. The main difference is that, instead of the electromagnetic vertex γ_μ, one has a vertex function Γ corresponding to a $\mathrm{q\bar{q}}$ bound state. A consequence of this, and of the particular arrangement of the projection operators, is that, even at lowest order, one cannot avoid the explicit appearance of the potential through (3.210). One ends with the expression found for the Cornell model in Chapter 2.

3.4.7 Center-of-Mass Motion and Internal Wave Functions

Another useful aspect of the BS formalism is the simple description that it gives for the internal wave functions of moving hadrons. For the general multitime amplitude, we have the relation (3.114) giving the amplitude for a moving hadron in terms of the rest frame one. In practice, it is interesting to know the instantaneous wave function to be used in the instantaneous approximation in the Mandelstam formalism, or in the Hamiltonian formalism. If we rely on (3.114) and take the equal-time limit in the frame where the particle has momentum $\boldsymbol{P}$, we get, after factorizing the center-of-mass wave function,

$$\begin{aligned}\chi_{\boldsymbol{P}}(\boldsymbol{x}_1-\boldsymbol{x}_2) &= \chi_{\boldsymbol{P}}(t_1-t_2=0, \boldsymbol{x}_{1\mathrm{T}}-\boldsymbol{x}_{2\mathrm{T}}, z_1-z_2) \\ &= S_1(\Lambda)S_2(\Lambda) \\ &\quad \chi_{\boldsymbol{P}=0}\left[-\frac{\beta}{(1-\beta^2)^{1/2}}(z_1-z_2), \boldsymbol{x}_{1\mathrm{T}}-\boldsymbol{x}_{2\mathrm{T}}, \frac{1}{(1-\beta^2)^{1/2}}(z_1-z_2)\right], \end{aligned} \tag{3.220}$$

$\boldsymbol{P}$ being assumed to lie in the z direction. *If we can neglect the dependence on the relative time in* $\chi_{\boldsymbol{P}=0}$, the relation means that $\chi_{\boldsymbol{P}}(\boldsymbol{x}_1-\boldsymbol{x}_2)$ can be obtained from the rest-frame wave function by the *usual spin boost and a Lorentz contraction in the z direction.* Anyway, the Lorentz-contraction effect is expected on general physical grounds (Licht and Pagnamenta, 1970). This prescription is useful for deducing higher-order effects due to the center-of-mass motion, which are not included in the Dirac equation. For instance, one finds, for the interaction with an electromagnetic field, in addition to the spin–orbit term, terms of a similar structure, but which are not additive, i.e. which are not a sum over the quarks (Brodsky and Primack, 1969; Close and Osborn, 1971; Le Yaouanc *et al.*, 1977c). It may also give a rough intuition concerning the structure functions of hadrons as suggested by us (Le Yaouanc *et al.*, 1975a,b). The boost of spin is important for two-body reactions at large momentum transfer (Lipkin, 1969), and has also been found to be crucial for helicity-flip reactions even at small momentum transfer (Krzywicki and Le Yaouanc, 1969). This type of approximation has been successfully used in lepto-production (Le Yaouanc *et al.*, 1972; Andreadis *et al.*, 1974; Alcock, Cottingham and Dunbar, 1980).

3.4.8 QCD Radiative Corrections to Current Vertices

In quark-model calculations, one introduces by hand QCD gluon-loop radiative corrections to the quark-current interaction vertex. This is analogous to the QED radiative correction generating the electron anomalous magnetic moment. Such a procedure has already been mentioned for the axial current at the end of Section 3.2.7. Such corrections are also important in the discussion of meson e^+e^- annihilation (see Section 4.5). In principle, the calculation makes sense only for heavy quarks, for which α_s may be considered as small; nevertheless, it may also give an indicative trend for light quarks. Such corrections appear to be quite sizable, even for heavy quarks, because the loop integral runs over large momenta. This is why they should be taken into account.

We will not explain here how one calculates Feynman graphs, and we refer to the original papers for the details of concrete calculations. What are more difficult to understand, of course, are the connection and combination of such calculations with the usual technology of quark models, as explained in Chapter 2 and the present chapter. In fact, the very distinction between perturbative and nonperturbative effects is not perfectly clear (for a brief discussion see Mackenzie and Lepage, 1981). We want to address a more modest question. We recall that a QCD Coulomb contribution α_s/r is introduced in the quark potential (Section 1.3.4). We have now introduced in this chapter corrections to the nonrelativistic quark model, which represent higher-order effects of the potential. It is then necessary to enquire whether these corrections include or forget part or the totality of such QCD radiative corrections. Since we can disregard gluon self-couplings at low orders, and for further simplicity we forget the confining potential, we can ask the following question: what part of the QED radiative corrections do we obtain by using an α/r instantaneous potential in the different frameworks of this chapter.

It is obvious that the Dirac formalism, where the potential is only an external field, does not produce vertex radiative corrections.

In contrast in the two-body Hamiltonian of Section 3.3, such radiative corrections are partly included through graphs analogous to Figure 3.5, but with instantaneous interaction acting on one quark line only. If we retain only the $\gamma^0\gamma^0$ component in the Hamiltonian, the answer to the above question is simple: since the exact Hamiltonian of QED is obtained by adding to the Coulomb interaction the transverse photon one (see Bjorken and Drell, 1964), we simply miss the contribution of transverse gluons. If we write a four-component $\gamma^\mu\gamma_\mu$ potential as in (1.112), we include transverse gluons in the instantaneous approximation; we miss retardation effects. Since, anyway, it is easier to use Feynman graphs to calculate radiative corrections, it is then more expedient to forget the contribution from the instantaneous potential and to add the full radiative correction. On the other hand, we should still include the instantaneous potential in the calculation of the nonrelativistic wave function as well as in other independent effects. Note that when one deals with annihilation into one photon one must take care to avoid double counting of Coulomb exchange.

In the Mandelstam formalism of Section 3.4 the situation is simpler. We are not limited *a priori* to an instantaneous interaction. We can include the full gluon exchange in the BS kernel. Then the radiative corrections to e^+e^- meson annihilation of Section 4.5 are automatically obtained by the approach of Section 3.4.2 (Poggio and Schnitzer, 1979, 1980; Bergström, Snellman and Tengstrand, 1980). In contrast, it is obvious that in the usual triangle-graph approximation used for one-particle matrix elements of currents (Section

3.4.2, Figure 3.7) no vertex correction is included, and the full radiative corrections should be added.

Finally, we must note that in the Wilson expansion of the second-order weak interaction (Section 4.6.3), a new type of QCD perturbative correction is introduced. It represents the contribution of "hard" gluons. The separation of hard and soft gluons, and the combination of such corrections with the calculations of local matrix elements by the standard methods of quark models, as explained in the present book, raise problems similar to the ones examined above, but still more complex.

CHAPTER 4

Phenomenological applications

... This strange eventful history ...

W. SHAKESPEARE, *As you like it*, II.7

Assurons-nous bien du fait, avant que de nous inquiéter de la cause.† *Il est vrai que cette méthode est bien lente pour la plupart des gens qui courent naturellement à la cause, et passent par-dessus la vérité du fait; mais enfin nous éviterons le ridicule d'avoir trouvé la cause de ce qui n'est point...*

De grands physiciens ont fort bien trouvé pourquoi les lieux souterrains sont chauds en hiver et froids en été. De plus grands physiciens ont trouvé depuis que cela n'était pas.

FONTENELLE, *Première Dissertation*, IV

4.1 GENERAL REMARKS

4.1.1 Basic Processes and Experimental Measurements

What is directly yielded by the quark model is the estimate of the basic processes that we have learned to calculate in Chapter 2: hadron couplings or vertices of the type $g_{NN\pi}$, $N^* \to N\gamma$, $\langle N|H_W^{pc}|\Lambda\rangle$, etc. Basic processes are simple in the quark model, but in general they are not directly measurable.

One of the most extensive applications concerns the hadronic resonances and their different decays, like $N^* \to N\gamma$ and $N^* \to N\pi$. As emphasized in Section 1.1.2, we can analyse most low- and medium-energy reactions, like $\gamma N \to \pi N$, in terms of these basic processes, the radiative or pionic transitions of resonances, but this deserves further theoretical work. In this case, what are directly measured are scattering amplitudes like $\gamma N \to \pi N$; in fact, if simplicity is gained in considering basic processes, one could be afraid that the gain is lost on returning to the full complexity of scattering amplitudes. Indeed, one can doubt whether hadronic physics can face such complex phenomena at a

† Let us first inquire about facts, and then about causes.

theoretical level. In particular, it may seem that the continuum always present in addition to resonances escapes any theoretical evaluation. In the example $\gamma N \to \pi N$ we have N* intermediate states, but also a nonresonant continuum. Happily, one can avoid the need for detailed theoretical knowledge of the scattering amplitude. Resonances appear as poles in the scattering amplitudes, considered as analytic functions of the energy–momentum variables. Certain methods allow these poles to be reached from the experimental scattering amplitudes; then the residues at the poles in various channels give the partial widths for various basic processes. For example, in $\gamma N \to N\pi$, the residues of the N* pole gives the partial widths of $N^* \to N\gamma$, $N^* \to N\pi$. Of course, the task of extracting the resonance poles is a difficult one, done by dispersion-relation methods, but it is quite standard.

An area where there is great simplification is that of hadrons built out of heavy quarks, because we have, apart from new stable particles like the D mesons, a number of states that, although decaying through the strong interaction, have a very narrow width, and then electromagnetic decays have a sizable branching ratio. Such is the case of charmonium: radiative cascades like $\psi(3685) \to \chi\gamma$, $\chi \to J/\psi\gamma$ are directly observable and have been measured with great precision.

Of course, there are directly observable transitions like weak decays of particles stable under strong and electromagnetic interactions, such as $\Lambda \to N\pi$ and $\Lambda \to pe^-\bar{\nu}_e$. There are also some electromagnetic decays that are directly observable, like $\pi^0 \to 2\gamma$. However, it must be noted that these electroweak transitions are not always basic processes in the sense considered here. The decay $\eta \to 3\pi$ and the nonleptonic weak decays, like $K \to 2\pi$ or $\Lambda \to p\pi^-$ are not simple in the sense of the quark model. In fact, their analysis raises difficult problems.

In the following, we assume that we have at our disposal the experimental estimation of the above basic processes, hadron vertices and electroweak matrix elements. There is one exception, however. In hyperon weak hadronic decays, the baryon-to-baryon matrix elements $\langle B'|H_W|B\rangle$ cannot be reached by unambiguous methods such as those used in scattering processes to obtain particle pole residues. Therefore, we shall have to perform a lengthy analysis to test the quark-model predictions; in fact, in Section 4.6 we shall have to use many methods of particle physics in addition to the quark model.

4.1.2 Parameters of the Nonrelativistic Quark Model

The expressions obtained in Chapter 2 for the nonrelativistic quark model may be calculated if we know the quark masses and the wave functions ψ. These wave functions will be derived from the Schrödinger equation, and we therefore need to know, in addition to the quark masses, the potential. It is not our aim here to reconsider in detail the whole discussion of spectroscopy, sketched in Chapter 1. It is enough to say that in principle the *spectroscopy* will allow us to fix the parameters of the potential. Then the quark model is fully predictive for some electromagnetic *transitions*, and more generally for hadron–hadron current-matrix elements. We can calculate these transitions with parameters that are independently determined from the fit of the mass spectrum. We can also completely predict strong decays in the Cornell model. On the other hand, more phenomenological models of strong decays, such as the elementary-meson-emission or the quark-pair-creation model still require the determination of another parameter, either the pion–quark coupling constant $g_{\pi qq}$, or the quark-pair-creation constant γ, to be discussed later.

An additional feature of hadron–hadron current matrix elements is that they do not depend very sensitively on the wave functions. More precisely, they look like a mean value of some power of the quark momenta, and depend essentially on the peripheral part of the wave function and of the potential. The same is true for strong decays. In this case it is not a bad approximation to use Gaussian wave functions, as would be derived from the harmonic-oscillator potential. Once the quark mass is known, the parameter of the Gaussian wave functions is determined by the excitation spectrum, at least roughly, through simple formulae. For a better determination of the parameter, we should use a variational principle starting from the actual binding Hamiltonian. For simplicity, we shall often make an even cruder approximation consisting in the use of harmonic-oscillator wave functions (a polynomial times a Gaussian).

For baryons, we write this Gaussian factor as

$$\psi \propto \exp\left[-\frac{1}{6R^2}\sum_{i<j}(\boldsymbol{r}_i-\boldsymbol{r}_j)^2\right]. \tag{4.1}$$

This leads to a universal Gaussian factor in the transition form factors, expected to be valid only for small $\boldsymbol{q}^2$:

$$F(\boldsymbol{q}) = \exp(-\tfrac{1}{6}\boldsymbol{q}^2R^2). \tag{4.2}$$

Equation (4.2) shows that in the most naive quark model the square of the wave-function radius R^2 should be identified with the square of the proton charge radius, since it will give all the $\boldsymbol{q}^2$ dependence for the elastic form factor $F_1(Q^2)$ of the ground state, $\langle r^2\rangle_{\text{em}} = -6F'_1(Q^2)|_{Q^2=0}$. The most naive model would be to identify the elastic form Factor F_1 (Close, 1979) with the form factor (4.2) (in the Breit frame $Q^2=-q^2=\boldsymbol{q}^2$), assuming the quarks to be point-like.

For mesons, we write

$$\psi \propto \exp\left[-\frac{(\boldsymbol{r}_1-\boldsymbol{r}_2)^2}{2R'^2}\right], \tag{4.3}$$

which gives a Gaussian,

$$F(\boldsymbol{q}) = \exp(-\tfrac{1}{16}\boldsymbol{q}^2 R'^2), \tag{4.4}$$

in the transition form factors.

Then, in the naive quark model, R'^2 is related to the square of the meson charge radius through

$$\langle r^2 \rangle = \tfrac{3}{8}R'^2. \tag{4.5}$$

It is also useful to have the relations between the squares of the wave-function radii R^2 and R'^2 and the excitation energy ΔE_1 from the ground state to the first orbital excitation. We have, respectively for baryons and for mesons

$$\Delta E_1 = 1/mR^2, \tag{4.6}$$

$$\Delta E_1 = 2/mR'^2, \tag{4.7}$$

m being the quark mass. These expressions are expected to be reasonable approximations. The energy shifts between the ground state and the first radial excitation are for mesons and baryons respectively

$$\Delta E' = 2/mR^2, \tag{4.8}$$

$$\Delta E' = 4/mR'^2, \tag{4.9}$$

These harmonic-oscillator expressions do not compare so well with experiment as (4.6) and (4.7).

With these parameters R^2, R'^2 and m, it is very simple to calculate a large number of processes, the simplicity being due to the easy handling of integrals over Gaussians. It must be said, however, that the criterion of simplicity must not be overestimated and the harmonic oscillator should be considered as only a crude approximation to more realistic potentials. At the same time, the naive quark model is also by itself an approximation to the real situation, where relativistic and strong-coupling effects are important.

More specifically, it must be noted that the relative insensitiveness of the decays to the detailed shape of the potential, which allows use of the convenient harmonic-oscillator wave functions, is not shared by what we have termed central interquark interactions (Sections 2.1.3, 2.1.4 and 2.2.2). In the processes considered, $q\bar{q}$ annihilation into leptons, nonleptonic matrix elements OZI-forbidden strong decays, the amplitude depends essentially on the ground-state wave function at *short distance*, which we denote concisely by

$$|\psi(0)|^2 \equiv \langle \psi_0 | \delta(\boldsymbol{r}_1-\boldsymbol{r}_2) | \psi_0 \rangle. \tag{4.10}$$

$|\psi(0)|^2$, in contrast with *peripheral* quantities, is not well described by the above harmonic approximation, if the true potential possesses, for instance, a Coulomb part. Moreover, it is doubtful that the relevant parts of the potential could be determined only from the hadron spectrum, unless rather detailed fits involving the fine structure of the spectrum are attempted. We shall later give some indications about $|\psi(0)|^2$, but let us just emphasize for now the general fact that $|\psi(0)|^2$ should be *larger* than expected from the harmonic-oscillator approximation for a given excitation energy ΔE_1.

Let us finally come to actual numbers. Although masses could be determined by the analysis of the spectrum, there are considerable uncertainties in the case of light quarks; the bounds obtained by the Feynman–Hellman theorem (Bertlmann and Ono, 1981; Ono, 1981b) are less and less constraining on moving to lighter quarks; it is possible to make a more direct determination using magnetic moments; it is then necessary to check the compatibility with these spectroscopic bounds. Using the magnetic moment of the proton (for the deduction see Section 4.3.1),

$$\mu_p = e/2m, \tag{4.11}$$

we find for the light quarks (u,d):

$$m \approx 330\,\text{MeV}. \tag{4.12}$$

The strange-quark mass can be extracted from hyperon magnetic moments or from the mass differences, leading to

$$m_s \approx 500\,\text{MeV}. \tag{4.13}$$

This is in agreement with a determination relying only on spectroscopy and fitting the ϕ, ψ, Υ systems (see e.g. Martin, 1981).

With (4.6) and (4.7), we can now determine the radius parameters R^2 and R'^2. For baryons, we can estimate from $\Delta E_1 \approx 0.5\,\text{GeV}$,

$$R^2 \approx 6\,\text{GeV}^{-2}. \tag{4.14}$$

It is essential to note that (4.14) is much smaller than the actual square of the proton charge radius, $R^2 \approx 19.5\,\text{GeV}^{-2}$. This is not worrying, in our opinion, although many authors have insisted on fitting the charge proton radius. We attribute the proton charge radius to a combination of the wave-function effect (4.2) and a cloud effect, a *quark structure*, which would represent higher-order strong-interaction effects (Licht and Pagnamenta, 1970; Le Yaouanc *et al.*, 1973). In fact, it is simply not possible in the naive approximation to get a coherent fit at the same time for the excitation energy, the magnetic moment and the proton charge radius. A choice then has to be made. If we insisted on a large R^2, we would have very bad consequences for the transition rates, and we could not understand the striking successes of the naive model in this area. We would be also in disagreement with spectroscopy. In contrast with other authors, we do not insist either on fitting the strong hyperfine structure, which could lead to rather different values; we think that the strong hyperfine structure is not very well understood and that it is much more important to fit the orbital-excitation spectrum.

As to mesons, there is a serious ambiguity in the estimate of ΔE_1 because of the very large ρ–π splitting. We cannot trust a nonrelativistic model for the π mass. It is reasonable to choose for the splitting between the ground state and the first orbitally excited state the difference between the mean value of the $L = 1$ mesons and the ρ mass, giving $\Delta E_1 = 0.5$–0.7 GeV, whence

$$R'^2 \approx 9\text{–}12\,\text{GeV}^{-2}. \tag{4.15}$$

We have here the same problem for the pion radius as we had for the proton: it is much smaller than the observed charge radius.

We must observe that the above parameters unavoidably lead to the conclusion that quarks have large internal velocities, as mentioned in Section 3.1.

So much for the light quarks. Concerning the heavy flavors, from the φ, ψ, ϒ analysis (Martin, 1981), the charmed and the bottom quarks have masses

$$m_c \approx 1.8\,\text{GeV}, \qquad m_b \approx 5.2\,\text{GeV}. \tag{4.16}$$

For the charmed quark mass, this is a high estimate, as opposed to a series of earlier estimates, but, as pointed out by Martin (1982), this high value is more reasonable from the point of view of *constituent* masses (as opposed to *current* masses). These heavy-quark masses are rather well constrained by the bounds resulting from a general potential (Bertlmann and Ono, 1981). From (4.16) and the mean value of the $\chi(L=1)$ masses, we can deduce (we have $\Delta E_1 \approx 0.43$ GeV),

$$R_c'^2 \approx 3\,\text{GeV}^{-2}. \tag{4.17}$$

In general, it must be emphasized that no coherent fit to all the hadronic properties is obtained, as exemplified by the proton charge radius. This is not only the effect of the harmonic-oscillator approximation. We can hope to get an overall coherent fit, especially for light quarks, only by taking into account relativistic and other high-order effects.

Another consequence of the approximate character of the model is that the above determined values only make sense in the naive approximation. Since the corrections are expected to be large, a fit with a more refined model could in principle yield rather different values of m, R^2 and R'^2. For instance, the bag model with approximately massless quarks can reproduce many results of the naive quark model, where quarks are massive.

Notwithstanding all these caveats, it is important to emphasize that, *within our definite framework*, we do not have so much freedom in choosing our parameters, and, also, that our definite choice has important consequences for almost all of the following discussions: quite different conclusions would come from a different choice.

4.1.3 Ambiguities of the Nonrelativistic Quark Model

Ambiguities and inconsistencies are encountered in confronting a nonrelativistic model with a reality that, as emphasized before, is always relativistic in some aspect. There is no logical way to solve these ambiguities without a more complete scheme. However, a sufficiently comprehensive treatment covering the various aspects of relativity is not available at present. Then, sticking to the nonrelativistic model, we must at least make definite choices after having explained where are the ambiguities.

A first category of ambiguities concerns the Lorentz-frame dependence. Even with a choice of normalizations corresponding to Lorentz-invariant phase space and amplitudes, the nonrelativistic approximation to this invariant amplitude would retain a frame dependence. After making the nonrelativistic approximation, the model is no longer covariant. Indeed, the amplitude depends on the external momenta, and since some hadrons at least are treated as composite and in a nonrelativistic way, they will possess only Galilean invariance. For simplicity, let us consider the case of radiative transitions. The Galilean invariance manifests itself through the fact that the amplitude depends only on $\boldsymbol{k}=\boldsymbol{P}'-\boldsymbol{P}$. It is usual to calculate $\boldsymbol{k}$ through its relativistic expression, which depends on the reference frame. For instance, let us consider three natural frames for the reaction $A \to B\gamma$; we have for the A rest frame, the equal-velocity frame (a frame where $\boldsymbol{v}_A = -\boldsymbol{v}_B$), and the B rest frame respectively

$$|\boldsymbol{k}_0| = \frac{m_A^2 - m_B^2}{2m_A}, \tag{4.18}$$

$$|\boldsymbol{k}_1|^2 = \frac{m_A}{m_B}|\boldsymbol{k}_0|^2, \tag{4.19}$$

$$|\boldsymbol{k}_2|^2 = \left(\frac{m_A}{m_B}\right)^2 |\boldsymbol{k}_0|^2. \tag{4.20}$$

For the example $\omega \to \pi\gamma$, dramatically different results are obtained since

$$m_\omega/m_\pi = 5.5. \tag{4.21}$$

In view of this, one might suspect that the nonrelativistic quark model is inconsistent when one considers such large mass differences. It is a fact, however, that the calculation in the A *rest frame* gives good results, even if the mass difference $m_A - m_B$ is not small, as it should be for the true nonrelativistic approximation.

Another related ambiguity is connected with the choice of normalization convention. Of course, results should be in principle independent of the convention; the change in the amplitude is compensated by a change in the phase space. Consider, for instance, the invariant choice

$$\langle p'|p\rangle = \frac{E}{m}\delta(p-p'). \tag{4.22}$$

With respect to this choice, the amplitudes will be multiplied by factors $(E/m)^{1/2}$ and the phase space by factors $(m/E)^{1/2}$. However, making the nonrelativistic approximation in the amplitude leads to setting $(E/m)^{1/2} \approx 1$, while the exact phase-space expression is usually retained, and so the results differ by factors $(m/E)^{1/2}$. Therefore a definite normalization convention has to be chosen to get definite physical results, and here there are not even pragmatic reasons for making the choice.

A second category of ambiguities concerns the treatment of the SU(6) $\times$ O(3) symmetry, the approximate invariance of the forces under space, flavor and spin rotations described in Chapter 1. Indeed, the Schrödinger equation is assumed to be spin and flavor SU(3)-independent, and rotation-invariant, and therefore the masses should be the same for all members of a SU(6) $\times$ O(3) multiplet, where O(3) labels the orbital angular momentum, and the spatial wave functions are the same for all the members of an SU(6) multiplet, and build a multiplet of O(3).

However, it is true that SU(6) $\times$ O(3) symmetry is actually broken; in particular, the masses are often widely different from the symmetry pattern. Having fixed the reference frame, the amplitudes depend on the mass through the momentum transfer k, which differs greatly for transitions between various members of the multiplets of states.

It is clear that we have no theoretically justified way of accounting for such differences, since they are generated by relativistic corrections to the Schrödinger equation, and we should then consistently introduce also the relativistic corrections to the transition operator and to the wave equation (in contrast, SU(3) breaking is easily accounted for in the naive approximation by just introducing a quark mass difference that plays a crucial role in some cases). What then is to be done if we remain in the nonrelativistic approximation? The transitions $\Delta \to N\gamma$ or $\rho \to \pi\pi$ would not exist as real decays in the symmetry limit (since Δ and N, ρ and π belong respectively to the same multiplet of SU(6) $\times$ O(3)), and this indicates that, at least for small k, the phase space must take into account SU(6) breaking by introducing the actual value of k. However, the effect of the variation of k cannot be confined to the phase space. It is certainly present in the amplitude. In particular, it seems that the l-partial waves must contain a factor k^l, k being the actual momentum. But in the case of radial excitations, as exemplified by the decays $\psi(4.03) \to D\bar{D}$, $D\bar{D}^* + \bar{D}D^*$, $D^*\bar{D}^*$, the dependence of the amplitude on the actual k, as given by the space integral, is not accounted for by the simple centrifugal barrier factor k^l, as we shall see in Section 4.4.6.

In conclusion, on the one hand, it is theoretically left completely ambiguous whether or not one must take the actual SU(6)-broken k or the symmetric one, and on the other hand, in practice, it seems that one must choose the actual value of k.

Finally, this ambiguity in the treatment of SU(6) breaking furthermore combines with the previous ones (choice of reference frame and of normalization) to generate new ambiguities, which we shall not discuss, but which are indeed present in many calculations.

All the above ambiguities are especially crucial when a π or a K is present in the reaction and treated as a composite of quarks, as examplified by the discussion of k. In this case, the nonrelativistic approximation is especially doubtful. In fact, it is possible that π and K cannot even be approximately treated in the naive way owing to their ultrarelativistic Goldstone-boson nature (Section 1.2.4). This drawback could be cured in the quark model only by a correct treatment of chiral symmetry.

What then is to be done practically? We repeat that in principle we should look for reactions where the ambiguities are minimized as much as possible. But we shall try to choose a prescription to make definite and correct predictions for the largest possible set of data. We choose to calculate in the *rest frame of the decaying resonance*, with the choice of *normalization (2.5b)*, and retaining the *relativistic expression for the phase space*, as well as the *relativistic expression for k* in the matrix elements, both with the physical value of the masses.

4.1.4 Algebraic Predictions and Symmetry Approaches

In the nonrelativistic quark-model expressions for the transition amplitudes, it is possible to distinguish a part that comes from the color, flavor, spin and orbital-momentum calculations. What is typical of this part of the calculations is that it can be done by standard techniques from group representation theory. We indeed have finite-dimensional representations of the groups $SU(3)_c$, SU(6) (or the spin symmetry group only for heavy quarks) and O(3). We can perform the calculation with the help of the standard coupling coefficients like the Clebsch–Gordan coefficients. A good example of this is found in Section 2.2.5, but we shall find the same structure in all the calculations of this Chapter. After factorization of all these algebraic coefficients, the remaining expressions involve only reduced matrix elements given by spatial integrals, and depend on three types of quantities: the quark masses, the parameters defining the potential—or equivalently the radii R^2, R'^2, etc.—and the external momenta (k in two-body decays). From the sole knowledge of the algebraic coefficients, we can predict certain ratios, for instance the well-known $\mu_p/\mu_n = -\frac{3}{2}$ (see section 4.3.1).

In the so-called *SU(6) × O(3) analysis*, an additional assumption is made that the reduced matrix elements are the same for all the members of an SU(6) × O(3) multiplet. This assumption is the logical consequence of the SU(6) × O(3) symmetry of the Schrödinger equation; indeed, this symmetry would imply that the parameters m, R^2 and the decay momentum are the same inside the multiplet. As we have just explained in the preceding section, we think that this assumption is not very good. First, it is necessary to take into account the SU(3) breaking of quark masses and its consequences for the spatial wave functions, i.e. on R^2. Secondly, and more crucially it is the physical value of k rather than its symmetric value that should be inserted in the amplitude. The SU(6) × O(3) analysis can be improved by including a factor $(k_{\text{phys}}/k_0)^l$, different for each member of the multiplet, and reflecting the observed centrifugal barrier effect for the l partial wave. This is l-broken SU(6) (Petersen and Rosner, 1972; Faiman and Plane, 1972; Faiman and Rosner, 1973). However, as we have said, it fails completely for radial excitations, as well as the equivalent prescription for the charmonium decays above $D\bar{D}$ threshold (see Section 4.4.6). The interest of SU(6) × O(3) analysis lies in the fact that it can include new types of operators not included in the nonrelativistic model, and shows up their presence in experimental data (see an example in Section 4.3.1).

SU(6) × O(3) *analysis* must not be confused with the *hypothesis* of $SU(6)_W \times O(2)$ symmetry. In the specific case of *strong decays*, an $SU(6)_W \times O(2)$ *invariance of the interaction* has been considered; W-spin (Lipkin, 1967) differs from the usual spin by the fact that the spin singlet $S=0$ becomes the $W=1$, $W_z=0$ triplet member, and the $S=1$, $S_z=0$ triplet member becomes the W-singlet, for mesons and for spin operators. The interest of $SU(6)_W$ lies in the fact that if the quark momenta were all *collinear* along the z axis, the strong-decay interaction operator would be *$SU(6)_W$-invariant*; indeed, in the elementary-pion-emission model the spin operator in the interaction is then σ_z, therefore a W-scalar; also the pair creation in models having the $^3P_0$ structure would be an $S_z=0$ triplet, therefore also a W-scalar. Simultaneously, the space part of the transition operator would be invariant by O(2), the rotation group around the z axis. This symmetry is badly broken owing to the large quark momenta transverse with respect to the z axis, and is in gross contradiction with experiment; it is only of historical interest.

4.1.5 General Status of the Naive Quark Model for Light Quarks

Before entering into particular discussions, it is still important to make precise what is to be expected from the calculations of the naive quark model in the case of light quarks. What must first be emphasized is the remarkable overall success of the picture. A huge number of decays of hadronic states and of hadronic vertices are described correctly by simple calculations.

However, the success is quantitative but far from accurate. Rates may be wrong by a factor of 2 or more. For instance, the neutron β-decay ratio $|g_1/f_1|$ is predicted to be $\frac{5}{3}$ instead of the experimental value of 1.22, as we shall see below. The success is often only statistical: it holds in average over a wide sample of similar processes, although in some few cases the individual predictions may be bad (for such a statistical analysis see e.g. Feynman, 1969).

The crudeness of the description is not unexpected in view of the uncertainties emphasized above, and of the basic fact that the expansion parameter v/c is not small. However, it would not be fair to blame only the nonrelativistic approximation. Efforts to reduce the discrepancies by more refined methods are often not rewarding. At best, the higher-order corrections only indicate a qualitative trend in the right direction. We shall consider these corrections in this qualitative spirit only and shall not present quantitative calculations.

Even so, the achievement of the nonrelativistic quark model is remarkable, especially when contrasted with the previous situation in hadron physics. Various competing approaches have been proposed, but no one can claim to offer such systematic predictions from a few basic parameters. For quite a while, there was a tendency to admit only relations based on general principles, like analyticity, unitarity, current algebra and partial conservation of axial current (PCAC). But then there is no estimate of absolute rates. For instance the Adler–Weisberger sum rule (Adler, 1965; Weisberger, 1966) gives a much better value for $|g_1/f_1|$ than the naive quark model estimate of $\frac{5}{3}$. But in fact the sum rule is only a relation between this ratio and other theoretically undetermined quantities like f_π or $\sigma_{tot}(\pi p)$. There are only a few cases like $\pi^+ \to \pi^0 e^+ \nu_e$ where an absolute rate is given. Although more systematic, the algebraic approaches suffer the same basic drawback of not predicting the absolute rates, but giving only relations within a multiplet.

More ambitious dynamical approaches competing with the naive quark model have recently been developed: QCD sum rules and lattice QCD calculations. Although less clearly grounded, the QCD sum rules are presently more useful than the latter in the prediction of transition rates. Yet they cannot claim to cover a range of processes comparable to that covered by the quark model. Anyway, we refer the reader to the original works and to our brief discussion in Section 1.1.3.

4.2 SEMILEPTONIC WEAK DECAYS

In semileptonic transitions the main body of data to be considered is given by the decays of strange or heavier-flavor baryons or mesons into the ground state. The transition between the ground state and excited baryons is not directly observable as a decay, since it competes with the much faster decay through the strong interactions. On the other hand, these couplings may be obtained in neutrino production of resonances, a subject that we shall not consider here. We therefore now concentrate on the decays of *ground-state* hadrons, excluding also the $L=0$ with spin excitation, Δ or ρ, which decay strongly. A notable exception is the $\Omega^-(\frac{3}{2}^+)$, for which the strong decay is kinematically forbidden.

4.2.1 Baryon Semileptonic Decays

As far as the *vector part* of the current is concerned, the quark-model expression (2.49) simply implements the CVC (conserved vector current) hypothesis at the quark level: it is essentially identical with the electromagnetic current except for its SU(3) structure; in addition, SU(3) itself completely relates the weak vector current to its electromagnetic counterpart, owing to the conservation property (Bell, 1965). Since CVC is well verified, up to SU(3)-breaking effects, the confrontation with experiment amounts in fact to the verification of the quark-model predictions for the electromagnetic current. Moreover, at small momentum transfer, the only nontrivial quantity parametrizing the electromagnetic current within the ground-state octet is the magnetic moment, and we have indeed chosen the quark masses so as to reproduce the proton magnetic moment. Note that the counterpart of the magnetic moment in the weak vector current is weak magnetism, seen in nuclear β decay.

Really new tests of the quark model come from consideration of the *axial current*. The quark expression of the current is a realization of the Cabibbo (1963) hypothesis that the weak current belongs to an SU(3) octet. It is well established that the Cabibbo theory is in excellent agreement with experiment. Note that we neglect small SU(3)-breaking effects that may come from momentum transfer. The following are left undetermined by Cabibbo theory:

(i) the ration F/D of the axial couplings, since here there are two independent octets (see Appendix);
(ii) the overall magnitude of the axial coupling, which may be defined, for instance, by the axial coupling constant g_1 in neutron β decay $n \to pe^- \bar{\nu}_e$.

We shall concentrate on the prediction of these two quantities. In detail, we write the weak-current matrix element between two $\frac{1}{2}^+$ baryons,

$$
\begin{aligned}
&(2\pi)^3\langle \mathrm{B}|\,J_\mu(0)\,|\mathrm{A}\rangle \\
&=\bar{u}_\mathrm{B}\{f_1(q^2)\gamma_\mu+\frac{\mathrm{i}}{m_\mathrm{A}+m_\mathrm{B}}f_2(q^2)\sigma_{\mu\nu}q^\nu+f_3(q^2)q_\mu \\
&\quad+g_1(q^2)\gamma_\mu\gamma_5+\frac{\mathrm{i}}{m_\mathrm{A}+m_\mathrm{B}}g_2(q^2)\sigma_{\mu\nu}q^\nu\gamma_5+g_3(q^2)q_\mu\gamma_5\}u_\mathrm{A},
\end{aligned}
\tag{4.23}
$$

where q^ν is the four-momentum transfer. In practice the transfer is small, which allows us to neglect the contribution of f_2 and g_2, and set $f_1(q^2)=f_1(0)$ and $g_1(q^2)=g_1(0)$. Anyway, as indicated, we shall refrain from discussing f_2, which is simply related to the nucleon anomalous magnetic moment. As to g_2 and f_3, they could be theoretically interesting. They vanish in the limit of SU(3) symmetry, but, allowing for SU(3) breaking by a quark mass, one predicts $g_2 \neq 0$ and $f_3 \neq 0$ (Halprin, Lee and Sorba, 1976; Gavela *et al.*, 1980b; Donoghue and Holstein, 1982). These couplings f_3 and g_2 that vanish in the equal-mass limit if CP is conserved are called *second class* (see Bell, 1965). The effect of the term g_3 is negligible in baryonic decay.

For the present investigation, we remain with f_1, which is completely determined by the SU(3) quantum numbers (always neglecting SU(3) breaking), and with g_1.

$g_1\bar{u}_\mathrm{B}\gamma_\mu\gamma_5 u_\mathrm{A}$ is the matrix element of the axial part of the weak current J_μ (see (2.47)), which, with our conventions, is in fact $-\mathrm{j}_\mu^5$. We shall disregard g_3, whose correct treatment is connected with PCAC. To be consistent with the nonrelativistic approximation of the quark model, these expressions should be reduced to their nonrelativistic limit. Moreover, q is small with respect to the hadron masses, and can be reasonably neglected considering the approximate character of the model. Finally, we shall not try to account for SU(3) breaking, and this allows us to set at the same time $m_\mathrm{A}=m_\mathrm{B}$ and $q=0$. Then,

$$(2\pi)^3\langle \mathrm{B}|\,J_0(0)\,|\mathrm{A}\rangle=f_1(0)\chi_\mathrm{B}^\dagger\chi_\mathrm{A}, \tag{4.24}$$

$$(2\pi)^3\langle \mathrm{B}|\,\boldsymbol{J}(0)\,|\mathrm{A}\rangle=g_1(0)\chi_\mathrm{B}^\dagger\boldsymbol{\sigma}\chi_\mathrm{A}, \tag{4.25}$$

where $\chi_{\mathrm{A,B}}$ are the baryon Pauli spinors.

At the quark level, we have similarly to retain only the operator j^0 for the vector part, with the simple substitution (2.50). For the sake of simplicity, we consider only the charge-raising part of the hadronic current, corresponding to electron emission. Therefore, according to (2.40) and (2.44),

$$f_1(0)\chi_\mathrm{B}^\dagger\chi_\mathrm{A}=3\int\prod_i \mathrm{d}\boldsymbol{r}_i\,\delta(\tfrac{1}{3}\sum_i \boldsymbol{r}_i)\,\psi_\mathrm{B}^\dagger(\boldsymbol{r}_i)[\cos\theta_\mathrm{C}\tau^{(+)}(3)+\sin\theta_\mathrm{C}v^{(+)}(3)]\psi_\mathrm{A}(\boldsymbol{r}_i). \tag{4.26}$$

Considering that the spatial wave function is the same for the initial and final state, we get straightforwardly

$$f_1(0) = \cos\theta_C \langle B|\sum_i \tau^{(+)}(i)|A\rangle + \sin\theta_C \langle B|\sum_i v^{(+)}(i)|A\rangle, \tag{4.27}$$

where the matrix element is understood only in SU(3) space, and where, in terms of standard notation for the isospin and V-spin operators at the baryon level (Carruthers, 1966)

$$\sum_i \tau^{(+)}(i) = T_+ = \mathrm{I}_+, \tag{4.28}$$

$$\sum_i v^{(+)}(i) = V_-. \tag{4.29}$$

Equation (4.27) expresses exactly the result of CVC plus the Cabibbo theory. The space component is given entirely by the axial-vector current $\boldsymbol{j}_5$, and we have, in a manner analogous to the deduction of (4.26),

$$g_1(0)\chi_B^\dagger\boldsymbol{\sigma}\chi_A$$
$$= 3\int \prod_i \mathrm{d}\boldsymbol{r}_i\,\delta(\tfrac{1}{3}\sum_i \boldsymbol{r}_i)\psi_B^\dagger(\boldsymbol{r}_i)[\cos\theta_C\tau^{(+)}(3) + \sin\theta_C v^{(+)}(3)][-\boldsymbol{\sigma}(3)]\psi_A(\boldsymbol{r}_i). \tag{4.30}$$

To complete the calculation, we now have to explicitly write the octet baryons wave functions

$$\Phi = \frac{\varphi'\chi' + \varphi''\chi''}{\sqrt{2}}, \tag{4.31}$$

$$\psi = \Phi\psi^s \tag{4.32}$$

introduced in Section 1.3.3. Once more, the integral over spatial wave functions will give unity, and we are left with purely algebraic computations in the spin–SU(3) space. To express the results, it has become conventional to define F and D, which are reduced matrix elements for SU(3). Indeed, $g_1(0)$ depends only on the SU(3) quantum numbers in simple way because the operators in (4.30) with coefficients $\cos\theta_C$ and $\sin\theta_C$ are two components of the same octet operator. The matrix elements of an octet operator between octet states may be entirely expressed in terms of two reduced matrix elements, the constants F and D, combined with coefficients that are determined by the SU(3) group only.

To fit with standard conventions, let us express $\tau^{(+)}(3)$ and $v^{(+)}(3)$ in terms of the Gell-Mann matrices:

$$\tau^{(+)}(3) = \tfrac{1}{2}[\lambda_1(3) + \mathrm{i}\lambda_2(3)], \tag{4.33}$$

$$v^{(+)}(3) = \tfrac{1}{2}[\lambda_4(3) + \mathrm{i}\lambda_5(3)], \tag{4.34}$$

Since hadron octets (mesons or baryons) are in the same representation as the

Gell-Mann matrices, the octet hadron states can be represented by a 3×3 SU(3) matrix. We denote by Φ_i the hadron state that corresponds to λ_i. For example, $\Phi_1 = \Sigma^+ + \Sigma^-$, etc. The baryon matrix, for instance, will be

$$\sum_{i=1}^{8} \Phi_i \frac{\lambda_i}{2} = \begin{pmatrix} \dfrac{\Sigma^0}{\sqrt{2}} + \dfrac{\Lambda}{\sqrt{6}} & \Sigma^+ & p \\ \Sigma^- & -\dfrac{\Sigma^0}{\sqrt{2}} + \dfrac{\Lambda}{\sqrt{6}} & n \\ \Xi^- & \Xi^0 & -\dfrac{2\Lambda}{\sqrt{6}} \end{pmatrix}. \tag{4.35}$$

Note that this corresponds to opposite signs for Σ^+ and Ξ^- with respect to de Swart's conventions, which we use elsewhere (see Appendix). Then, choosing the z space direction for the spin operator, and using up spin states, we define the SU(3) reduced matrix elements F and D of the axial-vector coupling:

$$3\Phi^\dagger_{i,+1/2}\, \lambda_j(3)\sigma_z(3)\Phi_{k,+1/2} = \mathrm{i} f_{ijk} F + d_{ijk} D. \tag{4.36}$$

We can calculate from this expression two hadron matrix elements and determine F and D. The SU(3) calculation gives

$$3\Phi^\dagger_{\mathrm{p},+1/2}\, \tau^{(+)}(3)\sigma_z(3)\Phi_{\mathrm{n},+1/2} = F + D, \tag{4.37}$$

$$3\Phi^\dagger_{\Sigma^+,+1/2}\tau^{(+)}(3)\sigma_z(3)\Phi_{\Sigma^0,+1/2} = -\sqrt{2}\,F. \tag{4.38}$$

Now using the baryon wave functions for the octet $\frac{1}{2}^+$ ground state listed in Section 1.3.3 (with a change of sign for Σ^+ and Ξ^- to agree with the conventions of (4.35)), we compute in the static quark model the preceding expressions. We obtain, using the symmetry properties of the spin and flavor wave functions,

$$\begin{aligned} &3\Phi^\dagger_{\mathrm{B_i},+1/2}\tau^{(+)}(3)\,\sigma_z(3)\Phi_{\mathrm{B_f},+1/2} \\ &= \tfrac{3}{2}[\langle \chi''_{+1/2}|\,\sigma_z(3)\,|\chi''_{+1/2}\rangle \langle \varphi''_{\mathrm{B_f}}|\,\tau^{(+)}(3)\,|\varphi''_{\mathrm{B_i}}\rangle \\ &+ \langle \chi'_{+1/2}|\,\sigma_z(3)\,|\chi'_{+1/2}\rangle\, \langle \varphi'_{\mathrm{B_f}}\,|\tau^{(+)}(3)\,|\varphi'_{\mathrm{B_i}}\rangle], \end{aligned} \tag{4.39}$$

and now using

$$\left.\begin{aligned} \langle \chi''_{+1/2}|\,\sigma_z(3)\,|\chi''_{+1/2}\rangle &= -\frac{1}{3}, \qquad \langle \chi'_{+1/2}|\,\sigma_z(3)\,|\chi'_{+1/2}\rangle = 1, \\ \langle \varphi''_{\mathrm{p}}|\,\tau^{(+)}(3)\,|\varphi''_{\mathrm{n}}\rangle &= -\frac{1}{3}, \qquad \langle \varphi'_{\mathrm{p}}|\,\tau^{(+)}(3)\,|\varphi'_{\mathrm{n}}\rangle = 1, \\ \langle \varphi''_{\Sigma^+}|\tau^{(+)}(3)|\,\varphi''_{\Sigma^0}\rangle &= -\frac{1}{3\sqrt{2}}, \qquad \langle \varphi'_{\Sigma^+}|\tau^{(+)}(3)\,|\,\varphi'_{\Sigma^0}\rangle = -\frac{1}{\sqrt{2}}, \end{aligned}\right\} \tag{4.40}$$

we obtain

$$F+D=\frac{5}{3}, \tag{4.41}$$

$$F/D=\frac{2}{3}. \tag{4.42}$$

The ratio F/D is close to the experimental fit 0.63 (Bourquin *et al.*, 1983). In fact the degree of accuracy of (4.42) is not well defined, since we may form various amplitude ratios that will present very different relative errors. For instance, the ratio

$$\left|\frac{M(\Sigma^0\to pe^-\bar{\nu}_e)}{M(\Lambda\to pe^-\bar{\nu}_e)}\right|^2=3\left(\frac{F/D-1}{3F/D-1}\right)^2 \tag{4.43}$$

is very sensitive to small deviations in F/D, ranging from $\frac{1}{3}$ for $F/D=\frac{2}{3}$ to 0.75 for $F/D=0.6$, i.e. for a 10% variation in F/D.

The agreement is much less satisfactory for the absolute value of the axial coupling, set for instance by neutron β decay. We find

$$g_1/f_1=-(F+D)=-\frac{5}{3}, \tag{4.44}$$

against the experimental value $g_1/f_1=-1.22$. This results in a discrepancy of a factor of about 2 in excess for the width. This discrepancy has been ascribed to two different origins. First, *relativistic effects* found with the Dirac equation lower the value of $|g_1/f_1|$ owing to the small components of the quark Dirac spinors, reflecting the relativistic motion of quarks within hadrons:

$$|g_1/f_1|=\frac{5}{3}(1-2\delta), \tag{4.45}$$

with δ given by (3.20) if, like Bogoliubov, one uses the Dirac equation without a nonrelativistic expansion. In the systematic v/c expansion of Section 3.2.6 that leads to the operator (3.60) at order $(v/c)^2$, the correction δ is given by

$$\delta=\left\langle\frac{p_T^2}{4m^2}\right\rangle, \tag{4.46}$$

where the mean value is understood as being between the ground-state nonrelativistic spatial wave functions. Another phenomenon also reduces the ratio $|g_1/f_1|$, namely *gluon radiative corrections* to the quark axial vertex (Figure 3.3), which reduce the bare-quark axial coupling. One obtains, at the lowest order (Gavela *et al.*, 1980a,b)

$$\left|\frac{g_1}{f_1}\right| = \frac{5}{3}\left(1 - \frac{4}{3}\frac{\alpha_s}{2\pi}\right). \tag{4.47}$$

Both effects have the correct tendency of reducing the naive result, although one cannot say much more in view of the uncertainties involved.

4.2.2 Meson Semileptonic Decays

The simplest 0^- meson decays are those leading also to 0^- states like $K^+ \to \pi^0 e^+ \nu_e$ or $\pi^+ \to \pi^0 e^+ \nu_e$. They receive a contribution only from the vector current, and in that case, as already indicated, the quark model does not have much to tell, except that an explicit model can take into account SU(3) or SU(4) breaking, specified by the simple prescription of a quark mass difference between the relevant flavors. One could also in principle calculate form factors, but the status of form-factor calculations is not very good in the quark model; this is unfortunate in the case of D or B decays, because experiment yields mainly integrated rates, and therefore requires the knowledge of form factors.

By Lorentz covariance, two form factors contribute to these transitions:

$$\langle 0^- | j_\mu(0) | 0^- \rangle = (p_\mu + p'_\mu) f_+(q^2) + (p'_\mu - p_\mu) f_-(q^2). \tag{4.48}$$

$f_-(q^2)$ does not contribute to the width in the limit of zero lepton mass, and it is of first order in the mass differences, the flavor-symmetry breaking. Moreover $f_+(0) = 1$ in the limit of no flavor-symmetry breaking: this is essentially the CVC hypothesis. This completely determines $\pi^+ \to \pi^0 e^+ \nu_e$. For $K^+ \to \pi^0 e^+ \nu_e$ the quark model has been used to evaluate the flavor-symmetry-breaking effects: it correctly predicts $f_-(0)$ (Isgur, 1975). The other nontrivial thing to predict is the q^2 dependence, but the naive quark model is not expected to give this correctly (see the remarks of Section 4.1.2 on the charge radius). In fact, in the transition $K^+ \to \pi^0 e^+ \nu_e$, the q^2 dependence of $f_+(q^2)$ is described correctly by a K* pole,

$$f_+(q^2) = \frac{1}{1 - q^2/m_{K^*}^2}. \tag{4.49}$$

The motivation for this pole expression cannot be found in the single wave-function overlap effects expected in the naive quark model. The same can be said of $D^+ \to \bar{K}^0 e^+ \nu_e$. Recent data (Schindler, 1985) indicate an F* pole-dominated form factor, without quoting a precise value at $t = 0$.

To illustrate the model, let us nevertheless apply the formulae (2.63), (2.65) and (2.71), with the same static approximation used in the previous section. We have,

$$|f^S|^2 = \frac{\boldsymbol{q}^2}{q^2} |\langle B | j_0 | A \rangle|^2, \tag{4.50}$$

whence the decay width (setting $\cos^2\theta_C = 1$)

$$\Gamma = \frac{G_F^2}{60\pi^3}|\boldsymbol{q}|^5_{\max} \times \begin{cases} 2 & (\text{for } \pi^+ \to \pi^0 e^+ \nu_e), \\ \sin^2\theta_C & (\text{for } K^0 \to \pi^- e^+ \nu_e), \\ 1 & (\text{for } D^+ \to \bar{K}^0 e^+ \nu_e), \end{cases} \tag{4.51}$$

For $\pi^+ \to \pi^0 e^+ \nu_e$ the result 2.5×10^{-25} GeV is in perfect agreement with experiment. For $K^0 \to \pi^- e^+ \nu_e$, it is not so good: 1.70×10^{-18} GeV against an experimental 2.4×10^{-18} GeV. For $D^+ \to \bar{K}^0 e^+ \nu_e$ the calculation gives 0.3×10^{-13} GeV, in comparison with the experimental rate of about 0.6×10^{-13} GeV, deduced from the data reported by Schindler. It should be pointed out that in the above calculations, as in the preceding section on baryons, we have neglected the $\boldsymbol{q}$ dependence of the current-matrix elements. Although this is not so justified here, it greatly simplifies the calculation, and is not of consequence for the purpose of the discussion. In fact, for such integrated rates, one cannot expect too much from such very naive approximations, owing to the badly determined form factor, and to the sensitivity induced by the three-body phase space. The discrepancy for K and D is not unexpected.

For the transitions to other states, we can apply the naive quark model if the final hadronic state is a resonance like ρ or K*. The naive quark model cannot estimate $K \to \pi\pi e\nu$, for which the final state is far from the ππ resonances, and which is better analysed in terms of current algebra and dispersion relations. However, heavy-quark decays offer the possibility of producing a number of resonances (e.g. $D \to K^* e\nu$). It is here that the quark model could display its typical predictive power. There are as yet very few data. $D \to K^* e\nu$ has been measured to be roughly equal to $D \to K e\nu$ (Schindler, 1985). A calculation in the naive approximation, with relativistic phase space, similar to that performed for $D \to K e\nu$, but with a less trivial integration, leads to a ratio that is somewhat too large:

$$\frac{\Gamma(D \to K^* e\nu)}{\Gamma(D \to K e\nu)} = 1.7. \tag{4.52}$$

The same remarks are in order here as for $D \to K e\nu$. In the absence of better models or more tractable data, we prefer, however, to make such somewhat unsatisfactory predictions rather than indulge in fitting the experiment with new parameters.

4.3 RADIATIVE TRANSITIONS

The description of radiative decays is certainly one of the most impressive achievements of the naive quark model. We have here an exceptionally good situation, with a large amount of experimental data and reasonable kinematics for a nonrelativistic approximation in baryon and charmonium decays. The data concern not only the $\frac{3}{2}^+ \Delta$ ground state, but also the spatially excited resonances and charmonium, allowing detailed tests of the quark model. Since the data are accurate and detailed, it is even possible to discuss the departure from the nonrelativistic approximation, with particular accuracy for charmonium transitions.

4.3.1 Radiative Transitions of Baryons

4.3.1(a) The nonrelativistic approximation

Before considering the radiative couplings $N^*N\gamma$ of baryon resonances, the main subject of this section, we review the predictions of the quark model for the nucleon magnetic moments, which have the same nature, except for the fact that the photon is necessarily virtual. In a parallel way to (4.23), we write down the matrix element of the electromagnetic current j^μ_{em} between nucleon states:

$$(2\pi)^3 \langle N | j^{em}_\mu | N \rangle = \bar{u}_N \left\{ F_1^N(q^2)\gamma_\mu + \frac{F_2^N(q^2)}{2m_N} i\sigma_{\mu\nu} q^\nu \right\} u_N, \tag{4.53}$$

where we have neglected the possibility of parity or CP violation, which would induce other terms in higher orders of the electroweak coupling. For $q^2 = 0$ we have $F_1^N(0) = Q_N$ (i.e. 0 or 1 for neutron or proton), and $F_2^N(0) = \kappa_N$, the anomalous magnetic moment. From the nonrelativistic reduction of (4.53), and considering only the spin-flip part, we obtain the following coupling to photons of polarization $\boldsymbol{\varepsilon}$ (for photon emission):

$$\chi_N^\dagger \mu_N [-i(\boldsymbol{\sigma}_N \times \boldsymbol{q})] \cdot \boldsymbol{\varepsilon}^* \chi_N,$$

with (4.54)

$$\mu_p = \frac{1+\kappa_p}{2m_N} \qquad \mu_n = \frac{\kappa_n}{2m_N}.$$

Empirically, the values are $1 + \kappa_p = 2.79$ and $\kappa_n = -1.91$. To compute now the nucleon magnetic moments in the quark model, we have to use (2.18b) at the quark level. We then have the following identity between spin-flip transition operators:

$$\langle N | \mu_N (-i) \boldsymbol{\varepsilon}^* \cdot (\boldsymbol{\sigma}_N \times \boldsymbol{k}) | N \rangle$$

$$= 3 \left\langle \frac{\varphi'\chi' + \varphi''\chi''}{\sqrt{2}} \right| -iQ(3)\, \boldsymbol{\varepsilon}^* \cdot \frac{\boldsymbol{\sigma}(3) \times \boldsymbol{k}}{2m} \left| \frac{\varphi'\chi' + \varphi''\chi''}{\sqrt{2}} \right\rangle, \tag{4.55}$$

or, with a convenient spin-quantization axis,

$$\mu_{\mathrm{N}}\langle \mathrm{N}|\sigma_{\mathrm{N},z}|\mathrm{N}\rangle$$

$$=\frac{3}{2m}\left\langle\frac{\varphi'\chi'+\varphi''\chi''}{\sqrt{2}}\middle|Q(3)\sigma_z(3)\middle|\frac{\varphi'\chi'+\varphi''\chi''}{\sqrt{2}}\right\rangle. \tag{4.56}$$

Choosing N↑ for instance, we have to compute the spin-matrix elements and the proton and neutron charge-matrix elements:

$$\left.\begin{aligned}
&\langle\chi'_{\uparrow}|\sigma_z(3)|\chi'_{\uparrow}\rangle=1,\quad \langle\chi''_{\uparrow}|\sigma_z(3)|\chi''_{\uparrow}\rangle=-\frac{1}{3},\\
&\langle\varphi'_{\mathrm{p}}|Q(3)|\varphi'_{\mathrm{p}}\rangle=\frac{2}{3},\quad \langle\varphi''_{\mathrm{p}}|Q(3)|\varphi''_{\mathrm{p}}\rangle=0,\\
&\langle\varphi'_{\mathrm{n}}|Q(3)|\varphi'_{\mathrm{n}}\rangle=-\frac{1}{3},\quad \langle\varphi''_{\mathrm{n}}|Q(3)|\varphi''_{\mathrm{n}}\rangle=\frac{1}{3}.
\end{aligned}\right\} \tag{4.57}$$

We finally obtain

$$\mu_{\mathrm{p}}=\frac{1}{2m},\qquad \mu_{\mathrm{n}}=-\frac{2}{3}\frac{1}{2m}. \tag{4.58}$$

The predicted ratio $\mu_{\mathrm{p}}/\mu_{\mathrm{n}}=-\frac{3}{2}$ is in excellent agreement with experiment. Taking an effective quark mass $m\sim 300$ MeV $\sim\frac{1}{3}m_{\mathrm{N}}$, we obtain the absolute value for the total proton magnetic moment, $\mu_{\mathrm{p}}\approx 3/2m_{\mathrm{N}}$, which has the right magnitude as compared with the experimental value $\mu_{\mathrm{p}}=2.79/2m_{\mathrm{N}}$. We recall that we finally choose $m=330$ MeV to fit the value of μ_{p}.

Let us now come to the resonance couplings. What is directly measured is the photoproduction amplitude $A(\gamma\mathrm{N}\to\mathrm{N}\pi)$ in the resonance region. It gives the product $M^*(\mathrm{N}^*\to\mathrm{N}\gamma)\times M(\mathrm{N}^*\to\mathrm{N}\pi)$. We can deduce from it the photodecay amplitude in *absolute value*, since $\Gamma(\mathrm{N}^*\to\mathrm{N}\pi)$ is also known from the elastic scattering $\pi\mathrm{N}\to\pi\mathrm{N}$ analysis. But we can also measure the *sign* of the product of the two couplings and compare it with the theoretical prediction, since the quark model predicts both amplitudes $\mathrm{N}^*\to\mathrm{N}\gamma$ and $\mathrm{N}^*\to\mathrm{N}\pi$ for a given excited baryon N*. The comparison of experimental and theoretical signs would thus in principle involve a discussion of the quark-model predictions for $\mathrm{N}^*\to\mathrm{N}\pi$. We refer the reader to Section 4.4 for this discussion, and, for the present discussion on radiative decays, we assume knowledge of the theoretical signs of $M(\mathrm{N}^*\to\mathrm{N}\pi)$. Finally, it must be observed that only the relative sign of various photoproduction amplitudes is measured, and therefore there remains an overall sign that is a matter of convention. We may, for instance, arbitrarily fix the sign of the nucleon pole contribution, and then measure the sign of the resonance contributions with respect to it. If we want to make a comparison with theoretical predictions, it is therefore necessary to multiply the relative experimental signs by the theoretical sign of the nucleon contribution.

Ultimately, what is compared is the experimental measurement and the theoretical prediction for an amplitude $M(\mathrm{N}^*\to\mathrm{N}\gamma)$, which conventionally has the total sign of the amplitude for π photoproduction through the particular N^* state, $\gamma\mathrm{N}\to\mathrm{N}^*\to\mathrm{N}\pi$.

There are two independent amplitudes. Although the multipole analysis proves very useful, especially in electroproduction, we shall use helicity amplitudes for the study of radiative decays, because they are more standard and simpler to define. Taking the quantization axis in the resonance rest frame along the photon momentum, and choosing the photon helicity to be $+1$, the two independent amplitudes M_λ may be chosen as those corresponding to a resonance spin projection $\lambda=\frac{3}{2},\frac{1}{2}$. It has also become conventional to define

$$A_\lambda=(2\pi)^{3/2}M_\lambda, \tag{4.59}$$

whence, (2.31b),

$$\Gamma=\frac{1}{\pi}\frac{E_\mathrm{B}}{m_\mathrm{A}}k^2\sum_\lambda|A_\lambda|^2. \tag{4.60}$$

For $A_{3/2}$ and $A_{1/2}$, we still have two independent amplitudes corresponding to proton or neutron final state; hence the superscripts p and n are used in Table 4.1. As a first example, let us consider the decay of the ground state $\Delta(1236)\to\mathrm{N}\gamma$, which gives by a quick calculation a good idea of the possibilities of the quark model. We have, from (2.45),

$$M=-\frac{1}{(2\pi)^{3/2}}\frac{1}{(2k^0)^{1/2}}e(2\pi)^3\langle\mathrm{N}|\boldsymbol{\varepsilon}^*\cdot\boldsymbol{j}(0)|\Delta\rangle, \tag{4.61}$$

$\boldsymbol{\varepsilon}^*$ accounting for the fact that the photon is emitted instead of being absorbed.

Since Δ and N have orthogonal spin wave functions, respectively χ^s and χ', χ'', only the magnetic interaction will contribute:

$$\begin{aligned}&(2\pi)^3\langle\mathrm{N}|\boldsymbol{\varepsilon}^*\cdot\boldsymbol{j}(0)|\Delta\rangle\\&=3\left\langle\frac{\varphi'\chi'+\varphi''\chi''}{\sqrt{2}}\right|-\mathrm{i}Q(3)\,\boldsymbol{\varepsilon}^*\cdot\frac{\boldsymbol{\sigma}(3)\times\boldsymbol{k}}{2m}\left|\varphi^\mathrm{s}\chi^\mathrm{s}\right\rangle\\&\int\prod_i\mathrm{d}\boldsymbol{r}_i\,\delta\Big(\tfrac{1}{3}\sum_i\boldsymbol{r}_i\Big)\psi_0(\boldsymbol{r}_i)^\dagger\mathrm{e}^{-\mathrm{i}\boldsymbol{k}\cdot\boldsymbol{r}_3}\psi_0(\boldsymbol{r}_i),\end{aligned} \tag{4.62}$$

where ψ_0 is the ground-state *spatial* wave function. The space integral gives

$$\int \prod_i \mathrm{d}\boldsymbol{r}_i\, \delta(\tfrac{1}{3}\sum_i \boldsymbol{r}_i)\, \mathrm{e}^{-i\boldsymbol{k}\cdot\boldsymbol{r}_3} |\psi_0(\boldsymbol{r}_i)|^2 = \mathrm{e}^{-k^2R^2/6} \tag{4.63}$$

for the harmonic-oscillator model, but this does not play a crucial role, since it is close to one:

$$\mathrm{e}^{-k^2R^2/6} = 0.93. \tag{4.64}$$

Since the spatial overlap can be approximated by 1, we see that the transition-matrix element $\Delta \to \mathrm{N}\gamma$, being of the magnetic type according to (4.62), will be proportional to the *nucleon magnetic moments*. The transition $\Delta^+ \to \mathrm{p}\gamma$ is purely magnetic dipole M_1; the electric quadrupole amplitude E_2, allowed in principle, vanishes in the quark model (Becchi and Morpurgo, 1965).

Writing the transition-matrix element $\Delta^+ \to \mathrm{p}\gamma$ in (4.62), quantizing the spin along Oz,

$$(2\pi)^3 \langle \mathrm{p}_\uparrow | \boldsymbol{\varepsilon}^* \cdot \boldsymbol{j}(0) | \Delta^+_\uparrow \rangle \equiv -\mathrm{i}(\boldsymbol{\varepsilon}^* \times \boldsymbol{k})_z \mu^*, \tag{4.65}$$

we obtain

$$\mu^* = \frac{3}{2m} \left\langle \frac{\varphi'_\mathrm{p}\chi'_\uparrow + \varphi''_\mathrm{p}\chi''_\uparrow}{\sqrt{2}} \middle| Q(3)\sigma_z(3) \middle| \varphi^\mathrm{s}_{\Delta^+}\chi^\mathrm{s}_\uparrow \right\rangle, \tag{4.66}$$

and from

$$\begin{aligned} \langle \chi'_\uparrow | \sigma_z(3) | \chi^\mathrm{s}_\uparrow \rangle &= 0, \qquad \langle \chi''_\uparrow | \sigma_z(3) | \chi^\mathrm{s}_\uparrow \rangle = -\tfrac{2}{3}\sqrt{2}, \\ \langle \varphi'_\mathrm{p} | Q(3) | \varphi^\mathrm{s}_{\Delta^+} \rangle &= 0, \qquad \langle \varphi''_\mathrm{p} | Q(3) | \varphi^\mathrm{s}_{\Delta^+} \rangle = -\tfrac{1}{3}\sqrt{2}, \end{aligned} \tag{4.67}$$

we finally get (Dalitz, 1966)

$$\mu^* = \frac{2\sqrt{2}}{3}\frac{1}{2m} = \frac{2\sqrt{2}}{3}\mu_\mathrm{p}. \tag{4.68}$$

The ratio μ^*/μ_p is of the right order of magnitude, but 30% smaller than the experimental number. In terms of the helicity amplitudes defined above, we find (for $I = \frac{3}{2}$ the proton and neutron amplitudes are not independent),

$$\left.\begin{aligned} A^\mathrm{p}_{3/2} &= -0.191\,\mathrm{GeV}^{-1/2}, \\ A^\mathrm{p}_{1/2} &= -0.110\,\mathrm{GeV}^{-1/2}, \end{aligned}\right\} \tag{4.69}$$

versus the experimental $A^\mathrm{p}_{3/2} = -0.244\,\mathrm{GeV}^{-1/2}$ and $A^\mathrm{p}_{1/2} = -0.138\,\mathrm{GeV}^{-1/2}$. Although the magnitude is too small, it is of the right order. The ratio $A_{3/2}/A_{1/2}$ is well described, a test of the classification of the N and the Δ in the $(\mathbf{56}, 0^+)$ ground state, and of the fact that the transition is purely magnetic.

Since we want to describe the full set of decay amplitudes of resonances known for $L \leqslant 2$ and some radial excitations, we must use an explicit model of baryon wave functions. The use of the harmonic-oscillator model, which leads to very simple calculations, is here a crucial approximation, which may well be questioned for radial excitations.

Although the application of the harmonic oscillator to radiative decays of baryon resonances was fully developed by Copley *et al.* (1969) (see also Faiman and Hendry, 1969), who pointed out the remarkable features predicted by theory and verified by experiment, in Table 4.1 we give the more complete results taken from a later paper (Feynman, Kislinger and Ravndal, 1971) compared with experimental data.

Table 4.1 Theoretical (Feynman *et al.*, 1971) versus experimental (Particle Data Group, 1982) decay amplitudes for $N^* \to N\gamma$ as defined in the text. The ampltidues A_λ^N are in units of $10^{-3}\,\mathrm{GeV}^{-1/2}$. The sign is the total-coupling sign in $\gamma N \to N^* \to N\pi$.

	$A^p_{3/2}$		$A^p_{1/2}$		$A^n_{3/2}$		$A^n_{1/2}$	
$P_{33}(1236)$	−178	−258±11	−103	−141±6				
$S_{11}(1535)$			160	67±15			−109	−78±2
$D_{13}(1520)$	112	166±7	−29	−17±11	−112	−136±14	−30	−69±14
$D_{15}(1670)$	0	22±12	0	13±8	−53	−54±24	−38	−37±24
$S_{31}(1650)$			47	23±38				
$D_{33}(1670)$	91	73±35	92	109±31				
$F_{15}(1688)$	70	132±15	−15	−13±10	0	−28±16	41	29±15
$P_{11}(1470)$			32	−70±9			−20	42±19

Apart from a striking general agreement, this table has two remarkable features. First, the amplitudes $A^p_{1/2}$ for $D_{13}(1520)$ and $F_{15}(1680)$ are very small, while $A^p_{3/2}$ is much larger, both theoretically and experimentally. Secondly, the A^p amplitudes for $D_{15}(1670)$ and $A^n_{3/2}$ for $F_{15}(1680)$ are zero theoretically and indeed small experimentally. To show how these results come out in the quark model, let us write explicitly the transition amplitude in the general case, using the expressions (2.18b) and (2.43):

$$A = -\frac{e}{(2k)^{1/2}}(2\pi)^3 \langle N|\boldsymbol{\varepsilon}^* \cdot \boldsymbol{j}(0)\,|N^*\rangle$$

$$= -\frac{e}{(2k)^{1/2}} 3\int \prod_i \mathrm{d}\boldsymbol{r}_i\, \delta(\tfrac{1}{3}\sum_i \boldsymbol{r}_i) \mathrm{e}^{-\mathrm{i}\boldsymbol{k}\cdot\boldsymbol{r}_3} \psi_N(\boldsymbol{r}_i)^\dagger Q(3)\left[-\mathrm{i}\boldsymbol{\varepsilon}^* \cdot \frac{\boldsymbol{\sigma}(3)\times\boldsymbol{k}}{2m} - \mathrm{i}\frac{\boldsymbol{\varepsilon}^*\cdot\boldsymbol{\nabla}_3}{m}\right]\psi_{N^*}(\boldsymbol{r}_i). \tag{4.70}$$

In this expression, ψ_N and ψ_{N^*} include the spin–isospin wave functions. We have taken $\boldsymbol{P}=0$, $\boldsymbol{P}' = -\boldsymbol{k}$ (we are working in the rest frame of the N^*), and we have used $\boldsymbol{\varepsilon}^* \cdot \boldsymbol{k} = 0$.

For $A_{3/2}$, i.e. $N^*_{+3/2} \to N_{+1/2} + \gamma_{+1}$, with indices indicating the spin projection, the magnetic term will not contribute if the N^* belongs to a **56** multiplet. Indeed, in that case, the spin–isospin wave function is the same as for the $\frac{1}{2}^+$ octet, namely

$$\Phi = \frac{1}{\sqrt{2}}(\varphi'\chi' + \varphi''\chi''). \tag{4.71}$$

The quark spin of this wave function is $S = \frac{1}{2}$, to be combined with an orbital angular momentum $L = 2$, to give spin $\frac{5}{2}^+$ as in the F_{15} case. Now the magnetic term necessarily lowers the quark spin. Indeed, with

$$\left.\begin{aligned} &\sigma^{(-)}\uparrow = \downarrow, \qquad \sigma^{(-)}\downarrow = 0, \\ &-i\varepsilon^*_{+1}\cdot\frac{\boldsymbol{\sigma}(3)\times \boldsymbol{k}}{2m} = -\sqrt{2}\,\sigma^{(-)}(3)\frac{k}{2m} \end{aligned}\right\} \tag{4.72}$$

it is impossible to go to nucleon quark spin $+\frac{1}{2}$ by *lowering* the spin of an $S = \frac{1}{2}$ quark-spin wave function. Since only a spin-independent part remains, the matrix element involves a factor

$$\tfrac{3}{2}[\langle\varphi'|\,Q(3)\,|\varphi'\rangle + \langle\varphi''|\,Q(3)\,|\varphi''\rangle], \tag{4.73}$$

which is zero for a neutron because of (4.57), giving $A^{\mathrm{n}}_{3/2} = 0$. This is the case of $F_{15}(1680)$.

The $D_{15}(1670)$ belongs to $(\mathbf{70}, 1^-)$. It has a quark spin $S = \frac{3}{2}$ (to construct a $J = \frac{5}{2}$), and its wave function is a sum of terms of the form

$$\psi = \frac{\varphi'\psi' + \varphi''\psi''}{\sqrt{2}}\chi^{\mathrm{s}} \tag{4.74}$$

coupled with appropriate Clebsch–Gordan coefficients. Here, once more, we do not care about the spin projection. It is sufficient to note that, since the spatial matrix element involves the totally symmetric spatial wave function of the nucleon, only terms of the form $\varphi''\chi^{\mathrm{s}}\psi''$ can contribute. Therefore the matrix element involves a factor

$$\langle\varphi''|\,Q(3)\,|\varphi''\rangle, \tag{4.75}$$

which vanishes for a proton according to (4.57). We therefore get $A^{\mathrm{p}}_{3/2} = 0$ for $D_{15}(1670)$. This is the so-called Moorhouse (1966) selection rule. In contrast with the preceding one, it does not involve the specific structure of the operator. The only thing that is necessary is the proportionality to $Q(3)$.

Finally, let us explain the smallness of $A^{\mathrm{p}}_{1/2}$ for $D_{13}(1520)$ and $F_{15}(1680)$ (Copley, Karl, Obryk, 1969). Here the argument is not algebraic, and the complete calculation of the transition amplitude is necessary. We find for $D_{13}(1520)$ and $F_{15}(1680)$ respectively

$$\frac{A^{\mathrm{p}}_{1/2}}{A^{\mathrm{p}}_{3/2}} = \frac{1}{\sqrt{3}}(k^2R^2 - 1), \tag{4.76}$$

$$\frac{A^{\mathrm{p}}_{1/2}}{A^{\mathrm{p}}_{3/2}} = \frac{1}{\sqrt{2}}(2k^2R^2-1). \tag{4.77}$$

The two terms in these formulae come respectively from the magnetic and the convective terms in (2.18b). They cancel almost exactly, if $R^2 = 6\,\mathrm{GeV}^{-2}$, *in both cases* as k^2 differs by a factor of 2 from one case to another. Of course, this cancellation may be questioned. It depends on choosing the rest-frame values of k and taking the actual masses of the resonance, not the mean value of the SU(6) multiplets. Still more crucial is the objection that $k^2R^2 \sim v^2/c^2$, and therefore higher-order terms are involved in the cancellation, which seems to contradict the assumed nonrelativistic approximation. We are not able to justify this on theoretical grounds, as we have discussed in Section 3.1.3. Nevertheless, this is the same prescription that leads to other detailed successes of the quark model, as for instance in applications of the QPC model. There must be some deeper reason behind these semiempirical successes.

Apart from these specific quantitative agreements, it is also observed that the signs of the amplitudes are remarkably well described. One notable exception is the well-known $P_{11}(1470)$ Roper resonance, for which the failure is complete for signs as well as magnitudes. This delicate problem will be considered at the end of the discussion.

4.3.1(b) SU(6) analysis

We now consider the algebraic SU(6) analysis that we have presented in Section 4.1.4, and which reveals the necessity of taking into account relativistic corrections in the more explicit quark model considered up to now. For details concerning the SU(6) analysis, as well as for the whole question of radiative transitions, we refer the reader to Close (1979) and to Alcock *et al.* (1980). We assume that, as in the preceding calculations,

$$\langle \mathrm{B}|\boldsymbol{j}|\mathrm{A}\rangle = 3(\mathrm{B}|\boldsymbol{j}(3)\,|\mathrm{A}), \tag{4.78}$$

where the second matrix element is understood to be between *internal* nonrelativistic wave functions and where $\boldsymbol{j}(3)$ operates only on the third quark.

Let us consider the absorption of a photon of helicity $+1$:

$$A = -\frac{3e}{(2k)^{1/2}}\langle \mathrm{N}^*|\,\boldsymbol{\varepsilon}_{+1}\cdot\boldsymbol{j}(3)\,|\mathrm{N}\rangle. \tag{4.79}$$

We can write, by a decomposition into standard Pauli matrices, using the notation of Alcock *et al.* (1980),

$$\frac{e}{(2k)^{1/2}}\boldsymbol{\varepsilon}_{+1}\cdot\boldsymbol{j}(3) = Q(3)[1F^+_\alpha + \boldsymbol{\varepsilon}_{+1}\cdot\boldsymbol{\sigma}F^0_\beta + \sigma_z F^+_\gamma + \boldsymbol{\varepsilon}_{-1}\cdot\boldsymbol{\sigma}F^{++}_\delta]. \tag{4.80}$$

In (4.80) we have assumed quite naturally that the current operator is proportional to the quark charge, and the Fs contain only coordinate or momentum operators.

The superscripts make explicit the fact that F changes the L_z orbital-angular-momentum projection by $+1$, 0 or $+2$. For all the members of the same SU(6) × O(3) multiplet, the amplitude A can be found from the following four matrix elements, which are then combined with purely group-theoretical factors:

$$\left.\begin{aligned} \alpha &= \langle \psi_{L_z=+1} | F_\alpha^+ | \psi_0^s \rangle, \\ \beta &= \langle \psi_{L_z=0} | F_\beta^0 | \psi_0^s \rangle, \\ \gamma &= \langle \psi_{L_z=+1} | F_\gamma^+ | \psi_0^s \rangle, \\ \delta &= \langle \psi_{L_z=+2} | F_\delta^{++} | \psi_0^s \rangle. \end{aligned}\right\} \tag{4.81}$$

The final wave function is the purely spatial part of the resonance wave function, which is moreover symmetric for 1↔2 exchange, i.e. it is ψ^s for a **56** multiplet and ψ'' for a **70**. For a (**70**, 1^-) it is obvious that $\delta=0$. It is possible to fit α, β and γ from the experimental data for all the (**70**, 1^-) states if the amplitudes are approximately SU(6) × O(3) symmetric. We hope that this happens, although we know that the states have masses that break that symmetry. A good phenomenological fit is found (Babcock *et al.*, 1977) for

$$\alpha = 0.16, \qquad \beta = 0.05, \qquad \gamma = 0.075. \tag{4.82}$$

In an explicit quark-model calculation one does not expect to have universal α, β and γ unless one assigns to the resonances a mean mass. This universality hypothesis is different from what we have assumed up to now, that is, to calculate k from the actual masses. It will be nevertheless instructive to follow this new recipe of taking a mean mass to define theoretical α, β and γ. According to Alcock *et al.* (1980), in the nonrelativistic quark model described above,

$$\alpha = 0.150, \qquad \beta = 0.139, \qquad \gamma = 0. \tag{4.83}$$

The agreement is poor between the nonrelativistic quark model in the SU(6) × O(3) limit and the data, expressed by the $\text{SU}(6)_W$ phenomenological analysis (4.82). The particular feature $\gamma=0$ is a direct consequence of the nonrelativistic expression of the current. Indeed, the only spin-dependent contribution is the magnetic term, and this will contain only $\sigma_+(3)$ ($\sigma_-(3)$ for emission), which corresponds to F_β^0. A γ term would be provided by the spin–orbit contribution, and at order v^3/c^3 in the v/c expansion of the Dirac equation in the presence of the radiation field. Indeed the general structure of (3.64) or (3.67) is

$$\boldsymbol{j}^{\text{so}} \sim \boldsymbol{\sigma} \times \boldsymbol{a}, \tag{4.84}$$

where $\boldsymbol{a}$ is a vector operator. Now

$$\varepsilon_{+1}\cdot(\boldsymbol{\sigma}\times\boldsymbol{a}) = [-(\boldsymbol{\sigma}\times\boldsymbol{a})_x + \mathrm{i}(\boldsymbol{\sigma}\times\boldsymbol{a})_y]/\sqrt{2}$$
$$= \mathrm{i}(\varepsilon_{+1}\cdot\boldsymbol{\sigma})a_z + \mathrm{i}\sigma_z(\varepsilon_{-1}\cdot\boldsymbol{a}). \qquad (4.85)$$

Therefore (4.84) will give contributions to both β and γ. By reexpressing a calculation of Kubota and Ohta (1976) in terms of SU(6) analysis, Alcock *et al.* (1980) found

$$\alpha = 0.150, \qquad \beta = 0.076, \qquad \gamma = 0.064. \qquad (4.86)$$

The spin–orbit term has the remarkable effect of both reducing β in the right direction and creating a nonzero γ of the correct sign and order of magnitude. All of this is very encouraging, since we are now *testing the relativistic corrections* to the naive quark model, which seem to go in the right direction. It must be noted that the bag-model calculations (Hey, Holstein and Sidhu, 1978) also give a nonvanishing γ—not surprisingly, since the bag wave functions are solutions of the Dirac equation in a potential well—but with the wrong sign.

4.3.1(c) The Roper-resonance radiative decay

A remaining problem is the Roper resonance $P_{11}(1440)$. Although the spin–orbit contribution of Kubota and Ohta has the good effect of reversing signs for $P_{11}(1440)$ with respect to the naive calculation, and therefore giving them in agreement with experiment, the magnitude is too small. However, as argued in Section 3.2.6, the usual spin–orbit contribution used by these authors is not correct, and one must, moreover, take into account terms of order $1/m^3$ and not only $1/m^2$. A correct calculation is given by Sucher for transitions between S states in the framework of the Dirac equation, calculated to the lowest order in v/c that contributes to such transitions involving radial excitations, namely the fourth order. It is found (Sucher, 1978; Sucher and Kang, 1978) that

$$\langle \mathrm{f}|\boldsymbol{j}\cdot\varepsilon|\mathrm{i}\rangle$$
$$= \langle\chi_\mathrm{f}|\mathrm{i}\varepsilon\cdot\frac{\boldsymbol{\sigma}\times\boldsymbol{k}}{2m}|\chi_i\rangle\langle\varphi_\mathrm{f}|\,1-\frac{k^2r^2}{6}-\frac{2}{3}\frac{p^2}{m^2}-\frac{V_\mathrm{S}}{m}+\frac{rV_\mathrm{V}'}{3m}|\varphi_\mathrm{i}\rangle. \qquad (4.87)$$

Suppose that we take the standard choice of a confining scalar potential,

$$V_\mathrm{S} = m\omega^2r^2, \qquad V_\mathrm{V} = 0. \qquad (4.88)$$

We also assume that, as given by the harmonic-oscillator potential, $k = 2/mR^2 = 2\omega$, which is rather different from the physical value 0.4 GeV. We then obtain for the ratio of the correct amplitude to the naive calculation

$$\frac{M_{\text{corr}}}{M_{\text{naive}}} = \frac{k^2-\omega^2}{k^2}, \tag{4.89}$$

where M_{corr} denotes the whole amplitude (4.87) with the choice (4.88), and M_{naive} denotes the contribution coming only from the overlap with $e^{ik\cdot r}$, i.e $\langle\varphi_f|(1-\frac{1}{6}k^2r^2)|\varphi_i\rangle$. For the theoretical $k=2\omega$, we find $+\frac{3}{4}$ for (4.89) and -0.56 for the actual value of k. This illustrates the sensitivity of the calculation to the different assumptions. Still other results are obtained by considering a vector confining potential, for which we obtain a more stable negative sign, but which seems at odds with spectroscopical analysis; anyway, it is also difficult to obtain the correct absolute magnitude. Finally, the harmonic-oscillator assumption may be questioned, but it is difficult to find a potential locating the Roper resonance in the right place (see Hogaasen and Richard, 1983).

In fact, the lesson of all this is that a transition like $P_{11}\to N\gamma$, being of high formal order v^4/c^4, is very sensitive to many corrections and cannot be safely calculated. This lesson also applies to the charmonium transition $\psi'\to\eta_c\gamma$, where ψ' is the first radial excitation. We shall return to $P_{11}\to N\gamma$ in the context of the vector-meson-dominance model in Section 4.4.6.

4.3.2 Meson Radiative Transitions

One of the early calculations of the quark model was the electromagnetic decay $\omega\to\pi^0\gamma$. An easy calculation, along the same line as $\Delta\to N\gamma$, gives

$$\Gamma(\omega\to\pi^0\gamma) = \frac{1}{3\pi}\left(\frac{e}{2m}\right)^2\frac{E_\pi}{m_\omega}k^3, \tag{4.90}$$

since the only contribution comes once more from the magnetic term as ω has spin 1, and π spin 0. For example, for polarization $+1$ γ emission, we have

$$M(\omega(+1)\to\pi^0\gamma(+1)) = \sqrt{2}\,\mu_q\langle\chi_0\varphi_1^0|\,\sigma^{(-)}(1)Q(1)+\sigma^{(-)}(2)Q(2)\,|\chi_1^{+1}\varphi_0\rangle, \tag{4.91}$$

where χ_0 and χ_1^{+1} represent the spin wave functions of π and ω, and φ_1^0 and φ_0 the corresponding isospin wave functions. Since

$$\left.\begin{aligned}\langle\chi_0|\,\sigma^{(-)}(1)\,|\chi_1^{+1}\rangle &= -\langle\chi_0|\,\sigma^{(-)}(2)\,|\chi_1^{+1}\rangle = -\frac{1}{\sqrt{2}},\\ \langle\varphi_1^0|\,Q(1)\,|\varphi_0\rangle &= -\langle\varphi_1^0|\,Q(2)\,|\varphi_0\rangle = \frac{1}{2},\end{aligned}\right\} \tag{4.92}$$

we find the right-hand side of (4.91) to be equal to the quark magnetic moment. Numerically, we get

$$\Gamma(\omega \to \pi^0 \gamma) = 626\,\mathrm{keV}, \tag{4.93}$$

which is not too far from the experimental 860 keV. However, as has been emphasized in Section 4.1.3, when one is trying to treat in a decay the π as a quark composite state, there are large uncertainties, and the result cannot be considered as really significant.

It is encouraging that the ration $\rho^0 \to \pi^0\gamma/\omega \to \pi^0\gamma$, much less ambiguous, is in very good agreement with experiment, but this is merely due to flavor symmetry:

$$\frac{\Gamma(\rho^0 \to \pi^0 \gamma)}{\Gamma(\omega \to \pi^0 \gamma)} = \frac{1}{9}, \tag{4.94}$$

versus an experimental 0.08.

A better kinematical situation is found in the charmed-meson radiative transitions $D^* \to D\gamma$. Unhappily, the widths of the D*s are not known, and therefore we do not know the absolute rates. However, the quark model gives a very natural explanation for the fact that $\mathrm{BR}(D^{*+} \to D^+\gamma) \ll \mathrm{BR}(D^{*0} \to D^0\gamma)$ (Ono, 1976) because the magnetic moments of the two quarks add in the D^0 and cancel in the D^+, and moreover the magnetic moment of the u quark is twice that of the d. Indeed, we have

$$\begin{aligned} &M(D^*(+1) \to D\gamma(+1)) \\ &= \tfrac{2}{3}\mu_c \langle \chi_0 | \sigma^{(-)}(1) | \chi_1^{+1} \rangle + \begin{Bmatrix} -\tfrac{2}{3} \\ +\tfrac{1}{3} \end{Bmatrix} \mu_q \langle \chi_0 | \sigma^{(-)}(2) | \chi_1^{+1} \rangle, \end{aligned} \tag{4.95}$$

where μ_q is the light-quark magnetic moment. Because of (4.92) we obtain for D^0 and D^+ respectively

$$-\frac{1}{\sqrt{2}}\left[\frac{2}{3}\mu_c + \begin{Bmatrix} +\tfrac{2}{3} \\ -\tfrac{1}{3} \end{Bmatrix} \mu_q\right]. \tag{4.96}$$

However, the best place for testing the quark model is among the charmonium radiative transitions $\psi' \to \chi\gamma$, $\chi \to J/\psi\gamma$, $J/\psi \to \eta_c\gamma$, $\psi' \to \eta_c\gamma$, $\eta'^c\gamma$, $\eta'_c \to J/\psi\gamma$, where ψ' and η'_c are the first radial excitations $\psi(3685)$ and $\eta_c(3590)$. What is interesting here is that we have a number of transitions involving radial excitations, whose assignment is unambiguous, because the meson spectrum is much simpler than the baryon spectrum. Also, we can hope that the situation is reasonably nonrelativistic, since, although $R_c'^2$ is smaller than for light hadrons, m_c is much larger, and we have

$$\frac{1}{m_c^2 R_c'^2} \sim \frac{1}{10}. \tag{4.97}$$

Since v/c is not large, the v/c expansion could be expected to be reasonable. It will appear that relativistic corrections are still sizable. There are even some cases where the relativistic corrections are crucial, because the lowest-order nonvanishing approximation already corresponds to a high formal order in v/c. This is the case encountered in the charmonium *forbidden* radiative transitions $\psi' \to \eta_c \gamma$ or $\eta_c' \to J/\psi\,\gamma$ between the radial excitations and the ground state. More surprisingly, the allowed E_1 transitions receive a very large correction. For all these reasons, the charmonium transitions are a good testing place for relativistic corrections. At the same time, they may also be sensitive to the potential, which, fortunately, happens to be rather well known. Therefore it becomes significant to go beyond the naive harmonic-oscillator calculations. In all these processes, one must note also that *coupled-channel effects* are important. As explained in Section 3.3, they are in principle high-order effects, but the naive v/c power-counting argument given there is misleading, owing to the possibility of light-quark pair creation (with a large value of the expansion parameter).

Because of the smallness of $1/m_c^2 R_c'^2$, it is possible to make a further approximation in the expression of the naive quark model. It is reasonable to set $e^{-i\boldsymbol{k}\cdot\boldsymbol{r}_i} \approx 1$, because we have

$$\boldsymbol{k}\cdot\boldsymbol{r}_i \sim \frac{1}{m_c R_c'}. \tag{4.98}$$

This new approximation greatly simplifies the discussion. We have two main types of transitions. The first are the transitions between S states, $J/\psi \to \eta_c \gamma$, $\psi' \to \eta_c' \gamma$, $\psi' \to \eta_c \gamma$, $\eta_c' \to J/\psi\,\gamma$. The second are the ones between P and S states $\psi' \to \chi\gamma$, $\chi \to J/\psi\gamma$. With the additional approximation $e^{-i\boldsymbol{k}\cdot\boldsymbol{r}} \approx 1$, the P$-$S transition only receives a contribution from the convective current. Indeed, the magnetic term can be factorized out of the space integral, leaving the scalar product of the spatial wave functions, which is zero since they are orthogonal. On the other hand, independently of the neglect of the factor $e^{-i\boldsymbol{k}\cdot\boldsymbol{r}}$, the convective term cannot contribute to S–S transitions, because it necessarily changes the projection of the orbital angular momentum by one unit. Therefore S–P and S–S transitions correspond respectively to the convective and magnetic terms. They are also called E_1 and M_1 transitions, in multipole notation, according to their dominant multipole structure.

Let us now proceed to actual calculations beginning with the S–S transitions. For $J/\psi \to \eta_c \gamma$ and $\psi' \to \eta_c' \gamma$, we find from a calculation similar to that for $\omega \to \pi^0 \gamma$, making the approximation $e^{i\boldsymbol{k}\cdot\boldsymbol{r}} \approx 1$,

$$\Gamma = \frac{16}{3}\alpha\left(\frac{Q_c}{2m_c}\right)^2 \frac{E_{\eta_c}}{m_\psi} k^3, \tag{4.99}$$

where Q_c is the charmed-quark charge.

This gives, with $m(\eta_c)=2.98$ GeV and $m(\eta_c')=3.59$ GeV, the branching ratios BR = 3.3% and 0.5%, in rough agreement with experiment, 1.27 ± 0.36% and 0.5–1.2%. The correction to (4.99) when one takes into account $e^{ik\cdot r_i}$ is a factor $e^{-k^2R'^2/8}$, which is indeed very close to unity.

In contrast it is obvious that setting $e^{ik\cdot r}\approx 1$ in $\psi'\to\eta_c\gamma$ would result into a vanishing amplitude (a *forbidden* transition). However, since $\boldsymbol{k}\cdot\boldsymbol{r}_i$ is still small, we can make a power expansion

$$e^{ik\cdot r_i}=1+i\boldsymbol{k}\cdot\boldsymbol{r}_i-\tfrac{1}{2}(\boldsymbol{k}\boldsymbol{r}_i)^2+\dots. \tag{4.100}$$

The relevant term for a transition like $\psi'\to\eta_c\gamma$ is the third one. It must not be forgetten moreover, that $\boldsymbol{r}_i=\pm\frac{1}{2}\boldsymbol{r}$. We then find

$$\langle\psi_1^{\text{space}}|e^{ik\cdot r_i}|\psi_0^{\text{space}}\rangle=\tfrac{-1}{24}k^2\langle r^2\rangle_{1,0}, \tag{4.101}$$

where 1 and 0 mean the first radially excited and the ground states respectively. To calculate (4.101) we must use explicit wave functions. The harmonic oscillator gives

$$\langle r^2\rangle_{1,0}=\sqrt{\tfrac{3}{2}}\,R_c'^2, \tag{4.102}$$

whence

$$\Gamma=\frac{16}{3}\alpha\left(\frac{Q_c}{2m_c}\right)^2\frac{E_{\eta_c}}{m_\psi}k^3\frac{3}{2}\left(\frac{k^2R_c'^2}{24}\right)^2. \tag{4.103}$$

This gives a branching ratio of 0.5% — only a factor of 2 off the experimental value of 0.28%. For a radial excitation, it might be feared that the harmonic-oscillator approximation is not good. However, the Cornell model (Porter, 1981) yields almost the same value, 0.45%.

Let us consider the P–S transitions. We have for the matrix element of the convective contribution

$$A=-\frac{e}{(2k)^{1/2}}Q_c\int d\boldsymbol{r}_1\,d\boldsymbol{r}_2\,\delta\left(\frac{\boldsymbol{r}_1+\boldsymbol{r}_2}{2}\right)\psi_f^\dagger\left(-i\frac{\boldsymbol{\varepsilon}\cdot\boldsymbol{\nabla}_1}{m_c}+i\,\frac{\boldsymbol{\varepsilon}\cdot\boldsymbol{\nabla}_2}{m_c}\right)\psi_i. \tag{4.104}$$

We have $\boldsymbol{\nabla}_1\psi_i=-\boldsymbol{\nabla}_2\psi_i=\boldsymbol{\nabla}_r\psi_i$ ($\boldsymbol{r}=\boldsymbol{r}_1-\boldsymbol{r}_2$), whence

$$A=-\frac{e}{(2k)^{1/2}}Q_c 2\int d\boldsymbol{r}\,\psi_f^\dagger\left(-i\frac{\boldsymbol{\varepsilon}\cdot\boldsymbol{\nabla}_r}{m_c}\right)\psi_i. \tag{4.105}$$

Here, attention must be paid to the fact that the χ_J states ($J=0,1,2$) are described by linear combinations of products of spin and spatial wave functions with definite orbital angular momentum J;

$$\psi_J^M=\sum_{m,\mu}\langle 1,1;m,\mu|J,M\rangle\psi_1^m\Phi^\mu, \tag{4.106}$$

where m denotes the projection of the orbital angular momentum and μ the quark-spin projection. Note that, to fit with the usual convention, the Clebsch–Gordan coefficients are taken in the L–S order. Choosing photons with helicity $+1$, we find for $\psi' \to \chi_J \gamma$,

$$A = \left(\frac{2\pi\alpha}{k}\right)^{1/2} Q_c \langle 1, 1; -1, S_z | J, S_z - 1\rangle \left(-i\frac{2}{\sqrt{3}}\frac{1}{m_c R'_c}\right), \tag{4.107}$$

where S_z is the spin projection of the ψ'. We have

$$\sum_{S_z} |\langle 1, 1; -1, S_z | J, S_z - 1\rangle|^2 = \tfrac{1}{3}(2J+1), \tag{4.108}$$

and finally

$$\Gamma(\psi' \to \chi_J \gamma) = (2J+1)\,\alpha \frac{E_\chi}{m_{\psi'}} k \frac{2^6}{3^5} \frac{1}{m_c^2 R_c'^2}. \tag{4.109}$$

Quite similarly, we find for $\chi_J \to J/\psi\, \gamma$

$$A = -\left(\frac{2\pi\alpha}{k}\right)^{1/2} Q_c \langle 1, 1; +1, J_z - 1 | J, J_z\rangle \left(-i\frac{2}{\sqrt{2}}\frac{1}{m_c R'_c}\right) \tag{4.110}$$

and

$$\Gamma(\chi_J \to J/\psi\, \gamma) = \alpha \frac{E_\psi}{m_\chi} k \frac{2^5}{3^3} \frac{1}{m_c^2 R_c'^2}. \tag{4.111}$$

The transitions to the ground state $\chi_J \to J/\psi\, \gamma$ are expected to be less sensitive to the potential than the transitions involving radial excitations like $\psi' \to \chi_J \psi$, but the widths $\Gamma(\chi_J)$ are not well known. Let us now discuss the $\psi' \to \chi_J \psi$ rates.

The numerical results, normalized to the total *experimental* width, are

$$\left.\begin{aligned} \mathrm{BR}(\psi' \to \chi_2 \gamma) &= 57\%, \\ \mathrm{BR}(\psi' \to \chi_1 \gamma) &= 45\%, \\ \mathrm{BR}(\psi' \to \chi_0 \gamma) &= 22\%. \end{aligned}\right\} \tag{4.112}$$

These results compare badly with experiment, which gives for the three reactions roughly the same number $\mathrm{BR}(\psi' \to \chi_J \gamma) = 8\%$; they add to more than 100% because the sum of the calculated partial widths exceeds the experimental total width. But the branching ratios $\psi' \to \chi_J \gamma$ are very sensitive to the potential, and it is clear that the radial excitation ψ' is not described well by a harmonic-oscillator potential. Choosing the more realistic Cornell potential, and calculating in the same way (to obtain these results, we do a rescaling, as explained later, from the original Cornell calculations (Eichten *et al.*, 1980)):

$$\left.\begin{aligned} BR(\psi'\to\chi_2\gamma) &= 21\%,\\ BR(\psi'\to\chi_1\gamma) &= 19\%,\\ BR(\psi'\to\chi_0\gamma) &= \ \ 9\%. \end{aligned}\right\} \tag{4.113}$$

This illustrates the sensitivity to the potential and shows that we are going into the right direction, although the agreement is still poor. But we have an ambiguity as to how we should calculate the matrix element. Indeed, the convective term (4.105) is essentially the velocity operator $-\mathrm{i}\boldsymbol{\nabla}_r/m=\boldsymbol{p}/m$. Often the formula

$$2\frac{\boldsymbol{p}}{m}=\mathrm{i}[H,\boldsymbol{r}] \tag{4.114}$$

where H is the Hamiltonian in the Schrödinger equation, is used to transform (4.105) into the matrix element of $\boldsymbol{r}$ times the energy difference $E_\mathrm{i}-E_\mathrm{f}=k$ (called the *dipole formula* for the width). It must be observed, however, that we only have an identity if we take $E_\mathrm{i}-E_\mathrm{f}$ equal to the difference of the Schrödinger equation levels. On the other hand, it is usual to take for the photon energy k the difference between the *actual* energies of the hadrons. With this prescription, the use of (4.114) will lead to a modified estimate. The corresponding values of k are 0.13, 0.17 and 0.26 GeV instead of theoretical $\Delta E=0.37$ GeV for the harmonic oscillator and 0.16 GeV for the Cornell model. With this latter procedure, the results are completely different for the harmonic oscillator,

$$\left.\begin{aligned} BR(\psi'\to\chi_2\gamma) &= \ \ 7\%,\\ BR(\psi'\to\chi_1\gamma) &= 10\%,\\ BR(\psi'\to\chi_0\gamma) &= 11\%. \end{aligned}\right\} \tag{4.115}$$

and for the Cornell model the difference is still sizable,

$$\left.\begin{aligned} BR(\psi'\to\chi_2\gamma) &= 13\%,\\ BR(\psi'\to\chi_1\gamma) &= 21\%,\\ BR(\psi'\to\chi_0\gamma) &= 23\%. \end{aligned}\right\} \tag{4.116}$$

but is not an improvement with respect to (4.113). This procedure, which is a new illustration of the ambiguities of the nonrelativistic approximation, emphasized in Section 4.1.3, can be partly justified in a treatment that accounts for relativistic corrections, taking into account the fine structure of the spectrum. However, it is then necessary to take account of other corrections, in particular to the wave functions.

Let us now come to the question of higher-order relativistic corrections. We begin with the S–S transitions, which have been treated by Sucher and collaborators (Sucher, 1978; Sucher and Kang, 1978), following the Hamiltonian method described in Section 3.3. The result is analogous to that obtained from the Dirac equation, but with differences owing to the treatment of center-of-mass motion. Moreover, it happens that if the confining potential has a structure similar to that found in one-photon exchange (Lorentz vector), it does not appear in the effective operator. For a $Q\bar{Q}$ quarkonium,

$$(2\pi)^3\langle \mathrm{f}\,|\,\boldsymbol{j}\cdot\boldsymbol{\varepsilon}\,|\,\mathrm{i}\rangle$$

$$= e\sum_i (\mathrm{f}\,|\,Q(\mathrm{i})\mathrm{i}\boldsymbol{\varepsilon}\cdot\frac{\boldsymbol{\sigma}(\mathrm{i})\times\boldsymbol{k}}{2m}\left(1-\frac{k^2r^2}{24}-\frac{2}{3}\frac{p^2}{m^2}-\frac{V_{\mathrm{S}}}{m}\right)|\,\mathrm{i}) \tag{4.117}$$

between internal nonrelativistic wave functions $|\mathrm{i})$ and $|\mathrm{f})$, with possible corrections of order v^2/c^2. In this expression the recoil corrections are still neglected. For $J/\psi\to\eta_c\gamma$ and $\psi'\to\eta_c'\gamma$ we get $(\mathrm{f}\,|\,1\,|\,\mathrm{i})=1$ for the spatial parts of the wave functions, because they are identical. Among the corrections, the $\frac{1}{24}k^2r^2$ contribution, which corresponds to the expansion of $\mathrm{e}^{\mathrm{i}\boldsymbol{k}\cdot\boldsymbol{r}/2}$, can be safely neglected since k is small even with respect to the Schrödinger level differences. If we calculate the remaining corrections in the harmonic-oscillator model, we find, in the case of $J/\psi\to\eta_c\gamma$ (ground state), a ratio between the naive calculation and the one taking $(v/c)^2$ corrections into account:

$$\frac{M_{\mathrm{corr}}}{M_{\mathrm{naive}}}=1-\frac{5}{2}\,\frac{1}{m_c^2R_c'^2}. \tag{4.118}$$

This gives a reduction factor of 0.5 for the width, yielding a branching ratio of 1.6%, in better agreement with experiment. A calculation with the Cornell model yields an analogous result of $\sim 1\%$ (Martin, 1982). It can therefore be seen that it is not justified to appeal to a negative quark anomalous moment, as has been claimed, to explain the discrepancy between experiment and the naive calculation.

However, in the preceding calculation, a delicate point has been omitted. Indeed, if one wants to fit the spectrum with a mass $m_c\approx 1.8\,\mathrm{GeV}$, and therefore $m_{J/\psi}-2m_c<0$, one cannot use a purely positive potential. In fact a large *negative* constant of order 1 GeV is necessary in both the harmonic-oscillator potential and the Cornell potential. This large negative constant would result in a *positive* additional contribution to (4.118), cancelling the reduction previously found. This problem has been avoided by Sucher and Kang (1978) by simply attributing this negative constant to the vector part of the potential. However, this hypothesis is rather unjustified. A more careful analysis of the spectrum equation (McClary, 1982) indeed shows that the negative constant must be attributed to the scalar part

$$V_S = \frac{r}{a^2} + C, \tag{4.119}$$

with $C = -0.80\,\text{GeV}$ and $a = 2.34\,\text{GeV}^{-1}$ (this value is that of the Cornell model). An approximate calculation then gives

$$\frac{M_{\text{corr}}}{M_{\text{naive}}} \sim 1.2, \tag{4.120}$$

a value > 1, contrary to (4.118). For $\psi' \to \eta_c' \gamma$, the corrections corresponding to the first calculation would be much larger,

$$\frac{M_{\text{corr}}}{M_{\text{naive}}} = 1 - \frac{35}{6} \frac{1}{m_c^2 R_c'^2}, \tag{4.121}$$

a trend not indicated by the data, which agree roughly with the naive calculation. Taking the negative constant C into account cancels the correction completely:

$$\frac{M_{\text{corr}}}{M_{\text{naive}}} \sim 1. \tag{4.122}$$

In conclusion, it can be seen that *the situation is rather disturbing for allowed transitions.*

For the forbidden transition $\psi' \to \eta_c \gamma$, the situation is no better. Equation (4.117) is still valid, but we must pay attention to the calculation of the first term, since the spatial overlap is now $\neq 1$ because the spatial wave functions are different for different total quark spin $S_i = 1$, $S_f = 0$ (due to the spin dependence induced by the relativistic effects), and the difference contributes an $O(v^2/c^2)$ term. A perturbative calculation shows that

$$\Delta_{\text{spin}} \equiv (\text{f}\,|\,1\,|\text{i})_{\text{space}} = \frac{\Delta E}{E_i - E_f} \frac{\psi_i(0)}{\psi_f(0)}, \tag{4.123}$$

where ΔE is the ground-state hyperfine splitting $\sim 100\,\text{MeV}$.

On the other hand, the constant C in the potential will contribute in the same way as this first term 1; its contribution can be accounted for through the replacement $1 \to 1 + |C|/m$. Anyway, it must be noted that C is formally $O(v^2/c^2)$ and must be consistently neglected here. It is also found that there is a cancellation between the second contribution, coming from the expansion of $e^{i\boldsymbol{k}\cdot\boldsymbol{r}_i}$ and the third one, of kinetic origin. Moreover,

$$(\text{f}\,| -\frac{V_S}{m} |\text{i})_{\text{space}} > 0. \tag{4.124}$$

All of this results in a large ratio for $\psi' \to \eta_c \gamma$ (M_{naive} is given for this case of a forbidden transition by the second term in (4.117)):

$$\frac{M_{\text{corr}}}{M_{\text{naive}}} \sim 2\text{–}3, \tag{4.125}$$

which is certainly not suggested by the data, which present the *opposite trend.* In the case of a *vector* confining potential the results would be more satisfactory, but this advantage would be negated by coupled-channel effects (Zambetakis, 1986).

The situation regarding the relativistic corrections is much more encouraging for the P–S transitions. A rather detailed discussion is given by McClary and Byers (1983). As we have seen, the transition amplitude is essentially given by

$$A = -\left(\frac{2\pi\alpha}{k}\right)^{1/2} Q_c\, 2k(-\mathrm{i})\,(\mathrm{f}\,|\boldsymbol{r}|\,\mathrm{i}), \tag{4.126}$$

with a neglect of effects of $e^{i\boldsymbol{k}\cdot\boldsymbol{r}_i}$, which are indeed small. The correction to this naive calculation then comes only from the corrections of order v^2/c^2 to the wave functions, which are given by the terms of order v^4/c^4 of the binding Hamiltonian, in particular those due to transverse gluon exchange. It is then legitimate to take into account the difference between the values of k corresponding to the various χ_J, since they are relevant to this order of approximation. It is found that these wave-function corrections reduce the value of $(\mathrm{f}|\boldsymbol{r}|\mathrm{i})$ with respect to its naive value. These corrections are particularly sizable for radial excitations.

A calculation by Henriques, Kellett and Moorhouse (1976)

$$\left.\begin{aligned} \mathrm{BR}(\psi' \to \chi_2\gamma) &= 7\%, \\ \mathrm{BR}(\psi' \to \chi_1\gamma) &= 10\%, \\ \mathrm{BR}(\psi' \to \chi_0\gamma) &= 10\%. \end{aligned}\right\} \tag{4.127}$$

in good agreement with experiment (Table 4.2).

Corrections due to the coupling to two charmed meson channels have also been calculated and give a minor 20% reduction.

Table 4.2 shows that the E_1 (electric dipole) transitions are in striking agreement with relativistic calculations. In the case of the bottonium E_1 radiative transitions, the nonrelativistic calculations are in fair agreement with experiment (Königsmann, 1985).

Table 4.2 Branching ratios for the E_1 transitions of charmonium. The theory is from Henriques *et al.* (1976), using the dipole formula (4.126) and relativistic wave functions from the Bethe–Salpeter equation. Experimental data is from Porter (1981). Errors come from experimental uncertainties on the total width.

	Theory	Experiment
$\psi' \to \chi_0 \gamma$	10%	10%
$\psi' \to \chi_1 \gamma$	10%	9%
$\psi' \to \chi_2 \gamma$	7%	8%
$\chi_0 \to \psi \gamma$	$(0.75 \pm 0.20)\%$	0.6%
$\chi_1 \to \psi \gamma$	$> 10\%$	27%
$\chi_2 \to \psi \gamma$	$(18 \pm 9)\%$	16%

4.4 STRONG DECAYS ALLOWED BY THE OKUBO–ZWEIG–IIZUKA RULE

4.4.1 Introduction

Strong decays offer a large number of experimental results to be confronted with theory. We have much more data here than for radiative or semileptonic decays, because all the resonances decay strongly while their radiative or weak decays generally have a very small and unobservable branching ratio. Moreover, the strong decays of resonances are mainly described by decays into two-body (possibly still resonances) channels, to which the quark model applies directly. To avoid complicated developments, we shall concentrate on a few selected cases. It must be understood that the omitted pieces of data could be described along the same lines. The most systematic developments concern the baryon excitations (up to $L=2$) and the heavy quarkonia — two cases where a large set of states has been unambiguously identified, in contrast with the less well-known excited mesons composed of light quarks, and the baryon composites of heavy quarks, for which data are very scarce.

There are two types of models, both presented in Section 2.2. On the one hand, there are the elementary-meson-emission models, mainly developed for pion emission. (To be brief, we shall omit the SU(3) partners of the pion, K, η and η'). On the other hand, there are the models that describe the decay process more completely, treating the two decay products as composed of quarks, and the decay itself as the creation of a quark–antiquark pair. This latter category includes the $^3P_0$ quark-pair-creation (QPC) model, the Cornell model and the flux-tube models. All these models have their own advantages and disadvantages. In the elementary-emission model it is of course quite unsatisfactory to treat one of the decay products as elementary; moreover, one needs one independent coupling constant for each type of emission, and this implies many new parameters. On the other hand, the advantage is that the emitted quantum can be treated, like the photon, in a relativistic way, and this is especially welcome in the case of the pion. In contrast fully composite models are only simple in a completely nonrelativistic treatment, and therefore accurate results cannot be expected. But they are definitely superior, as they describe in a unified way any type of meson emission. In addition, the study of radial excitations reveals the necessity of taking the spatial extension of the emitted meson into account.

4.4.2 Strong Decays by Pion Emission. Direct and Recoil Terms

We have to start from the expressions given in Section 2.2. Here we shall use the momentum-space representation. Starting from the expressions (2.153,154) for the Hamiltonian operator, and expressing the matrix elements in terms of internal wave functions, we end up with the following formulae, respectively for pseudoscalar and axial-vector coupling for baryon decay $B_i \to B_f + \pi^\alpha$:

$$M = \frac{i}{(2\pi)^{3/2}} \frac{1}{(2\omega_\pi)^{1/2}} \frac{g_{\pi qq}}{2m}$$
$$\times \frac{1}{\sqrt{3}} \int d\boldsymbol{p}_\rho \, d\boldsymbol{p}_\lambda \, \psi_f(\boldsymbol{p}_\rho, \boldsymbol{p}'_\lambda)^\dagger \, \tau^\alpha(3) \, \boldsymbol{\sigma}(3) \cdot \boldsymbol{k} \, \psi_i(\boldsymbol{p}_\rho, \boldsymbol{p}_\lambda), \tag{4.128}$$

$$M = \frac{i}{(2\pi)^{3/2}} \frac{1}{(2\omega_\pi)^{1/2}} \frac{g_{\pi qq}}{2m}$$
$$\times \frac{1}{\sqrt{3}} \int d\boldsymbol{p}_\rho \, d\boldsymbol{p}_\lambda \, \psi_f(\boldsymbol{p}_\rho, \boldsymbol{p}'_\lambda)^\dagger \, \tau^\alpha(3) \left[\boldsymbol{\sigma}(3) \cdot \boldsymbol{k} - \omega_\pi \boldsymbol{\sigma}(3) \cdot \frac{\boldsymbol{p}_3 + \boldsymbol{p}'_3}{2m} \right] \psi_i(\boldsymbol{p}_\rho, \boldsymbol{p}_\lambda), \tag{4.129}$$

$\boldsymbol{k}$ is the pion momentum, $\boldsymbol{p}'_\lambda = \boldsymbol{p}_\lambda + \sqrt{\tfrac{2}{3}}\,\boldsymbol{k}$, $\boldsymbol{p}_3 = -\sqrt{\tfrac{2}{3}}\,\boldsymbol{p}_\lambda$ and $\boldsymbol{p}'_3 = \boldsymbol{p}_3 - \boldsymbol{k}$. We have worked in the momentum representation, choosing the rest frame $\boldsymbol{P} = 0$ of the decaying resonance. We have adapted the calculations of Section 2.2, devised for absorption, to an emission process by making $\boldsymbol{k} \to -\boldsymbol{k}$, and we have made some translations of variables and passed to the relative coordinates $\boldsymbol{p}_\rho$ and $\boldsymbol{p}_\lambda$.

For the quark-pair-creation model a parallel calculation leads to

$$M = -\frac{\gamma}{2(6\pi)^{1/2}}$$
$$\times \frac{1}{\sqrt{3}} \int d\boldsymbol{p}_\rho \, d\boldsymbol{p}_\lambda \, \psi_f(\boldsymbol{p}_\rho, \boldsymbol{p}'_\lambda)^\dagger \, \tau^\alpha(3) \, \boldsymbol{\sigma}(3) \cdot \boldsymbol{p}'_3 \, \psi_0(2\boldsymbol{p}_3 - \boldsymbol{k}) \, \psi_i(\boldsymbol{p}_\rho, \boldsymbol{p}_\lambda). \tag{4.130}$$

We recall that γ is a dimensionless phenomenological parameter, which is found to be $\gamma \approx 3$.

At least for the elementary-emission models, the discussion may be conducted in analogy with that of radiative decays by isolating different terms and considering their respective contributions to the different types of transitions. In the radiative decays these were the convective and the magnetic terms. For pion emission, one distinguishes the *direct* term, which has the $\boldsymbol{\sigma} \cdot \boldsymbol{k}$ structure, and the *recoil* term, which has the $\boldsymbol{\sigma} \cdot (\boldsymbol{p} + \boldsymbol{p}')$ structure. According to the expressions (4.128) and (4.129), the direct term is common to both the pseudoscalar and the axial-vector couplings, while the recoil term is only present in the axial-vector case. However, a recoil term is also obtained in the pseudoscalar case on going to the next v/c approximation (see Section 3.2.4). In fact, in a certain approximation, the two couplings are identical and contain direct and recoil terms. We shall therefore use henceforth the expression (4.129), which describes a better approximation, and we shall see below that this better approximation is needed, for example, for $L=1$ decays.

To discuss the respective effects of the direct and the recoil terms, it is useful to consider the partial waves of the decay. For the decay of a resonance of quark orbital angular momentum L into ground-state hadrons, it can be shown that there are two allowed partial waves $l = L \pm 1$. This result derives straightforwardly from the Wigner–Eckart theorem applied to the orbital angular momentum. Indeed, let us completely put aside the quark spin and consider only the quark space variables. We consider a decay $A \rightarrow B + C$, where B and C are ground-state hadrons, with quark orbital angular momentum $L=0$ each. The final $B+C$ state has only a relative angular momentum l, since the decay products have no quark orbital angular momentum. The transition operator is obviously a vector in configuration space, both in the elementary-emission models and in the $^3P_0$ and Cornell models. From the Wigner–Eckart theorem, we then have $l = L+1$, $L-1$ or L. Finally $l = L$ is excluded by parity conservation. Indeed,

$$P = (-1)^L = (-1)^{l+1} \tag{4.131a}$$

for the decay of a baryon (parity $P = (-1)^L$) into ground-state meson ($P = -1$) and baryon ($P = +1$), and

$$P = (-1)^{L+1} = (-1)^l \tag{4.131b}$$

for the decay of a meson $q\bar{q}$ system of parity $P = (-1)^{L+1}$ into two ground-state mesons (both of parity -1). From (4.131a,b) we get the expected result

$$l = L \pm 1 \tag{4.132}$$

Now, the interesting point is that the upper partial wave $l = L+1$ is essentially given by the direct term and receives only a small correction from the recoil term, while the lower partial wave $l=L-1$ receives its essential contribution from the recoil term. Although this may be demonstrated generally with the methods used in Section 2.2.5, we shall give the proof only for the case $L=1$, where $l=0$, 2, and shall restrict ourselves for definiteness to baryon decay and harmonic-oscillator wave functions.

We drop constants and normalization factors, and also omit the SU(3) quantum numbers and the quark-spin wave functions of the baryons, as they are irrelevant for the present analysis. We then have a typical expression for the amplitude, for the elementary-emission model, calculated in momentum space:

$$\begin{aligned} M \propto \int \mathrm{d}\boldsymbol{p}_\rho \boldsymbol{p}_\lambda \exp[-\tfrac{1}{2}(\boldsymbol{p}_\rho^2+\boldsymbol{p}_\lambda^2)R^2] \\ \times \mathscr{Y}_1^{L_z}(\boldsymbol{p}_\lambda)\mathscr{Y}_1^m\left(\boldsymbol{k}-\omega\frac{\boldsymbol{p}_3+\boldsymbol{p}_3'}{2m}\right)\exp[-\tfrac{1}{2}(\boldsymbol{p}_\rho^2+\boldsymbol{p}_\lambda'^2)R^2], \end{aligned} \tag{4.133}$$

where, again, $\boldsymbol{p}_3'=\boldsymbol{p}_3-\boldsymbol{k}$ and $\boldsymbol{p}_\lambda'=\boldsymbol{p}_\lambda+(\tfrac{2}{3})^{1/2}\boldsymbol{k}$.

It is necessary for the orbital excitation to be of the variable-λ type, since the ρ excitation would give zero by orthogonality, because $\boldsymbol{p}_\rho$ is not in the transition operator.

Defining $\boldsymbol{p}=\tfrac{1}{2}(\boldsymbol{p}_\lambda+\boldsymbol{p}_\lambda')$ and performing the trivial integration over $\boldsymbol{p}_\rho$, which gives simply a multiplicative constant, we are left with

$$\begin{aligned} M \propto \int \mathrm{d}\boldsymbol{p} \exp(-\boldsymbol{p}^2R^2-\tfrac{1}{6}\boldsymbol{k}^2R^2) \\ \times \mathscr{Y}_1^{L_z}\left(\boldsymbol{p}-\sqrt{\frac{1}{6}}\boldsymbol{k}\right)\mathscr{Y}_1^m\left[\boldsymbol{k}\left(1+\frac{\omega}{6m}\right)+\frac{\omega}{m}\sqrt{\frac{2}{3}}\boldsymbol{p}\right]. \end{aligned} \tag{4.134}$$

It is now possible to extract the partial waves by using the formula

$$\begin{aligned} \mathscr{Y}_{l_1}^{m_1}(\boldsymbol{v})\,\mathscr{Y}_{l_2}^{m_2}(\boldsymbol{v}) \\ = \sum_{l,m}\left[\frac{(2l_1+1)(2l_2+1)}{4\pi(2l+1)}\right]^{1/2}\langle l_1,l_2;m_1,m_2|l,m\rangle \\ \times\langle l_1,l_2;0,0|l,0\rangle\, v^{l_1+l_2-1}\mathscr{Y}_l^m(\boldsymbol{v}). \end{aligned} \tag{4.135}$$

We use this formula for $l_1=l_2=1$ and $\boldsymbol{v}=\boldsymbol{k}$ or $\boldsymbol{p}$. l represents the partial-wave angular momentum. From the Clebsch–Gordan coefficient $\langle 1,1;0,0|0,0\rangle=0$, we immediately conclude, as before, that $l=0$, 2. Furthermore, because of the integration over $\boldsymbol{p}$ with a spherically symmetric function $\exp(-\boldsymbol{p}^2R^2)$, $\mathscr{Y}_1^{L_z}(\boldsymbol{p})$ will give zero. Therefore the $l=2$ contribution is given by

$$M_2 \propto -\exp(-\tfrac{1}{6}k^2R^2)\left(1+\frac{\omega}{6m}\right)\mathscr{Y}_2^0(\boldsymbol{k}). \tag{4.136}$$

The respective contributions of the direct and recoil terms are clearly distinguished here, the latter being characterized by the presence of the factor ω. For $l=2$ the effect of the recoil term is simply to multiply the direct-term contribution by a factor close to 1. In contrast, for $l=0$ we find

$$M_0 \propto -\exp(-\tfrac{1}{6}k^2R^2)\left[\left(1+\frac{\omega}{6m}\right)k^2-\frac{3\omega}{mR^2}\right]\mathscr{Y}_2^0(\boldsymbol{k}). \tag{4.137}$$

Here the dominant contribution within the brackets is the term $3\omega/mR^2$, which comes from the recoil term and has a *sign opposite* to the direct contribution. These conclusions hold more generally for the upper and lower waves.

For the QPC model we cannot make such a natural distinction between direct and recoil terms. However, in (4.130) we can write

$$\boldsymbol{p}_3' = \tfrac{1}{2}[-\boldsymbol{k}+(\boldsymbol{p}_3+\boldsymbol{p}_3')], \tag{4.138}$$

and say that this corresponds to a combination of direct and recoil terms, with the substitution $\omega/2m \to 1$. In (4.136) and (4.137) this corresponds to the replacements

$$\left.\begin{aligned} 1+\frac{\omega}{6m} &\to \frac{4}{3}, \\ \left(1+\frac{\omega}{6m}\right)k^2-\frac{3\omega}{mR^2} &\to \frac{4}{3}k^2-\frac{6}{R^2}, \end{aligned}\right\} \tag{4.139}$$

if we neglect the effect of the pion wave function $\psi_0(2\boldsymbol{p}_3-\boldsymbol{k})$. With this assumption, it is true that, with respect to a simple $\boldsymbol{\sigma}\cdot\boldsymbol{k}$ interaction, the upper wave is only changed by a positive factor close to 1, while the lower wave is dominated by the contribution of $\boldsymbol{p}_3+\boldsymbol{p}_3'$, which has the opposite sign. We shall study the phenomenological implications of these features in Section 4.4.4.

4.4.3 The Decays of the Δ Trajectory into Nπ

The baryons show very striking towers of excitations, corresponding to the orbital excitations of the nucleon and the Δ, usually called Regge trajectories. As there are many possible combinations of total quark spin S and total quark orbital angular momentum L, and also several types of SU(6) multiplets, there are many possible trajectories (Dalitz, 1975). A well-established trajectory, which includes many states, is that of the Δ, starting from $J^P=\frac{3}{2}^+$ and ranging up to $\frac{15}{2}^+$. It is easily interpreted as the set of baryons with total quark spin $S=\frac{3}{2}$ and quark orbital angular momentum L, giving a total angular momentum $J=L+\frac{3}{2}$. L is even, as the states belong to the decuplet of a **56** multiplet. For odd L we have a different pattern, corresponding to a **70** supermultiplet, which does not contain a decuplet $S=\frac{3}{2}$.

In the decay of these Δ states into Nπ, only the upper wave $l=L+1$ is present. Therefore it is essentially given by the direct term, and we retain only the $\boldsymbol{\sigma}\cdot\boldsymbol{k}$ interaction. To see that $l=L+1$, we point out that $J=l\pm\frac{1}{2}$ by total-angular-momentum conservation and since $J=L+\frac{3}{2}$ we have $l=L+2$ or $L+1$. Since $l=L\pm1$ owing to parity conservation (4.131), (4.132), only $l=L+1$ is possible.

We now calculate the decay widths. We need the wave functions of the excited Δ. They are a combination of a spatial wave function with pure orbital excitation L and a spin wave function $S=\frac{3}{2}$:

$$\Psi=\sum_m\langle L,\tfrac{3}{2};m,J_z-m|L+\tfrac{3}{2},J_z\rangle\varphi\chi_{3/2}^{J_z-m}\psi_L^m. \tag{4.140}$$

The spatial wave function must be properly symmetrized. For $L\geqslant 6$ there are several orbital wave functions that are totally symmetric for the exchange of spatial coordinates and have the same energy in the harmonic-oscillator potential. This degeneracy has been carefully studied by Dalitz (1977) and Dalitz and Reinders (1979). We choose the one that has the strongest coupling to the Nπ channel, which is, for all L even,

$$\psi_L^m(\boldsymbol{r}_1,\boldsymbol{r}_2,\boldsymbol{r}_3)$$
$$=N_L^S\exp\left(-\frac{\rho^2+\lambda^2}{2R^2}\right)\frac{1}{\sqrt{3}}\left[\mathscr{Y}_L^m(\boldsymbol{\lambda})+\mathscr{Y}_L^m\left(\frac{\sqrt{3}\,\boldsymbol{\rho}-\boldsymbol{\lambda}}{2}\right)+\mathscr{Y}_L^m\left(-\frac{\sqrt{3}\,\boldsymbol{\rho}+\boldsymbol{\lambda}}{2}\right)\right], \tag{4.141}$$

with

$$N_L^S=\left(\frac{1}{3\sqrt{3}}\right)^{1/2}\frac{1}{R^{L+3}}\frac{1}{[(1+2^{L-1})(2L+1)!!]^{1/2}}\frac{2^{L+1/2}}{\pi}. \tag{4.142}$$

Using the usual ground-state wave function of the nucleon, we get the following widths in the elementary-pion-emission model:

$$\Gamma(\Delta_L^*\to\mathrm{N}\pi)=\frac{8}{3\pi}\left(\frac{g_{\pi\mathrm{qq}}}{2m_\mathrm{q}}\right)^2\frac{E_\mathrm{N}}{m_{\Delta^*}}k^3\frac{(L+1)(1+2^{L-1})}{(2L+3)!!}\left(\frac{kR}{\sqrt{6}}\right)^{2L}\exp(-\tfrac{1}{3}k^2R^2). \tag{4.143}$$

For $L=0$, (4.143) gives the width of the $\Delta(1236)$:

$$\Gamma(\Delta\to\mathrm{N}\pi)=\frac{4}{3\pi}\left(\frac{g_{\pi\mathrm{qq}}}{2m_\mathrm{q}}\right)^2\frac{E_\mathrm{N}}{m_\Delta}k^3. \tag{4.144}$$

Dalitz (1977) and Dalitz and Reinders (1979) have calculated the decay widths of the Δ trajectory. They use a spatial wave function that is somewhat different from (4.141) and gives for $L \geqslant 6$ and at the same R a value slightly smaller than (4.143). Remember that L is even in this formula. The coupling constant is to be fitted from one of the decays or from $g_{\pi NN}$, according to the relation

$$\frac{g_{\pi NN}}{2m_N} = \frac{5}{3}\frac{g_{\pi qq}}{2m_q}. \qquad (4.145)$$

The factor $\frac{5}{3}$ follows by the same algebra as for the nucleon axial-vector coupling, (4.44), because the operator for pion emission (2.146) has the same symmetry structure as the axial-vector current (2.51b). Taking the value given by this formula (4.145), we have the values given in Table 4.3 with $R^2 = 6\,\text{GeV}^{-2}$ (see Section 4.14). We see that $\Delta^* \to N\pi$ is too small. The calculations by Dalitz and Dalitz and Reinders use a larger radius ($R^2 = 11\,\text{GeV}^{-2}$) and they find theoretical widths in fair agreement with experiment. Note that our radius of $R^2 = 6\,\text{GeV}^{-2}$ was fitted in relation with the lowest baryon states ($L = 0,1$ as in Section 4.14). It is likely that the equivalent harmonic-oscillator radius increases with L as is the case in a linear potential. This is presumably the reason why the widths of high-L states are better described with larger radii.

Table 4.3 Widths of baryonic resonances into Nπ (in MeV). Experimental values are those of Höhler (1979).

	L	Theory	Experiment
$P_{3\,3}(1232)$	0	70	116
$F_{3\,7}(1913)$	2	27	85
$H_{3\,11}(2420)$	4	4.2	30
$K_{3\,15}(2950)$	6	1.2	15

4.4.4 Examples of Pion Emission With Two Partial Waves: $N^* \to \Delta\pi$, $B \to \omega\pi$

It is easy to see that any decay into two ground-state hadrons of spin $\frac{1}{2}^+$ or 0^- like $N^* \to N\pi$ has only one allowed partial wave among $l = L \pm 1$, (4.132). For a two-pseudoscalar final state this is obvious, because $l = J$. For a decay into $\frac{1}{2}^+ + 0^-$, $J = l \pm \frac{1}{2}$, which shows that the two possible values of l differ by one unit, while, according to $l = L \pm 1$, they differ by two units; therefore only one l can be present.

For two partial waves to be present (still restricting ourselves to pion emission, the possible channels are ρπ and ωπ for mesons, and Δπ for baryons (we exclude strange quarks for simplicity). For example, the meson B ($S=0$, $L=1$, $J^{PC}=1^{+-}$) decays into $\omega\pi$ through $l=0,2$ (S, D waves), and the excited baryon D_{13} ($L=1$, $S=\frac{1}{2}$, $J^P=\frac{3}{2}^-$) decays into $\Delta\pi$ through $l=0,2$ (S,D waves). Such decays where two partial waves are present are especially interesting, because the *relative sign* of these two waves can be predicted, and this relative sign is measurable.

To formulate this prediction, we note first that, if only the direct term $\boldsymbol{\sigma}\cdot\boldsymbol{k}$ were present, the amplitude relative signs would be completely determined by the Clebsch–Gordan coefficients of SU(6) × O(3), and would not depend on unknown spatial matrix elements. In fact, the dependence on quantities not determined by group-theoretical considerations would be reduced to a single common factor for all the amplitudes. The relative signs corresponding to the $\boldsymbol{\sigma}\cdot\boldsymbol{k}$ interaction are called *SU(6)$_W$ signs*, as this is the prediction of the symmetry scheme called SU(6)$_W$ already discussed in Section 4.1.4. As we have emphasized, this symmetry SU(6)$_W$ amounts to neglecting the transverse quark motion relative to the line of flight of the final hadrons.

Now, when we introduce the recoil term, and therefore the *transverse* quark motion, the remarkable fact is that the relative sign of the waves $l=L\pm 1$ is opposite to the SU(6)$_W$ sign and is therefore called *anti-SU(6)$_W$ sign*. The reason can be seen from the expressions (4.136) and (4.137) for the case $L=1$. In the upper wave $l=2$, the direct and recoil terms contribute with the same sign, while for $l=0$ the contributions are opposite, and that of the recoil term is dominant. Indeed, $-3\omega/mR^2$ is always greater in magnitude than $(1+\omega/6m)k^2$.

Of course this is not a general theorem, but rather the result of a concrete balance between two independent contributions: ω and k are taken at their physical value and are not simply connected with mR^2. Let us take the ideal SU(6)-symmetric case and $m_\pi=0$, whence $\omega=k=1/mR^2$. Even then, the balance of the two terms will depend on the sign of the dimensionless number $-2+1/6m^2R^2$, which, from our estimates (4.12) and (4.14), is definitely negative. Moreover, although the coefficients are model-dependent, the conclusion is still the same for the QPC model, as we have seen in the explicit calculation of Section 2.2.5. This prediction of *anti-SU(6)$_W$ signs* is experimentally successful for the well-known (**70**, 1^-) baryon multiplet decaying into $\Delta\pi$ (Le Yaouanc *et al.*, 1975b). In contrast, the best known $\Delta\pi$ decay in the (**56**, 2^+) multiplet, $F_{15}\to\Delta\pi$, does not satisfy the anti-SU(6)$_W$ prediction. There is no explanation for this isolated failure (see the discussion at the end of this section).

For mesons, the only well-known case is $B\to\omega\pi$, which gives a confirmation of the theory (see the QPC calculation in Section 2.2.3(e)). The B meson is a $J^{PC}=1^{+-}$, $L=1$, $S=0$, $q\bar{q}$ state. The two waves are $l=0,2$. In standard notation, the SU(6)$_W$-like ratio corresponding to a $\boldsymbol{\sigma}\cdot\boldsymbol{k}$ interaction would be

$D/S = -\sqrt{2}$, and experimentally it is $+0.3$ (see Particle Data Group tables). In this reaction, the presence of the recoil term is easily seen in another way, by considering helicity amplitudes. With the direct term only, the helicity of the ω in the B rest frame should be $\lambda_\omega = 0$. Indeed, the quark spin inside the B and along the π momentum is $S_z = 0$. Since the direct term is a σ_z operator, $S_z = 0$ also inside the ω, and since $L=0$ for the ω, this implies $J_z(\omega) = \lambda_\omega = 0$. In contrast the recoil term contains transverse operators $\sigma_\pm$, allowing $J_z(\omega) = \pm 1$. Experimentally, the transverse helicity amplitude $M[\mathrm{B} \to \omega(\lambda_\omega = \pm 1)\pi]$ is as important as the longitudinal one, $M[\mathrm{B} \to \omega(\lambda_\omega = 0)\pi]$.

The main conclusion from this analysis is that it is crucial to take into account the recoil term, which plays a dominant role in the lower wave.

Let us now discuss the v/c orders of the matrix elements. By setting $m_\pi = 0$, $k = \omega$ is of second order in v/c, and, in general, as operators, the direct and recoil terms are respectively of orders two and three. However, on taking matrix elements, these formal orders of the operators may be changed into higher orders. Indeed, this is what happens for the lower $(l = L-1)$ wave. For a $L = 1$ decay the direct-term matrix element is really of order three in v/c and not of order two, because the transition form factor, which is exactly factorized, is of order one in v/c $(\boldsymbol{k} \cdot \boldsymbol{r} = O(v/c))$. Thus the two terms are contributing to the same (third) order, and neither is in principle negligible: it is necessary to take the recoil term into account even in the lowest-order approximation, and this justifies the results presented in this section concerning the $L = 1$ decays.

The same line of argument is to be followed for $L = 2$ decays. But here the transition form factor that is factorized in the direct term contribution is of order two in $v/c\,((\boldsymbol{k} \cdot \boldsymbol{r})^2 = O(v^2/c^2))$. Therefore this contribution is of order *four*, and it is not only necessary to take into account the recoil term, but also terms of order four in the expansion of the operator. But these terms of order four are not included in the previous calculations, and there is some inconsistency for the $L = 2$ decays.

Could this be the source of the failure of the model for $F_{15} \to \Delta\pi$? It is difficult to calculate such fourth-order terms according to the complete method presented in Section 3.3. It is easier to use the more tractable Dirac equation, which may be reasonable for baryons; in this framework, we have calculated the second-order terms in the axial current, which correspond to the fourth-order terms in the pion-emission operator. The result is given by (3.51). The calculation of the matrix element in the harmonic oscillator shows that the amplitude signs are still predicted to be anti-$SU(6)_W$ (Gavela *et al.*, 1982c), in contradiction with experiment. However, as exemplified by the radiative decay of the Roper resonance, this type of calculation is very sensitive to certain approximations, especially to the choice of k, which may be considered as either the harmonic-oscillator theoretical value $k = 2/mR^2$ or the physical value. The calculation should be done with a proper treatment of the center-of-mass motion and with more realistic two-body potentials. A point to realize is that this discussion about the v/c expansion is only possible for elementary-emission models. For the QPC model, we do not yet have any derivation that would allow discussion of corrections.

4.4.5 $N^* \to \pi\pi N$ in the Quark-Pair-Creation Model

For the processes considered up to now, there is no essential difference between the elementary-emission and the QPC models. The first important difference arises when one has to consider, in addition to pion emission, channels involving the emission of other mesons like the vector mesons or possibly the orbital excitations, like the 0^+ ε meson. As noted before, if one insists on considering elementary emission models, one has to devise a model with new parameters for each type of mesons. In contrast a fully composite model like the QPC model will predict all the possible decay channels. One particularly interesting application of this feature of the QPC model is the prediction of all the relative signs of the various two-body decay channels in the process $N^* \to N\pi\pi$, i.e. not only $N^* \to \Delta\pi$, but also Nρ, Nε, etc. We have presented in Section 2.2.5 the full formalism that allows treatment of all these decays and the general features of the model: centrifugal barrier factors and anti-$SU(6)_W$ signs. We now proceed to a phenomenological discussion of the results (Le Yaouanc *et al.*, 1975b).

Considering first $N^* \to N\rho$, we note that such decays, like $N^* \to \Delta\pi$, are in general decays with two partial waves $l = L \pm 1$. The difference is that in addition there are two possibilities, $S = \frac{1}{2}$ or $S = \frac{3}{2}$, for the total Nρ spin $\boldsymbol{S} = \boldsymbol{J}_N + \boldsymbol{J}_\rho$. The amplitudes will thus be labelled $N\rho_1$ or $N\rho_3$.

However, in spite of this difference, the treatment of Nρ in the QPC model quite parallels that of Δπ and leads to the same conclusion of anti-$SU(6)_W$ signs for the partial waves. Indeed, the additional spin label S concerns only the quark spin ($S = \frac{1}{2}$ for the nucleon, $S = 1$ for the vector mesons) and not the orbital angular momentum. But the orbital-angular-momentum structure is given by (2.251)–(2.254) exactly as for a Δπ decay, since they are channels of two ground-state hadrons of zero quark orbital angular momentum.

We drop constants, normalization factors and omit SU(3) and quark-spin quantum numbers in the spirit of the expression given in (4.133) for pion elementary emission. We then have for the decay amplitude of $L = 1$ baryons into any ground-state baryon plus ground-state meson, in the QPC model,

$$M \propto \int \mathrm{d}\boldsymbol{p}_\rho \, \mathrm{d}\boldsymbol{p}_\lambda \exp\left[-\tfrac{1}{2}(\boldsymbol{p}_\rho^2 + \boldsymbol{p}_\lambda^2)R^2\right]$$

$$\times \mathscr{Y}_1^{L_z}(\boldsymbol{p}_\lambda)\, \mathscr{Y}_1^{m}(\boldsymbol{k} - \boldsymbol{p}_3)\, \psi_0(\boldsymbol{k} - 2\boldsymbol{p}_3) \exp\left[-\tfrac{1}{2}(\boldsymbol{p}_\rho^2 + \boldsymbol{p}_\lambda'^2)R^2\right], \quad (4.146)$$

where ψ_0 is the meson ground state spatial wave function, and $\boldsymbol{p}'_\lambda = \boldsymbol{p}_\lambda + (\frac{2}{3})^{1/2}\boldsymbol{k}$.

It is possible to extract from this formula the partial waves M_0 and M_2 by a procedure entirely similar to that followed in Section 4.4.2. The SU(6)$_W$-like relative signs are obtained by making the substitution

$$\mathscr{Y}_1^m(\boldsymbol{k}-\boldsymbol{p}_3) \rightarrow \mathscr{Y}_1^m(\boldsymbol{k}), \tag{4.147}$$

i.e. by neglecting the term in the transition operator that explicitly depends on the quark momentum. In this case the spatial dependence is factorized entirely into one spatial integral, and the amplitude signs are entirely derived from knowledge of the SU(6) Clebsch–Gordan coefficients. When calculating with the correct expression (4.146), it is found, just as for the pion emission examined in Section 4.4.2, that the relative sign M_2/M_0 is exactly opposite—whence the *anti-SU(6)$_W$ signs* in the prediction of the QPC model. This general feature of the model is made explicit by (2.254) appropriately modified for application to mesons (Sections 2.2.5(b) and (c)).

This prediction of the QPC model for Nρ signs is not shared by elementary-emission models, in contrast with Δπ. Indeed, taking the natural expression for the emission of an elementary vector meson, in analogy with electromagnetic emission (Moorhouse and Parsons, 1973), leads to two terms, the convective and the magnetic terms, for which the spatial integral analogous to (4.146) has a different structure. Dropping as before spin and SU(3) quantum numbers and retaining only the space variables, we have

$$\begin{aligned} M^{\text{conv}} \propto & \int \mathrm{d}\boldsymbol{p}_\rho \, \mathrm{d}\boldsymbol{p}_\lambda \exp[-\tfrac{1}{2}(\boldsymbol{p}_\rho^2+\boldsymbol{p}_\lambda^2)R^2] \\ & \times \mathscr{Y}_1^{L_z}(\boldsymbol{p}_\lambda)\mathscr{Y}_1^m\left(\frac{\boldsymbol{p}_3+\boldsymbol{p}'_3}{2m}\right)\exp[-\tfrac{1}{2}(\boldsymbol{p}_\rho^2+\boldsymbol{p}'^2_\lambda)R^2], \end{aligned} \tag{4.148}$$

$$\begin{aligned} M^{\text{mag}} \propto & \int \mathrm{d}\boldsymbol{p}_\rho \, \mathrm{d}\boldsymbol{p}_\lambda \exp[-\tfrac{1}{2}(\boldsymbol{p}_\rho^2+\boldsymbol{p}_\lambda^2)R^2] \\ & \times \mathscr{Y}_1^{L_z}(\boldsymbol{p}_\lambda)\exp[-\tfrac{1}{2}(\boldsymbol{p}_\rho^2+\boldsymbol{p}'^2_\lambda)R^2], \end{aligned} \tag{4.149}$$

where, again $\boldsymbol{p}'_3 = \boldsymbol{p}_3 - \boldsymbol{k}$ and $\boldsymbol{p}'_\lambda = \boldsymbol{p}_\lambda + (\frac{2}{3})^{1/2}\boldsymbol{k}$. There is therefore no general reason to have anti-SU(6)$_W$ signs: if M^{mag} dominates, the signs will be SU(6)$_W$-like, and if M^{conv} dominates, they will be anti-SU(6)$_W$. In the case of the decay of a resonance with total quark spin $S=\frac{3}{2}$ into Nρ, since the nucleon has total spin $S=\frac{1}{2}$, there is a contribution from M^{mag} only. Therefore the elementary-emission model predicts SU(6)$_W$ signs, in contrast with the QPC model. This is the case, for example, for $F_{35} \rightarrow N\rho_3$. Experiment confirms the anti-SU(6)$_W$ prediction of the QPC model. It may be that the elementary-emission-model predictions are changed when higher-order corrections (e.g. spin–orbit contributions) are taken into account. Up to now, no concrete calculations have been made.

As to the simultaneous prediction of all the quasi-two-body channels in N* →Nππ, Δπ, Nρ, Nε, which is made by the QPC model, the following situation holds: the relative signs of these channels are well described, except for Nε, where there is an almost complete failure, for which no explanation is known (though one can invoke the unclear status of the 0^+ mesons). However, on the whole, the agreement of theory and experiment is rather good (Table 4.4).

Table 4.4 Theoretical (upper) versus experimental (lower) coupling signs of baryonic resonances in πN→πΔ, ρN; a 0 means a small experimental amplitude with undetermined sign. A ? signals an uncalculated theoretical sign or an uncertain experimental sign. Experimental data is from Manley *et al.* (1984).

		Δπ		$N\rho_1$		$N\rho_3$	
		$l=L-1$	$l=L+1$	$l=L-1$	$l=L+1$	$l=L-1$	$l=L+1$
	$S_{11}(1535)$		+/0	+/−			+/0
	$S_{11}(1650)$		+/+	−/−			−/+?
	$D_{13}(1520)$	−/−	−/−		+/0	−/−	−/0
$(\mathbf{70},1^-)$	$D_{13}(1700)$	−/−?	+/+		−/0	+/0	+/0
	$D_{15}(1675)$		+/+		−/0		−/−
	$S_{31}(1620)$		−/−	+/+			+/−
	$D_{33}(1700)$	+/+	+/+		+/0	+/+	?/0
	$P_{13}(1720)$	−/0	−/0	+/0		+/0	+/0
$(\mathbf{56},2^+)$	$F_{15}(1680)$	+/−	+/+		+/0	−/−	−/−
	$F_{35}(1905)$	−/+	+/+		−/0	+/+	−/0
	$F_{37}(1950)$		+/+		−/0		−/0

For the absolute magnitudes the agreement is undoubtedly rough. The widths can disagree by factors of two or more. However, statistically, if we plot the ratio between the theoretical and experimental numbers, most of the points lie within a factor of two in excess or defect. If we take averages, we find striking agreements. For instance, the ratio of decay amplitudes to Δπ from positive-parity ($L=2$) baryons to negative-parity ($L=1$) baryons is correct on average to within 10%. We must emphasize that the same average pair-creation constant $\gamma \approx 3$ works for baryons and mesons, and, as we shall see, the same value also works for charmonium decays. It is everywhere understood here that this value is used in the calculation of absolute magnitudes.

4.4.6 The Decays of Radial Excitations

Up to now we have examined only decays of purely orbital excitations. In that case, the discussion essentially involves the angular momenta L and S and the mean-square momentum $\langle p^2 \rangle \sim 1/R^2$ and not the details of spatial wave functions. When radial excitations are considered, new features arise because there are nodes in the radial wave function. In any explicit quark model the amplitude is expressed as a kind of convolution product of spatial wave functions, and the nodes in the decaying-state wave function induce *zeros in the amplitude as a function of k*—which may be considered as an independent variable if the energy-conservation constraint is relaxed. Then, according to the position of these zeros with respect to the physical values $k = k_{\mathrm{phys}}$, as determined by energy conservation, there is either an amplitude comparable in magnitude to that of a purely orbital excitation or, near to a zero, a strong suppression. We thus have a possible striking consequence of the nodal structure of radial excitations.

Moreover, it is found that the position of the amplitude zeros is strongly model-dependent. The expression for the amplitude is different for the elementary-emission model and for models in which the emitted meson is considered as composite (QPC and Cornell models). In this latter case the decay amplitude is proportional to a spatial overlap expressed as a convolution product that also contains the wave function of the emitted meson. In contrast, the elementary-emission model involves only the overlap of the two systems considered as composite. The decays of radial excitations give a test not only of the wave functions, but also of the emission model. In practice, however, the test is hampered by two things. First, the positions of the nodes of the radial wave function are rather sensitive to the potential. The harmonic approximation may, for example, give bad results. Only potentials that fit the mass spectrum give reasonable values for the nodes. Secondly the zeros of the amplitude are sensitive to higher-order corrections.

The decays with only one independent amplitude are especially interesting, because there is only one spatial integral, and we can therefore isolate the zeros due to the radial wave functions from other possible cancellation effects. As examples, we shall consider the ρ' decay into $\pi\pi$, the decays of excited charmonium states above $D\bar{D}$ threshold and the Roper resonance $N^*(1440)$ decays (this resonance is also often denoted by $P_{11}(1470)$).

4.4.6(a) ρ′(1600) decays

To illustrate the nature of the effects that are encountered in the decays of radial excitations, let us first consider the decays of ρ′(1600), a $J^{PC}=1^{--}$ state that is naturally classified as the first radial excitation in the light $q\bar{q}$ system (if we exclude a possible lower radial excitation). Possible SU(6) partners of this state are the π′(1300) and ϕ′(1680). The Particle Data Group tables reveal the disturbing feature that ρ′ decays mostly into 4π(∼60%) and that the 2π mode, strongly favored by phase space, is suppressed (∼20%).

The nodes present in the ρ′ wave function, plus the form of the amplitude in the QPC or Cornell models, as the convolution product of the spatial wave functions of all hadrons involved in the decay, provide an explanation of this striking feature. Leaving aside detailed experimental features of the decay $\rho'\to 4\pi$, we shall show that the model naturally gives a suppression (Kaufman and Jacob, 1974)

$$\Gamma(\rho'\to 2\pi) \ll \Gamma(\rho'\to \rho\rho),\ \Gamma(\rho'\to \omega\pi). \tag{4.150}$$

In a different model, this has also been emphasized by Böhm *et al.* (1973a,b). To see this qualitatively, we shall use the harmonic-oscillator potential. As we have seen in Section 2.2.3(a), the QPC model for a decay of a meson A→B+C expresses the spatial part of the amplitude in the form

$$\int\prod_i d\boldsymbol{p}_i\, \psi_B(\boldsymbol{p}_1-\boldsymbol{p}_3)^*\psi_C(\boldsymbol{p}_4-\boldsymbol{p}_2)^*\psi_A(\boldsymbol{p}_1-\boldsymbol{p}_2)$$
$$\times\, \mathscr{Y}_1^m(\boldsymbol{p}_4-\boldsymbol{p}_3)\delta(\boldsymbol{p}_1+\boldsymbol{p}_2)\delta(\boldsymbol{p}_1+\boldsymbol{p}_3-\boldsymbol{k}_B)\delta(\boldsymbol{p}_2+\boldsymbol{p}_4-\boldsymbol{k}_C). \tag{4.151}$$

We see here a convolution product of the three spatial wave functions involved with the P-wave solid spherical harmonic of the created pair. For the decay $\rho\to 2\pi$, where all hadrons are the ground state, the wave functions are all equal to a Gaussian:

$$\psi_0(\boldsymbol{p}) = N_0 \exp\left(-\tfrac{1}{8}R'^2\boldsymbol{p}^2\right), \tag{4.152}$$

where R'^2 is the meson radius (4.3) and $\boldsymbol{p}=\boldsymbol{p}_q-\boldsymbol{p}_{\bar{q}}$.

In contrast, for the decay $\rho'\to\pi\pi$, ρρ, ωπ, the wave function of the initial state, corresponding to the first radial excitation, is

$$\psi_1(\boldsymbol{p}) = \sqrt{\tfrac{2}{3}}\,N_0\,L_1^{1/2}\left(\tfrac{1}{4}R'^2\boldsymbol{p}^2\right)\exp\left(-\tfrac{1}{8}R'^2\boldsymbol{p}^2\right), \tag{4.153}$$

where

$$L_1^{\alpha}(x) = (1+\alpha) - x \tag{4.154}$$

is a Laguerre polynomial. We see that the wave function (4.153) has a zero at $R'^2\boldsymbol{p}^2=6$ or, with $R'^2=12\,\mathrm{GeV}^{-2}$ (4.15), at $\boldsymbol{p}^2=0.5\,\mathrm{GeV}^2$. From these wave functions we get for the rates of two-body decays the following k dependence

$$\Gamma \propto Ck^{2l}\rho(k)F(\beta), \tag{4.155}$$

where C is an algebraic spin–flavor factor and $\rho(k)$ is the phase-space factor (2.31a). $F(\beta)k^{2l}$ comes from the square of the spatial overlap (4.151) ($\beta = \frac{1}{12}R'^2k^2$). For example, for $\rho \to \pi\pi$ we obtain $C=1$, and

$$F(\beta) = e^{-2\beta}, \tag{4.156}$$

and for $\rho' \to \pi\pi$, $\omega\pi$, $\rho\rho$ we obtain $C=1, 2, 7$ respectively and

$$F(\beta) = \frac{2}{27}[L_1^{3/2}(4\beta)]^2 e^{-2\beta}, \tag{4.157}$$

or, using (4.154),

$$F(\beta) = \frac{2}{27}\left(\frac{5}{2} - 4\beta\right)^2 e^{-2\beta}. \tag{4.158}$$

We see that this expression has a zero at $\beta = \frac{5}{8}$, or $R'^2k^2 = \frac{15}{2}$. It is clear that the value of k will be widely different for the various modes $\rho' \to \pi\pi$, $\omega\pi$, $\rho\rho$, decreasing in this order. In fact, for $R'^2 = 12\,\mathrm{GeV}^{-2}$, we find that the mode $\rho' \to \pi\pi$ is completely suppressed, having $R'^2k^2 \approx \frac{15}{2}$. We find

$$\Gamma(\rho' \to 2\pi) \ll \Gamma(\rho' \to \omega\pi) < \Gamma(\rho' \to \rho\rho). \tag{4.159}$$

We do not want to go beyond this qualitative statement, because the actual detailed structure of the 4π modes of ρ' (the ratio of $\pi^+\pi^-\pi^+\pi^-$ to $\pi^+\pi^-\pi^0\pi^0$) is poorly understood (Le Yaouanc *et al.*, 1978a; Busetto and Oliver, 1983). Finally, we must emphasize that in elementary-emission models the zeros of the amplitude $M(k)$ are located at rather different places from those in the QPC model.

4.4.6(b) Charmonium decays above the $D\bar{D}$ threshold

We now turn to the OZI-allowed decays of radial excitations of charmonium, i.e. above the $D\bar{D}$ threshold. The QPC model seems particularly natural for several reasons. First, the situation really does become nonrelativistic because the decay products $D\bar{D}$, $D\bar{D}^* + \bar{D}D^*$, $D^*\bar{D}^*$ are all heavy, with the sum of masses not far from that of the decaying particle $\psi(4.030)$, $\psi(4.160)$, etc. This is at least true for the lowest excitations, for which there exists a detailed analysis of exclusive decays. Secondly, the internal velocities of the heavy quarks are smaller. Finally, it seems absolutely necessary to treat the two decay products on the same footing; since no difference of status can be reasonably introduced between them, the elementary-emission model consequently loses its main justifications, while the quark-pair-creation model offers straightforwardly a symmetric treatment of the two charmed mesons. It is important to be aware that there are now quarks with different masses and that the treatment outlined before for $\rho'(1600)$ must be accordingly modified.

In the frame of the model, we still need to choose the potential. In fact, the harmonic-oscillator model describes the *qualitative* features very well, and the picture was first developed with this potential (Le Yaouanc *et al.*, 1977a). However, we do not think that the harmonic oscillator is sufficient; to faithfully describe the nodes of radial excitations like the second radial excitation $\psi(4.030)$, we must use a potential that is realistic as regards the mass spectrum. We do not want at the moment to be involved in a critical discussion of the numerous charmonium potentials. This is more crucial for annihilation processes and will be considered in Section 4.5. We shall simply quote the results in Ono (1981a,b), who has made a systematic study of charmonia and bottonia decays, and some results of the Cornell model (Eichten *et al.*, 1978), which, however, cannot be easily compared, because they include coupled-channel corrections.

What is to be explained? The first radial excitation $\psi(3.68)$ is below threshold and decays strongly via OZI-suppressed decays. Above threshold, the simplest process is the decay of the $L=2$ orbital excitation, $\psi(3.77)\to D\bar{D}$, which gives the total rate. The $\psi(4.03)$ presents an outstanding case, because it is the second radial excitation 3S, whose decays into $D\bar{D}$, $D\bar{D}^* + \bar{D}D^*$, $D^*\bar{D}^*$ have been measured. These decay modes give a crucial and extensive test of the QPC model: a nontrivial spatial wave function, a large variety of spin structures of the final states, and several possible values of k, allowing a test of the predicted dependence of the spatial matrix element on k. Next, the $\psi(4.16)$ assignment is controversial (Ono, 1983) because the usual assignment as a 2D orbital–radial excitation gives a branching into e^+e^- that is much too small. The $\psi(4.41)$ has a safer assignment as a 4S radial excitation, but no detailed analysis of decay products is available. Therefore the only test consists in predicting the total width.

It is found that the QPC or the Cornell models, which seem practically equivalent in this area, work strikingly well. First, it is verified that, with the same $\gamma \approx 3$ as used for light quarks, and with the potential and masses of Ono (1981), the $\psi''(3.77)$ width is in perfect agreement with experiment:

$$\Gamma(\psi''(3.77)\to D\bar{D}) \approx 22\,\text{MeV}, \tag{4.160}$$

versus an experimental 25 ± 3 MeV. The Cornell model, where the transition amplitude is completely determined by the potential and the masses, gives

$$\Gamma(\psi''(3.77)\to D\bar{D}) \approx 30\,\text{MeV}, \tag{4.161}$$

It must be noted that the Cornell result includes coupled-channel effects. The achievement is rather impressive, considering that the magnitude is entirely deduced from parameters of spectroscopy.

For the decays $\psi(4.03) \to D\bar{D}$, $D\bar{D}^* + \bar{D}D^*$, $D^*\bar{D}^*$, the main phenomenon to be explained is the rate of these transitions relative to the naive expectations (analogous to the SU(6) semialgebraic predictions discussed in Section 4.1.4), which take into account the $l=1$ (P-wave) centrifugal barrier, the different spin counting (SC) for each mode, and the phase space. Let us define

$$r = \Gamma / SPk^2, \tag{4.162}$$

where $S=1$, 4, 7 is the spin-counting coefficient respectively for $D\bar{D}$, $D\bar{D}^* + \bar{D}D^*$ and $D^*\bar{D}^*$, $k^{2l} = k^2$ $(l=1)$ stands for P-wave decay, and P is the phase-space factor (2.31a). The naive expectation is then

$$r = \text{constant}. \tag{4.163}$$

Empirically, the ratio r is, for the different modes,

$$r(D\bar{D}) : r(D\bar{D}^* + \bar{D}D^*) : r(D^*\bar{D}^*) = 1 : 5 : 100. \tag{4.164}$$

This corresponds to the experimental width ratios

$$\Gamma(D\bar{D}) : \Gamma(D\bar{D}^* + \bar{D}D^*) : \Gamma(D^*\bar{D}^*) = 1 : 7 : 9, \tag{4.165}$$

versus the spin-counting expectation

$$\Gamma(D\bar{D}) : \Gamma(D\bar{D}^* + \bar{D}D^*) : \Gamma(D^*\bar{D}^*) = 1 : 1.56 : 0.11, \tag{4.166}$$

We see that the data strongly contradict the naive expectation.

In view of our discussion of the $\rho'(1600)$ decays, we can suspect that this is a phenomenon linked to the nodes of the radial excitation $\psi(4.03)$. Here again the harmonic oscillator gives a very nice qualitative illustration of the phenomenon. Being a 3S state, the spatial wave function of $\psi(4.03)$ will be

$$\psi(r) = N L_2^{1/2}\left(\frac{r^2}{R'^2}\right) \exp\left(-\frac{r^2}{2R'^2}\right) \tag{4.167}$$

where

$$L_2^{1/2}(x) = \frac{15}{8} - \frac{5}{2}x + \frac{1}{2}x^2. \tag{4.168}$$

We assume, for illustrative purposes only, that the quark masses are all equal, as well as the meson radii. We obtain a theoretical ratio r (see (4.162)) that, far from being constant, is proportional to a widely varying function of k, analogous to (4.158):

$$r \propto |F(k)|^2 = \left(\frac{35}{12} - \frac{7}{9}R'^2k^2 + \frac{1}{27}R'^4k^4\right)^2 \exp\left(-\tfrac{1}{6}R'^2k^2\right). \tag{4.169}$$

The zeros in the polynomial (4.169) are the reflections, in the decay amplitude, of the zeros of the wave function (4.167) of the decaying system and explain qualitatively the ratios (4.164) (Le Yaouanc *et al.*, 1977a).

Extensive calculations have been performed by Ono (1981a), using realistic quark masses and potential models that fit the hadron masses. The dependence on k is factorized into one single function, the same for the three channels; this function has very different values in the three cases. In Figure 4.1 we plot the function $M(k)$ proportional to the amplitude with spin factors dropped, but without removing the centrifugal barrier factor k^l, $l=1$, unlike in expressions (4.162) and (4.169). We see that the three physical k are located in very different regions of the curve. In particular, the k corresponding to $D\bar{D}$ is located near a zero, which explains the very small value of $\psi(4.03)\to D\bar{D}$.

We emphasize again that the predictions are completely different from what could be expected from the naive phenomenological treatment (4.163), which has been otherwise successful. The crucial disagreement lies in the k dependence. The k^{2l} dependence, often used in phenomenology, is the so-called *centrifugal-barrier* factor. Although for small k the behavior of the amplitude is indeed $|M(k)|^2 \sim k^{2l}$, it is on the whole very different as soon as a radial excitation is considered instead of a pure orbital excitation (Figure 4.1). In fact, k^{2l} is multiplied by an oscillating and decreasing factor $|F(k)|^2$, like (4.169). The first striking effect of this behavior of $M(k)$ concerns the *ratios* between different decays having the same partial wave l. They are completely different from the naive expectation (4.163). The second effect is that for the typical values of k corresponding to the decay of radial excitations like the $\psi(4.028)$ or $\psi(4.414)$, $|M(k)|^2$ is *much smaller* than what would be indicated by the extrapolation from the behavior at the origin $|M(k)|^2 \sim k^{2l}$, as clearly seen in Figure 4.1 for $\psi(4.028)$. This explains that, although k is already quite sizable, especially for $\psi(4.414)$, the width remains quite *narrow* with respect to a ground-state decay with similar k (compare for instance $\psi(4.414)\to D\bar{D}$ with $\rho \to \pi\pi$). The smallness of $|M(k)|^2$ in the *intermediate region* of k is of course due to the nodes. $|M(k)|^2$ remains relatively small between two nodes, and the fact that k is located at a node just adds an additional suppression.

With the QPC model it is then found that

$$\Gamma(D\bar{D}),\ \Gamma(D\bar{D}^* + \bar{D}D^*),\ \Gamma(D^*\bar{D}^*) = 2.5,\ 30,\ 35\ \text{MeV} \tag{4.170}$$

and

$$\Gamma_{tot}(\psi(4.03)) \approx 70\ \text{MeV}, \tag{4.171}$$

which are in reasonable agreement with the data quoted above.

The success is remarkable. These ratios are especially interesting, since they test the assumption that the amplitude $M(k)$ is to be calculated at the physical k, taking into account the mass differences between D and D*.

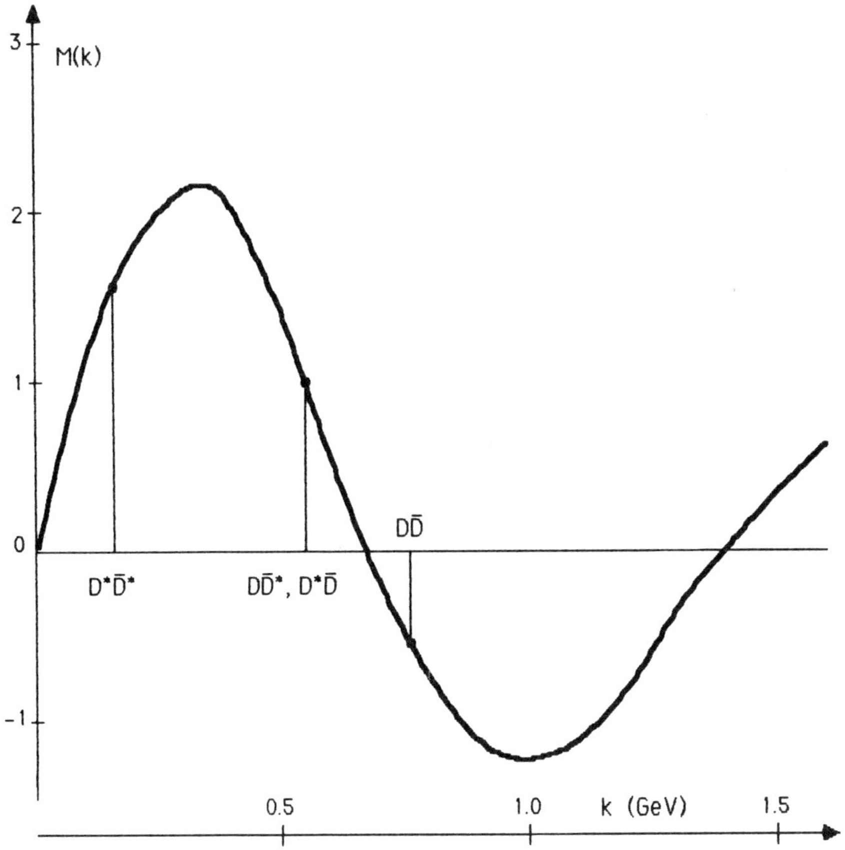

Figure 4.1 A plot of the spatial part $M(k)$ of the QPC amplitude for the decay of the $\psi(4.028)$, from Ono (1983). The node structure originates in the analogous structure of the 3S radial wave function of the decaying state.

Finally the total width of the $\psi(4.414)$ is well explained by the same model, $\Gamma(\psi(4.414)) = 30$ MeV, versus an experimental 43 ± 20 MeV. This width seems surprisingly narrow in view of the many channels opened at such a high mass: $D\bar{D}$, $D\bar{D}^* + \bar{D}D^*$, $D^*\bar{D}^*$, $F\bar{F}$, $F\bar{F}^* + \bar{F}F^*$, $F^*\bar{F}^*$ and possibly channels with P-wave mesons. The explanation is simple however: the nodes of the decay amplitude, considered as a function of k (Le Yaouanc *et al.*, 1977a), imply a general depression of the amplitude for any value of k. We compare the model with experiment in Table 4.5.

Table 4.5 Decay widths in MeV. Theory is model II of Ono (1981). Experimental data are taken from Particle Data Group (1984).

	Theory	Experiment
$\psi(3772)\to D\bar{D}$	22.5	25 ± 3
$\psi(4028)\to D\bar{D}$	2.44	3
$D^*\bar{D}+D\bar{D}^*$	29.9	21.4
$D^*\bar{D}^*$	34.8	27.5
total	67.2	52 ± 10
$\psi(4414)\to$ total	28.5	43 ± 20

4.4.6(c) Roper-resonance decays

Let us now consider a radially excited baryon, the Roper resonance, the $P_{11}(1470)$ baryon state. Although there has been much controversy about the quark-model identification of this resonance, and there are still open problems, we shall support the standard identification as a radial $(\mathbf{56}, 0^+)$ excitation, by showing that it allows a satisfactory description of the transitions. The study of the decays of this resonance will clarify the differences between the elementary and composite models, as the prediction of the positions of the nodes in the decay amplitude varies widely from one another. We first discuss the elementary-pion-emission model for $P_{11}\to N\pi$ decay. The width is

$$\Gamma = \frac{1}{2\pi}\frac{m_N}{m_{N^*}}k^3\frac{1}{f_\pi^2}\sum_\lambda \frac{|\langle N|X|N_\lambda^*\rangle|^2}{2J+1}, \tag{4.172}$$

where we have set $m_\pi = 0$ for the sake of simplicity, $k=\omega_\pi$, and X is given by (2.154):

$$X = \sum_i X(i) = \sum_i \tau^{(+)}(i)\{\sigma_z(i)\,e^{-ikz_i} - \frac{1}{2m}\{\boldsymbol{\sigma}(i)\cdot\boldsymbol{p}_i, e^{-ikz_i}\}] \tag{4.173}$$

for a positively charged pion moving along the Oz direction. We shall expand this expression up to second order in v/c, which is the lowest nonvanishing order for the transition between the ground state and a radial excitation. We shall, however, keep k as a free variable instead of setting $k=\omega_\pi$ immediately, in order to display the zeros in k of the decay amplitude. Furthermore, to simplify the notation, we drop the isospin operator, the index i and the sum over i; we then have

$$\begin{aligned} X' &\equiv \sigma_z e^{-ikz} - \frac{1}{2m}\{\boldsymbol{\sigma}\cdot\boldsymbol{p}, e^{-ikz}\} \\ &\approx \sigma_z - ikz\sigma_z - \frac{\boldsymbol{\sigma}\cdot\boldsymbol{p}}{m} - \frac{1}{2}k^2z^2\sigma_z + \frac{1}{2m}\boldsymbol{\sigma}\cdot\{\boldsymbol{p}, ikz\}. \end{aligned} \tag{4.174}$$

In a transition between $L=0$ states we can drop the transverse operators, and retain only

$$\begin{aligned} X'' &\approx \sigma_z - \left(ikz + \frac{p_z}{m}\right)\sigma_z + \frac{ik}{2}\sigma_z\left[ikz^2 + \frac{1}{m}\{p_z, z\}\right] \\ &= \sigma_z - \left(ikz + \frac{p_z}{m}\right)\sigma_z + \frac{ik}{2}\sigma_z\left\{z, \left(\frac{ikz}{2} + \frac{p_z}{m}\right)\right\}. \end{aligned} \tag{4.175}$$

The first two terms give zero—the first because of the orthogonality of the wave functions, the second because of parity. In the third term we can use the equation of motion to write

$$\left\langle \left\{ z, \frac{p_z}{m} \right\} \right\rangle = \langle -ik_0 z^2 \rangle, \tag{4.176}$$

where $k_0 = E_i - E_f$. We obtain finally the effective operator

$$X''' = -\frac{1}{2}k\sigma_z(k-k_0)z^2. \tag{4.177}$$

It must be noted that k_0 is the theoretical energy difference coming from the Schrödinger equation. For example, for the harmonic-oscillator potential, $k_0 = 2/mR^2$. It is now clear that there is a zero in the decay amplitude at $k=k_0$. In a consistent nonrelativistic approximation, we would set $k=k_0$, and the amplitude would exactly vanish. According to the procedure that we are following, the amplitude will not vanish exactly because of two reasons. First $k = (M_{N^*}^2 - M_N^2)/2M_{N^*}$ in the center of mass of the N* differs from $M_{N^*} - M_N$. Secondly, the potential does not give exactly the physical masses. For example, the harmonic oscillator gives a very high location for the Roper resonance (about 2 GeV). Other improved potentials, increasing more slowly, are unable to predict the low Roper-resonance mass (Hogaasen and Richard, 1983).

A calculation with the harmonic-oscillator model gives

$$\Gamma(P_{11} \to N\pi) \sim 20\,\text{MeV}, \tag{4.178}$$

where $g_{\pi qq}$ has been adjusted to give a correct $g_{\pi NN}$. This is a poor result, the experimental width being about 100 MeV. We have here a signal of failure of the standard elementary-emission model. We expect that a better potential will give different results, because the matrix element of z^2 is sensitive to the position of the node of the radial wave function. However, the qualitative difficulty is not changed, because, on the other hand, $k-k_0$ will be closer to zero.

This difficulty is not an isolated one. The failure is systematic for all the decays of the radial excitations into $N\pi$ and $\Delta\pi$, including the $P_{33}(1690)$, which is the decuplet partner of the Roper resonance in the **56** multiplet, as emphasized by Burkhardt and Pulido (1978). With respect to the experimental values, the predictions of the Feynman, Kislinger and Ravndal (1971) (FKR) model qualitatively similar to the nonrelativistic model, are down by a factor of about 10.

A possible way out of this bad prediction of the elementary-emission would be to consider higher-order relativistic corrections. This is a logical possibility for positive-parity excitations. We have already emphasized for $L=2$ decays that, since we begin with second-order terms in v/c in the matrix element, the expansion of X must be consistently extended up to this order. Usually, only *first-order* operators are kept, except for the expansion of the factor $\exp(-ikz)$. The correct expression for X up to second order has been given in the framework of the Dirac equation in (3.2.3). This expression is valid independently of this equation, since only the free-particle Foldy–Wouthuysen transformation is needed at this order. With respect to (4.177), the term (3.55) has to be added. We retain only the part relevant for $L=0$ transitions, and consider averages over space wave functions,

$$X'^{(2)} = -\frac{1}{2m^2} p_{\mathrm{T}}^2 \sigma_z = -\frac{p_z^2}{m^2} \sigma_z. \tag{4.179}$$

The total effective operator is, therefore,

$$X''' = -\left[k(k-k_0)\frac{z^2}{2} + \frac{p_z^2}{m^2} \right] \sigma_z. \tag{4.180}$$

A simple calculation for the harmonic oscillator gives the following conclusion: the sign of the amplitude is changed relatively to the naive model, but the magnitude is the same, and the width is therefore still small. However, it is not excluded that a more realistic potential may change this situation. We see once more that transitions of radial excitations are sensitive quantities, which could be drastically changed by various types of corrections.

In contrast with the elementary-emission model, the QPC model offers a very natural explanation. We have seen that for decays of orbital excitations the two models are practically equivalent, with a suitable choice of the strength constants $g_{\pi qq}$ and γ. Here the quark-pair-creation model gives definitely different predictions. In fact, the QPC model gives a large $\mathrm{P}_{11} \to \mathrm{N}\pi$ width. The point is that the position of the zero in k in the decay amplitude is markedly different from that found in the elementary-pion-emission model, as it lies far from the physical region. To illustrate this point, we will write the space integral (2.202) with the δ-function removed. Using harmonic-oscillator wave functions, we find

$$M \propto |\boldsymbol{k}| \exp\left[-\frac{12R^2+5R'^2}{3R^2+R'^2}\frac{R^2\boldsymbol{k}^2}{24}\right]$$
$$\times\frac{3}{4}\left[\frac{4R^2+R'^2}{3R^2+R'^2}\frac{R'^2}{R^2}+2\frac{2R^2+R'^2}{3R^2+R'^2}\right.$$
$$\left.-\frac{1}{4}\left(\frac{2R^2+R'^2}{3R^2+R'^2}\right)^2(4R^2+R'^2)\boldsymbol{k}^2\right]. \qquad (4.181)$$

Taking $R'^2=8\,\mathrm{GeV}^{-2}$, which is a lower bound for the meson-radius, we have a zero at $|\boldsymbol{k}|=0.82\,\mathrm{GeV}$, which is indeed far from the physical $|\boldsymbol{k}|=0.41\,\mathrm{GeV}$. This zero is furthermore located at a negative $k^2=k_0^2-\boldsymbol{k}^2$. For $m_\pi=140\,\mathrm{MeV}$ the conclusion drawn for $m_\pi^2=k^2=0$ is strengthened. This zero can be reached only in electroproduction. It seems indeed that such a zero is observed in electroproduction at the right place (M. Davenport, private communication, 1980). We end up with a width of the right order of magnitude:

$$\Gamma(\mathrm{P}_{11}\to\mathrm{N}\pi)=80\,\mathrm{MeV}. \qquad (4.182)$$

The other rates, $\mathrm{P}_{11}(1470)\to\Delta\pi$ and $\mathrm{P}_{33}(1690)\to\mathrm{N}\pi$, $\Delta\pi$ are also in good agreement with the data. The success of the QPC model in contrast with the FKR elementary-emission model is impressive, as we see in Table 4.6 (Gavela *et al.*, 1980a,b).

Table 4.6 Decay widths of the radial excitations N*(1470), Δ*(1690) and N*(1780) into Nπ or Δπ. Data are from Höhler (1979) and Burkhardt and Pulido (1978).

	Nπ			Δπ		
	FKR	QPC	Expt	FKR	QPC	Expt
$\mathrm{P}_{11}(1470)$	8	80	70	0.8	13	20
$\mathrm{P}_{33}(1690)$	0.4	33	30	4	80	80
$\mathrm{P}_{11}(1780)$	?	26	14			

It can be seen that the presence of the pion wave function in the space integral is very important in obtaining such large widths. To illustrate this point, we can compare this with the ideal case $R'^2=0$, which corresponds to a pointlike pion. To make a valid comparison, we must also rescale γ in order to get correct $\Delta\to\mathrm{N}\pi$ or $L=1$ decay widths in the case $R'^2=0$. Having done this rescaling, we find a value close to (4.178), about a factor 4 smaller than (4.182). This clearly confirms that the difference with the elementary-emission model consists essentially in the nonlocal effect of the pion spatial extension.

We do not know how to combine the relativistic corrections with the effects of the spatial extension of the emitted meson, described by the QPC model. In light-quark physics there is no reason why internal velocities or the meson spatial extension should be negligible. The quantitative success of the QPC nonrelativistic calculations therefore remains empirical, but it is nonetheless striking.

These QPC model strong interaction coupling calculations can also be used, via vector dominance ideas, to compute *radiative* couplings. This may be an alternative treatment of the strong interaction corrections to radiative decays, different from the one presented in Section 3.3. and used in 4.3 (Petersen, 1975). It gives a natural solution to the difficulties of $P_{11}(1470)\to N\gamma$ (Gavela *et al.*, 1980a).

4.5 MESON $q\bar{q}$ ANNIHILATION

4.5.1 Introduction

We now confront with experiment the processes that have been described in Sections 2.1.3 and 2.2.2. Their characteristic feature is the annihilation of a $q\bar{q}$ pair into gluons or into lepton pairs via photons or $W^{\pm}$, Z^0 bosons. The theoretical expression is simple in the sense that the strong-interaction dynamics of bound quarks is involved only through the wave function at the centre, $\psi(0)$, present as a universal factor in all processes, multiplied by the amplitude of annihilation of *free* quarks.

However, this simplicity does not allow for a very straightforward calculation. The quantity $|\psi(0)|^2$ depends sensitively on the potential. We cannot expect good results here from the harmonic oscillator. First, we must know accurately the potential and then solve the Schrödinger equation numerically. We just sketch the procedure. To find the potential, we have both theoretical principles and experimental data. On the one hand, we have suggestions from QCD concerning the short- and long-distance behavior of the potential (see Section 1.3.4). The potential should be linear at long distance and should be given at short distances by perturbative QCD:

$$V(r) \sim \begin{cases} \lambda r & \text{(large } r\text{)} \\ -\dfrac{1}{r\log(r\Lambda)} & \text{(small } r\text{)} \end{cases} \tag{4.183}$$

At short distances the potential will be the Coulomb potential modified by the logarithmic corrections of asymptotic freedom, the effective coupling decreasing as $r\to 0$. Although the theory gives the small- and large-r behavior, it does not give the detailed potential. Ultimately, there should be only one parameter, the scale parameter Λ of QCD, which could be measured independently in hard (large-Q^2) processes. However, the connection between the confining strength λ and Λ is not very direct, and although numerical calculations for lattice QCD now establish a relation between them, this quantitative relation is not yet completely reliable. Moreover, the Λ parameter needed through (4.183) to describe the empirical spectrum does not coincide with that deduced from hard processes. A smaller value of Λ is found in this latter case. Therefore, we have to appeal to some empirical fitting. This can be done by trying to fit the meson spectrum, and especially the radially excited states. Several radial excitations are well known for heavy quarks, in the charmonium $c\bar{c}$ and bottonium $b\bar{b}$ spectra. The fit of the spectrum determines well the intermediate region of the potential, fixes λ and gives indications on Λ. We can then proceed to the calculation of $|\psi(0)|^2$, if the quark mass is also known.

It is surprising, in view of the preceding considerations, that there is yet no consensus about the value of $|\psi(0)|^2$, especially for the case of charmonium, which is of the greatest interest. A tabulation of the values given by different models can be found in a study by Ono (1981b), together with a critical discussion. For the J/ψ, it can be seen that $|\psi(0)|^2$ may vary by a factor of 3. The main factor explaining the theoretical discrepancies appears to be the mass of the charmed quark. There are approximately three groups of rather different values for m_c: around 1 GeV, 1.5 GeV, and 1.8 GeV. This wide range of values for m_c is understandable since the spectrum is not very sensitive to the quark mass. In contrast $|\psi(0)|^2$ is strongly dependent on m_c. For example, for a potential $V \propto r^2$, for *fixed energy eigenvalue*, the dependence of $|\psi(0)|^2$ on the mass is

$$|\psi(0)|^2 \propto (m_q)^{3/2}, \tag{4.184}$$

which implies a variation by a factor of 2–2.5, according to the value of m_c. For the b quark the variation is not so important because of the smaller relative uncertainty on the b quark mass.

We want to emphasize that in our opinion there is no such freedom for the choice of m_c if the *nonrelativistic* approximation is used consistently for the spectrum and also for the decays. In the case of light quarks, m_q is constrained to be 0.33 GeV by the nonrelativistic expression for the proton magnetic moment. Analogously, m_c should be constrained by the rate of the M_1 transition $J/\psi \to \eta_c \gamma$. Although the accuracy is not yet very good, this branching ratio compels the choice of a large value $m_c \sim 2$ GeV (see Section 4.3.2). Moreover, by discussing the Schrödinger equation and the spectrum for various flavours, an inequality can be established on $m_c - m_s$ that also implies $m_c \sim 2$ GeV (Martin, 1981). It seems that there is no possibility of accomodating the smaller values — hence our choice stated in Section 4.1.2. It is only by taking into account *relativistic* corrections that one could consider smaller values of m_c, but this is another story.

Another source of theoretical uncertainty concerning the estimate of $|\psi(0)|^2$ is the disagreement about the behavior of the potential at the origin. In particular, the Cornell model assumes a very strong singularity at small distances, $\sim -\alpha_s/r$, while the Martin potential $\sim r^{0.1}$ (Martin, 1980) is regular. This explains why $|\psi(0)|^2$ is larger for the Cornell than for the Martin model. The behavior of the potential at the origin has a considerable effect, and this becomes more and more the case with increasing quark mass. The difference may be seen by passing from charmonium to bottonium. For a fixed r^α potential, $|\psi(0)|^2$ varies with the quark mass according to the scaling law (Quigg and Rosner, 1979)

$$|\psi(0)|^2 \propto m_q^{3/(2+\alpha)}. \qquad (4.185)$$

It can be seen that $|\psi(0)|^2$ should increase much more for the Cornell potential than for the Martin one.

It seems to us that, theoretically speaking, there is no real choice: the short-distance behavior given by asymptotic freedom (4.183) should be assumed. We therefore adopt the predictions of the model of Ono (1979), which has $m_c = 1.9$ GeV, and the Richardson (1979) potential, which behaves like (4.183) as $r \to 0$. However, the Martin potential is not very different at intermediate distance, and it can be very useful because of its simplicity. It allows the scaling relation (4.185) to be used to extrapolate down to light quarks.

We make a selection among all the applications. We first consider e^+e^- annihilation, the best testing ground for the calculation of the wave function at the center. We then consider total rates of OZI-forbidden processes. $\pi^0 \to \gamma\gamma$ decay has already been treated in Section 3.4.5.

4.5.2 Meson Annihilation Into Lepton Pairs

Let us first recall the basic formulae. In Section 2.1.3 we found (2.109)

$$\Gamma(P \to \mu\nu) = \frac{1}{\pi}\left(\frac{G_F}{\sqrt{2}}\right)^2 m_\mu^2 \frac{(m_P^2 - m_\mu^2)^2}{m_P^4} 3[\mathrm{Tr}(F\Phi_P)]^2 |\psi(0)|^2 \qquad (4.186)$$

for a pseudoscalar meson P. We recall that Φ_P is the normalized SU(3) wave function, and that F is the relevant SU(3)-flavor operator. By the same method, we establish the formula for the decay of a vector meson V into e^+e^-:

$$\Gamma(V \to e^+e^-) = 16\pi \frac{\alpha^2}{m_V^2}[\mathrm{Tr}(Q\Phi_V)]^2 |\psi(0)|^2. \qquad (4.187)$$

Under the SU(6) (flavor–spin) symmetry, one would expect the same wave function $\psi(0)$ to govern the π and K leptonic weak decays, and the ρ, ω and ϕ electromagnetic decays into e^+e^-. However, in order to fit (4.186) and (4.187) with experiment, the π, K, ρ, ω and ϕ wave function at the origin should scale approximately like

$$|\psi_M(0)|^2 \propto m_M, \qquad (4.188)$$

as first noticed by Van Royen and Weisskopf (1964).

But this empirical fact is misleading. It does not make sense, in the nonrelativistic quark model, to have a wave function depending on the hadron mass: it can only depend on the quark masses, and should be the same, at least for π, ρ and ω, which are composed of the same light quarks. The relation (4.188) must be considered as indicating a failure of the approximation.

This failure of the nonrelativistic approximation is not unexpected for π and K annihilation, since they are quasi-Goldstone bosons. In fact, these processes should be considered in the light of chiral symmetry, which represents a complementary approach: the ratio $f_K/f_\pi \approx 1$ is satisfactorily understood in the chiral-symmetry limit of vanishing *current* masses $m_s = m_u = m_d = 0$.

If we discard π and K decays, we remain with the decays of the vector mesons ρ, ω and ϕ into e^+e^-, for which no such difficulties are expected in the nonrelativistic approximation. Indeed, (4.187) now gives satisfactory results. Using the experimental rates, we find

$$|\psi_\rho(0)|^2_{\text{exp}} = |\psi_\omega(0)|^2_{\text{exp}} = 2.9 \times 10^{-3}\,\text{GeV}^3, \tag{4.189}$$

$$|\psi_\phi(0)|^2_{\text{exp}} = 4.5 \times 10^{-3}\,\text{GeV}^3. \tag{4.190}$$

We first note that the qualitative ratios are as expected: the wave functions should be the same for ρ and ω, composed of u and d quarks, and $|\psi(0)|^2$ should be larger for ϕ, as it is composed of s quarks. According to (4.185), the ratio between $|\psi_\phi(0)|^2$ and $|\psi_\rho(0)|^2$ should be about 2. There is also agreement with the expectation for the absolute magnitude at least in a qualitative sense. Considering, for instance, the ϕ with $m_s = 0.5$ GeV, we get, using the Martin-potential results for the charmonium ground state (Martin, 1980) and the scaling relation (4.185),

$$|\psi_\phi(0)|^2 \sim 10^{-2}\,\text{GeV}^3. \tag{4.191}$$

The agreement is only qualitative, but we cannot expect more from the first-order nonrelativistic approximation for light quarks. Better agreement could be expected for heavy quarks. For J/ψ, the charmonium 1^3S_1 state, the prediction of Ono (1981b) is

$$|\psi(0)|^2 = 7 \times 10^{-2}\,\text{GeV}^3, \tag{4.192}$$

which leads to

$$\Gamma_{e^+e^-}(J/\psi) = 8.6\,\text{keV}, \tag{4.193}$$

to be compared with the experimental 4.7 keV. This is again encouraging, but it is also disappointing to note that the discrepancy is still large. We are led to state that we have not yet obtained a satisfactory nonrelativistic approximation or a situation where the potential-model description is valid.

For the much heavier b quark, the agreement becomes much better (Ono, 1981b):

$$|\psi(0)|^2 = 0.42\,\text{GeV}^3, \tag{4.194}$$

which leads to

$$\Gamma_{e^+e^-}(\Upsilon) = 1.41\,\text{keV}, \tag{4.195}$$

against an experimental 1.17 keV.

It is encouraging to see that the agreement improves as the mass increases. This observation coincides with what we have noted for the E_1 radiative decays.

It is also encouraging to observe that the *ratios* of the various charmonium or bottonium decay rates are in rather good agreement with experiment. Since there are many such states, we have an extensive test of the quark model, which we give in Table 4.7.

It is instructive to compare these predictions of potentials that fit the spectrum rather well with those of the harmonic oscillator. For instance, the ratio of the first excitation to the ground-state wave function at the centre is, for the harmonic oscillator,

$$\psi_{2S}(0)/\psi_{1S}(0) = \sqrt{\tfrac{3}{2}}. \tag{4.196}$$

This is in strong contradiction with the data, and indicates that the wave functions at the center really do test the potential. This is to be compared with other processes that depend on the peripheral part of the wave function, and are less sensitive to the potential.

Table 4.7 Decay rates into e^+e^- of radial excitations, normalized to the ground-state decay rate (Martin, 1980; Ono, 1981b).

$c\bar{c}$	Ono	Martin	Experiment
2S	0.51	0.40	0.44 ± 0.6
3S	0.35	0.25	0.16 ± 0.10
4S	0.27	0.16	0.09 ± 0.06
$b\bar{b}$			
2S	0.51	0.51	0.46 ± 0.03
3S	0.36	0.35	0.33 ± 0.03
4S	0.29	0.27	0.23 ± 0.03
5S	0.23	0.21	
6S	0.19	0.17	

Of course, the agreement for ratios still leaves as unexplained the discrepancies in the absolute values (4.193) and (4.195). Generally speaking, higher-order corrections should be considered. But, since the ratios are good, the discrepancy consists in a constant factor for all the ψs or all the ϒs, namely, a factor of 0.5 in the ψ system, and a factor of 0.8 for the ϒ states. Such factors are indeed found by considering perturbative QCD corrections, if, instead of the systematic methods of Chapter 3, a heuristic prescription is followed. The nonrelativistic formula for the annihilation rate (Sections 2.1.3 and 2.2.2) is retained:

$$\Gamma = |\psi(0)|^2\,|v_1 - v_2|\,\sigma[(q\bar{q})_C \to X], \tag{4.197}$$

with the $q\bar{q} \to X$ annihilation amplitude up to first order in α_s. A deduction of this heuristic procedure is given by Mackenzie and Lepage (1981). It is correct only for heavy quarks and for the lowest-order correction in α_s. The correction factor, which depends on the process $q\bar{q} \to X$, is found to be, for e^+e^- annihilation (Barbieri *et al.*, 1975a,b),

$$\Gamma = \Gamma_0\left(1 - \frac{16}{3\pi}\alpha_s\right), \tag{4.198}$$

where Γ_0 denotes the right-hand side of (4.187). A more complete discussion, including the v^2/c^2 corrections, neglected here, can be found in Poggio and Schnitzer (1979, 1980).

In trying to estimate the correction numerically, we are confronted with the delicate question of the value of α_s. Ambiguities are both experimental—how large is the scale parameter Λ of QCD?—and theoretical—what is the scale μ to be considered in computing $\alpha_s(\mu^2)$? The reduction factor is unfortunately very sensitive to the value of α_s, as the coefficient is very large. For the scale $\mu = m_\psi$ and $\Lambda = 0.1$ GeV, we have $\alpha_s = 0.2$, and for $\Lambda = 0.5$ GeV, $\alpha_s = 0.4$. These two values lead to the correction factors

$$1 - \frac{16}{3\pi}\alpha_s \approx 0.66 \quad \text{or} \quad 0.32. \tag{4.199}$$

For the scale m_Υ we have respectively

$$1 - \frac{16}{3\pi}\alpha_s \approx 0.75 \quad \text{or} \quad 0.6. \tag{4.200}$$

In the presence of such uncertainties, the best we can say is that there is a reduction factor in the right direction, in qualitative agreement with the data. Moreover, we can point out that the correction is smaller for the ϒ than for the ψ states.

In conclusion, the situation is satisfactory, at least for the model considered so far. As to the other models, it can be noted (see Ono, 1981b, Table 3) that some of them may give results closer to experiment for the absolute magnitude of the e^+e^- rates of the ground states J/ψ and ϒ. For instance, the model of

Kraseman and Ono (1979) gives a good result for both the J/ψ and the Υ, although at the cost of a very low charm mass $m_c \sim 1$ GeV, which would give unreasonable results for $J/\psi \rightarrow \eta_c \gamma$. The Martin potential gives a good fit for the Υ but not such a good one for the J/ψ although he uses a realistic value $m_c \approx 1.8$ Ge V; as observed before, the behavior at the origin of the potential is at odds with the QCD expectations. In contrast, the Cornell potential gives values that are much too high, owing to the strong singularity of its potential at small distances, in $-1/r$. In fact, it is the potentials that are on better theoretical grounds, with a behaviour $1/(r \ln r)$, that give results in better agreement with experiment.

Up to now we have considered only S states. There are other states with $J^P = 1^-$, which can therefore decay into e^+e^-, the $^3D_1$ states, like $\psi(3770)$, which is the lowest state of this type in charmonium. For these orbitally excited states, the wave function at the origin vanishes in the nonrelativistic limit, $\psi(0) = 0$, and therefore the lowest-order approximation vanishes. We are in a situation, analogous to forbidden transitions, where the calculation is not reliable because the amplitude is sensitive to various small effects, contributing with different signs. We list the effects that have been considered: (i) the tensor force generated by one-gluon exchange gives an S–D mixing; (ii) the relativistic corrections give a nonvanishing value proportional to the second derivative $\psi''(0)$ instead of $\psi(0)$; (iii) coupled-channel effects also given an S–D mixing. The latter effect seems to be the most important (Eichten *et al.*, 1980). The result is roughly satisfactory, but still small for $\psi(3770) \rightarrow e^+e^-$, a real problem (Ono, 1983).

4.5.3 Hadronic OZI-Forbidden Rates

As we have indicated in Section 2.2.2, it is possible to describe the OZI-forbidden decays, at least for heavy quarkonia, by annihilation into a few gluons. We have already presented the calculation for annihilation into two gluons. However, for the case of 1^- states ψ and Υ, we must have at least three gluons, since two-gluon decay is forbidden by C conservation (Landau and Lifshitz, 1972). The calculation is a little more complicated, but analogous to the positronium decay into three photons. We give only the final result:

$$\Gamma(C \rightarrow 3g) = \frac{40}{81}(\pi^2 - 9)\frac{\alpha_s^3}{m_q^2}|\psi(0)|^2. \tag{4.201}$$

This rate is the dominant contribution and can be considered as an estimate of the total width. In contrast with the e^+e^- decay, since α_s is renormalization-point-dependent and the expression is very sensitive to α_s, this leads only to qualitative predictions.

The first point to be made is that one gets a correct description with a fixed value of $\alpha_s \approx 0.2$. Indeed, we find

$$\Gamma(J/\psi) = 66\,\text{keV}, \tag{4.202}$$

$$\Gamma(\Upsilon) = 52\,\text{keV}, \tag{4.203}$$

versus the experimental 63 keV and 48 keV respectively. This success must be considered as striking, in view of the complexity of the physical channels that give Γ. It is a confirmation not only of the quark model, but specifically of perturbative QCD, which gives a simple picture in terms of annihilation into gluons.

However, as we apply perturbative QCD, we should also observe the logarithmic evolution of α_s from the ψ to the Υ mass scale: α_s should be larger for ψ than for Υ. To be quantitative, let us fit α_s from the width of Υ, since perturbative QCD can be expected to work better at this larger scale. We find

$$\alpha_s(\Upsilon) \approx 0.195, \tag{4.204}$$

and from the relation

$$\frac{m_\Upsilon}{m_\psi} = \exp\left[\frac{6\pi}{27}\left(\frac{1}{\alpha_s(\Upsilon)} - \frac{1}{\alpha_s(\psi)}\right)\right] \tag{4.205}$$

we deduce

$$\alpha_s(\psi) = 0.283. \tag{4.206}$$

This value leads to a total rate that is much too large for charmonium:

$$\Gamma(J/\psi) \approx 187\,\text{keV}. \tag{4.207}$$

We must conclude that the running coupling constant does not present the expected logarithmic evolution. The situation is not improved by the introduction of radiative corrections. Indeed, for annihilation into three gluons, the QCD correction, analogous to (4.198), is given by

$$\Gamma = \Gamma_0(3g)\left(1 + \frac{4}{\pi}\alpha_s\right), \tag{4.208}$$

where $\Gamma_0(3g)$ is the lowest-order approximation, given by (4.201), and the renormalization point is m_Υ. Introducing this correction, we find, instead of (4.204) and (4.206),

$$\alpha_s(\Upsilon) \approx 0.182, \tag{4.209}$$

$$\alpha_s(\psi) = 0.256, \tag{4.210}$$

whence

$$\Gamma(\psi) \approx 185\,\text{keV} \tag{4.211}$$

This result is close to (4.207), a total hadronic decay width for the J/ψ that is much too large. Accounting for the log (log) corrections to the evolution of α_s does not improve the conclusion. The solution of this problem is probably not to be found in a modification of the wave functions $\psi(0)$, because if $\psi(0)$ decreases for charmonium, the agreement for $J/\psi \to e^+e^-$ will be destroyed.

It must be stressed that this difficulty does not compel us to abandon the three-gluon picture, supported moreover by the observation of three-jet events at the Υ. It only sheds some doubt on the use of perturbative QCD at the ψ. It could happen that the three-gluon picture is approximately correct, while perturbative QCD is not yet valid at this scale. The disagreement might also be due to v^2/c^2 relativistic effects.

4.6 NONLEPTONIC WEAK DECAYS

4.6.1 General Framework

Nonleptonic weak decays are decays involving only hadrons, mediated by the weak interaction, like the hyperon decays $\Lambda \to N\pi$ and $\Sigma \to N\pi$ and the kaon decays $K \to 3\pi$ and $K \to 2\pi$. With the advent of heavy-quark spectroscopy, one has to consider also the decays like $D \to K\pi$ and $B \to D\pi$ and baryon decays such as $\Lambda_c \to NK$ or $\Lambda_c \to \Lambda\pi$. Moreover, in addition to the strict weak decays, there are also weak radiative decays like $\Sigma \to N\gamma$. For definiteness, we shall first confine ourselves to hyperon decays like $\Lambda \to p\pi^-$. They provide the largest sample of available data that is at the same time rather well understood. In contrast, there are still fundamental difficulties for $K \to 2\pi$. Moreover, the D decay modes are also controversial.

Let us first assume that the weak interaction through W exchange has been reduced, owing to the large W mass, to a point-like $(V-A)(V-A)$ four-quark interaction. The $(V-A)(V+A)$ contribution that enters through radiative corrections (penguin diagrams) will be discussed in Section 4.6.3. The simplest process leading to the decay $B' \to B + \pi$ is the direct weak emission of the π by a quark via this four-quark interaction (Figure 4.2), leaving the other quarks as spectators. There is also another possibility, a baryon-to-baryon transition induced by the weak interaction (Figure 4.3a), followed or preceded by strong pion emission. This process could be called *internal baryon conversion* (Figure 4.3b).

Of course, it has always been the principle of quark models to consider the

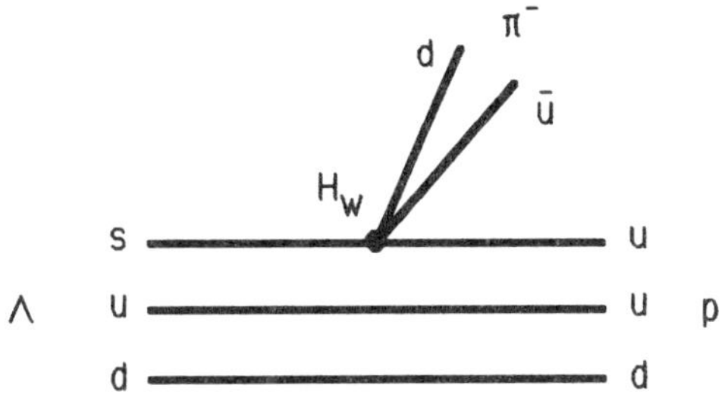

Figure 4.2 Direct weak pion emission in the decay $\Lambda \to p\pi^-$.

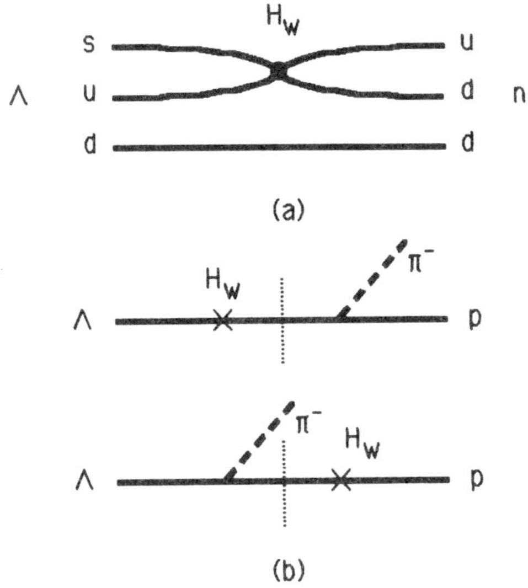

Figure 4.3 Internal baryon conversion in the decay $\Lambda \to p\pi^-$. A cross in (b) denotes a weak transition at the hadron level. The transition $\Lambda \to n$ in the first diagram in (b) is detailed in (a).

simplest possible process, which is expected to give the largest contribution. But it happens that, at least for hyperon decays (in contrast with Λ_c decays or b-flavored baryon decays (Guberina, Tadić and Trampetić, 1981)), the direct-emission process gives only a small contribution to the decay amplitude, and that the internal-baryon-conversion process — although it is a two-step process involving intermediate states — is much larger. This can be easily seen by calculating the decay amplitude M corresponding to the direct emission process $\Lambda \to p\pi^-$. From the matrix element of the Hamiltonian

$$H_W = \frac{G_F}{\sqrt{2}} \int d\boldsymbol{x} \frac{1}{2} \{ j_\mu^-(\boldsymbol{x}), j^{+\mu}(\boldsymbol{x}) \} = \int d\boldsymbol{x}\, \mathscr{H}_W(\boldsymbol{x}) \qquad (4.212)$$

the general expression for M is deduced by using translation operators:

$$j_\mu(\boldsymbol{x}) = \mathrm{e}^{-\mathrm{i}\boldsymbol{P}\cdot\boldsymbol{x}} j_\mu(0)\,\mathrm{e}^{\mathrm{i}\boldsymbol{P}\cdot\boldsymbol{x}},$$

where $\boldsymbol{P}$ is the momentum operator. M is then given by

$$M = \frac{G_\mathrm{F}}{\sqrt{2}}(2\pi)^3 \langle \mathrm{f} | \{ j_\mu^-(0), j^{+\mu}(0) \} | \mathrm{i} \rangle. \tag{4.213}$$

Hyperon nonleptonic decay has two possible partial waves because of the possibility of parity violation. The weak Hamiltonian can be divided into parity-conserving and parity-violating parts:

$$H_\mathrm{W} = H_\mathrm{W}^\mathrm{pc} + H_\mathrm{W}^\mathrm{pv}, \tag{4.214}$$

which correspond respectively to the parts with even or odd numbers of γ_5 matrices. (γ_5 is chosen according to the convention of Bjorken and Drell, 1964). For decays like $\Lambda(\frac{1}{2}^+) \to \mathrm{p}(\frac{1}{2}^+) + \pi(0^-)$, H^pc and H^pv generate respectively the P and the S waves. In the rate the squared contributions of P and S amplitudes must be added:

$$\Gamma = 8\pi^2 \frac{E_\mathrm{B} E_\pi}{m_\mathrm{A}} k \left[|M^\mathrm{pc}|^2 + |M^\mathrm{pv}|^2 \right], \tag{4.215}$$

the fermions being conventionally chosen with spin up along $\boldsymbol{k}$. Other physical quantities, like asymmetries in decay angular distributions that involve interference terms, allow measurement of the phase between M^pc and M^pv and their respective magnitudes.

Let us consider first the direct-emission process. To illustrate the methods, we shall only calculate, for the sake of simplicity, the S wave contribution to $\Lambda \to \mathrm{p}\pi^-$. To calculate the diagram of Figure 4.2, we factorize the amplitude M into a product of physical current-matrix elements:

$$M = -\frac{G_\mathrm{F}}{\sqrt{2}} \cos\theta_\mathrm{C} \sin\theta_\mathrm{C} (2\pi)^3 \langle \pi^- | \bar{d}\gamma^\mu\gamma_5 u | 0 \rangle \langle \mathrm{p} | \bar{u}\gamma_\mu s | \Lambda \rangle. \tag{4.216}$$

This additive piece is thus often called the *separable* or *factorizable* contribution.

Although the two current-matrix elements could be evaluated in the naive approximation, it is still simpler to use the matrix element, independently measured in $\pi \to \mu\nu$ decay,

$$\langle \pi^- | \bar{d}\gamma^\mu\gamma_5 u | 0 \rangle = -\frac{1}{(2\pi)^{3/2}} f_\pi k^\mu \frac{1}{(2\omega_\pi)^{1/2}}, \tag{4.217}$$

where k_μ is the pion momentum. Then

$$M=\frac{G_{\mathrm{F}}}{\sqrt{2}}\cos\theta_{\mathrm{C}}\sin\theta_{\mathrm{C}}\frac{1}{(2\pi)^{3/2}}\frac{f_\pi}{(2\omega_\pi)^{1/2}}(2\pi)^3 k^\mu\langle \mathrm{p}|\,\bar{u}\gamma_\mu s\,|\Lambda\rangle. \quad (4.218)$$

Assuming

$$(2\pi)^3\langle \mathrm{p}|\,\bar{u}\gamma_\mu s\,|\Lambda\rangle\approx\sqrt{\tfrac{3}{2}}\,\bar{u}_{\mathrm{p}}\gamma_\mu u_\Lambda, \quad (4.219)$$

since $k_\mu=p_\mu(\Lambda)-p_\mu(\mathrm{p})$, we obtain

$$k^\mu(2\pi)^3\langle \mathrm{p}|\,\bar{u}\gamma_\mu s\,|\Lambda\rangle\approx\sqrt{\tfrac{3}{2}}(m_\Lambda-m_{\mathrm{p}})\,\bar{u}_{\mathrm{p}}u_\Lambda\approx\sqrt{\tfrac{3}{2}}(m_\Lambda-m_{\mathrm{p}}). \quad (4.220)$$

We finally have

$$M\approx\frac{G_{\mathrm{F}}}{\sqrt{2}}\cos\theta_{\mathrm{C}}\sin\theta_{\mathrm{C}}\frac{1}{(2\pi)^{3/2}}\frac{f_\pi}{(2\omega_\pi)^{1/2}}\sqrt{\tfrac{3}{2}}(m_\Lambda-m_{\mathrm{p}}), \quad (4.221)$$

from which the rate is found to be

$$\Gamma^{\mathrm{pv}}(\Lambda\to\mathrm{p}\pi^-)\approx\frac{3G_{\mathrm{F}}^2 f_\pi^2}{8\pi}\cos^2\theta_{\mathrm{C}}\sin^2\theta_{\mathrm{C}}\,k(m_\Lambda-m_{\mathrm{p}})^2, \quad (4.222)$$

Numerically, this gives $\Gamma^{\mathrm{pv}}(\Lambda\to\mathrm{p}\pi^-)=4\times10^{-17}$ GeV versus the experimental value 1.4×10^{-15} GeV. This is convincing evidence against the direct-emission process for the explanation of nonleptonic decay, since we did not have to make any detailed dynamical assumption to evaluate M. This direct-emission process has the same structure as the semileptonic decay, with the leptons replaced by the quarks inside the pion. The fact that hadronic weak decays are much larger than predicted by the mechanism operating in semileptonic decays is known as *nonleptonic enhancement*.

Another failure of the direct-emission process concerns the isospin structure. The operator (4.212) has the form $(\bar{u}s)(\bar{d}u)$ in terms of the quark fields. The initial state us has $I=\frac{1}{2}$, and the final state ud has $I_3=0$ and $I=0$ or 1. Therefore the operator is a combination of $\Delta I=\frac{1}{2}$ and $\Delta I=\frac{3}{2}$. Experimentally, the $\Delta I=\frac{1}{2}$ piece is enhanced by a factor of 20 over the $\Delta I=\frac{3}{2}$ piece. This is the so-called $\Delta I=\frac{1}{2}$ rule. It is easy to verify that the direct-emission amplitude has a comparable amount of $\Delta I=\frac{1}{2}$ and $\Delta I=\frac{3}{2}$, both being simply related by SU(3) algebraic factors. We are led to suspect that not only the quantitative estimate is to be questioned, but indeed the whole mechanism. We have to face the more difficult task of evaluating the contribution of the baryon-internal-conversion process (Figure 4.3). We need to calculate the transition amplitude for pion emission between a baryon state and any possible intermediate state $|n\rangle$, to estimate the matrix element of the weak Hamiltonian between a baryon state and $|n\rangle$, and finally to sum over the intermediate states $|n\rangle$.

Moreover, the $|n\rangle$ states are not only the usual baryonic three-quark states, but possibly also multiquark intermediate states generated by Z graphs (Figure 4.4). In fact it happens that these Z graphs are crucial.

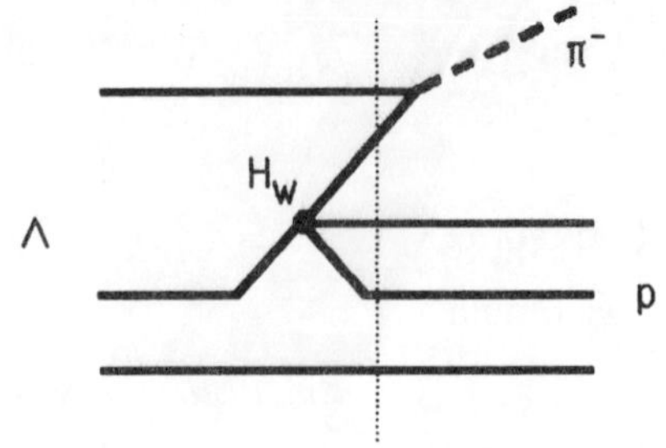

Figure 4.4 Z-graph contribution to the $\Lambda \to p\pi^-$ weak transition.

On the whole, this seems very complicated. It is in fact comparable in complexity to the calculation of a process like $\gamma N \to N\pi$ or $\pi N \to N$ at low energy. In the latter case it is possible to experimentally reach the resonance poles by partial-wave analysis, and therefore to confront the quark-model predictions with experiment for transitions like $N^* \to N\gamma$ or $N^* \to N\pi$. However, in the present case of weak decays it is no longer possible to reach the resonance poles, because in $Y \to N\pi$ there is no free kinematical variable that could be varied.

However, it is still possible to obtain a great simplification by methods similar to those used for scattering processes. We can consider the amplitude

$$M = (2\pi)^3 \langle N\pi | \mathscr{H}_W(0) | Y \rangle \tag{4.223}$$

for momenta that do not satisfy energy–momentum conservation and that are only constrained to be on shell ($p_i^2 = m_i^2$). Then $s = (p_N + p_\pi)^2$ plays the same role as the energy in a scattering amplitude. As we shall now see, it is possible to separate the amplitude into two parts from their respective behavior as functions of s: a slowly varying part and a part dominated only by baryon poles close to the physical region.

This can be done in practice with the help of the identity between the divergence of the axial current and the pion field (partial conservation of axial current, PCAC). The amplitude $B_1 \to B_2 + \pi^i$ may be written, with the help of the reduction formula for the pion state, as

$$\langle B_2 \pi^i | \mathscr{H}_W(0) | B_1 \rangle$$
$$= i \int d^4x \frac{e^{ik\cdot x}}{(2\pi)^{3/2}(2\omega_\pi)^{1/2}} (\overrightarrow{\Box_x + m_\pi^2}) \langle B_2 | T(\varphi_R^i(x) \mathscr{H}_W(0)) | B_1 \rangle, \tag{4.224}$$

where $\varphi_R^i(x)$ is a pion interpolating field, normalized through

$$\langle 0 | \varphi_R^i(x) | \pi^i \rangle = \frac{e^{-iP\cdot x}}{(2\pi)^{3/2}(2\omega_\pi)^{1/2}}. \tag{4.225}$$

To apply PCAC, we choose as the pion interpolating field

$$\varphi_{\mathrm{R}}^{i}=\frac{\partial_{\mu}j_{5}^{i\mu}}{m_{\pi}^{2}f_{\pi}}. \tag{4.226}$$

The expression of the coefficient in (4.226) has been chosen to have

$$\langle 0|j_{5}^{i\mu}|\pi^{i}\rangle=\frac{\mathrm{i}f_{\pi}k^{\mu}}{(2\pi)^{3/2}(2\omega_{\pi})^{1/2}}, \tag{4.227}$$

where $f_{\pi}=130$ MeV; $j_5^{i\mu}$ is defined as the *same* flavor combination of quark *fields* that defines the π^i state in terms of quark *states*. For example, for an outgoing $\pi^+=+\mathrm{u}\bar{\mathrm{d}}$,

$$j_{5}^{(\pi+)\mu}=\bar{d}\gamma^{\mu}\gamma_{5}u. \tag{4.228}$$

Of course, for an ingoing pion we must use the Hermitian conjugate, obtained by exchanging the quark and antiquark flavors.

It must be remembered that, in addition to the standard conventions for color and spin wave functions (see Appendix), we have chosen the configuration-space wave function of the meson to be real, with $\psi(0)>0$. The comparison of (4.227) with (2.94) and (2.95) shows that the pion states used in (4.225) and (4.227) are multiplied by a phase i compared with our standard conventions. We shall from now on return to the standard conventions. The definition (4.228) of the axial current is more convenient than the usual one based on Gell-Mann SU(3) matrices (differing by a factor of $\sqrt{2}$ and signs) when one needs to reduce pion states defined in the standard way by the quark model.

Through standard steps (Dashen and Weinstein, 1969), starting from (4.224) and using the relation (4.226), we end up after having neglected corrections $O(m_\pi^2)$, with

$$\langle \mathrm{B}_2\pi^i|\mathscr{H}_{\mathrm{W}}(0)|\mathrm{B}_1\rangle=\frac{1}{(2\pi)^{3/2}}\frac{1}{(2\omega_{\pi})^{1/2}}\frac{1}{f_{\pi}}$$
$$\times\Big[\langle \mathrm{B}_2|\big[\textstyle\int \mathrm{d}\boldsymbol{x}\,\mathrm{e}^{-\mathrm{i}\boldsymbol{k}\cdot\boldsymbol{x}}\,j_5^{i0}(\boldsymbol{x}),\,\mathscr{H}_{\mathrm{W}}(0)\big]|\mathrm{B}_1\rangle$$
$$+\,\mathrm{i}k_{\mu}\textstyle\int \mathrm{d}^4x\,\mathrm{e}^{\mathrm{i}k\cdot x}\langle \mathrm{B}_2|\,\theta(x^0)\,\overline{j_5^{i\mu}(x)}\,\mathscr{H}_{\mathrm{W}}(0)+\theta(-x^0)\,\mathscr{H}_{\mathrm{W}}(0)\overline{j_5^{i\mu}(x)}\,|\mathrm{B}_1\rangle\Big], \tag{4.229}$$

where a bar over $j_5^{i\mu}$ means that the pion pole contributions have been taken out of the T product. Of the two terms inside the brackets, the first is slowly varying as a function of k_μ and can be approximated by its $\boldsymbol{k}=0$ value

$$\langle \mathrm{B}_2|\big[\textstyle\int \mathrm{d}\boldsymbol{x}\,\mathrm{e}^{-\mathrm{i}\boldsymbol{k}\cdot\boldsymbol{x}}\,j_5^{i0}(\boldsymbol{x}),\mathscr{H}_{\mathrm{W}}(0)\big]|\mathrm{B}_1\rangle\approx\langle \mathrm{B}_2|\big[Q_5^i,\mathscr{H}_{\mathrm{W}}(0)\big]|\mathrm{B}_1\rangle, \tag{4.230}$$

where

$$Q_5^i\equiv\textstyle\int \mathrm{d}\boldsymbol{x}\,j_5^{i0}(\boldsymbol{x}). \tag{4.231}$$

In contrast, the second term contains baryon poles in k^0, which can be exhibited by introducing a set of intermediate baryon states:

$$ik_\mu \int d^4x\, e^{ik\cdot x}\, \theta(x^0) \langle B_2 | j_5^{i\mu}(x)\, \mathscr{H}_W(0)\, | B_1 \rangle$$
$$= -k_\mu (2\pi)^3 \sum_n \frac{\langle B_2 | j_5^{i\mu}(0)\, | n\rangle \langle n | \mathscr{H}_W(0)\, | B_1\rangle}{k^0 + p_2^0 - p_n^0 + i\varepsilon}\Bigg|_{\boldsymbol{p}_n = \boldsymbol{p}_2 + \boldsymbol{k}}, \tag{4.232}$$

$$ik_\mu \int d^4x\, e^{ik\cdot x}\, \theta(-x^0) \langle B_2 |\, \mathscr{H}_W(0)\, j_5^{i\mu}(x)\, | B_1 \rangle$$
$$= k_\mu (2\pi)^3 \sum_n \frac{\langle B_2 | \mathscr{H}_W(0) | n\rangle \langle n | j_5^{i\mu}(0)\, | B_1\rangle}{k^0 - p_1^0 + p_n^0 - i\varepsilon}\Bigg|_{\boldsymbol{p}_n = \boldsymbol{p}_1 + \boldsymbol{k}}. \tag{4.233}$$

Note that on shell $\boldsymbol{p}_1 = \boldsymbol{p}_2 + \boldsymbol{k}$, giving respectively in (4.232) and (4.233) $\boldsymbol{p}_n = \boldsymbol{p}_1$ and $\boldsymbol{p}_n = \boldsymbol{p}_2$. Since the physical k_μ is small, we hope that these contributions are negligible, *unless* $p_2^0 - p_n^0$ or p_1^0 are small, which leads to small denominators. This corresponds to intermediate baryonic states, low excitations of the three-quark system, lying not very far from the nucleon or hyperon masses.

In the standard analysis of dispersion relations, the contributions (4.232) and (4.233) correspond to the *s*- and *u*-channel poles (Landau and Lifshitz, 1973). We shall exploit the fact that (4.224) extrapolates to momenta not constrained by $k = p_1 - p_2$ (still keeping $k^2 = m_\pi^2 \approx 0$). The kinematical variables of a scattering amplitude can be defined by introducing a "spurion" of momentum $q \equiv p_2 + k - p_1$ with $s \equiv (p_2 + k)^2$, $u \equiv (p_2 - q)^2 = (p_1 - k)^2$, $t \equiv (k - q)^2 = (p_2 - p_1)^2$. We easily find that the poles in (4.232) and (4.233) correspond respectively to $s = m_n^2$ and $u = m_n^2$. The corresponding pole contributions can equivalently be found by considering Figures 4.3(b) as Feynman diagrams. The Feynman diagrams are covariant, while the terms (4.232) and (4.233) are not separately for each *n*. In fact, Feynman diagrams associate two poles in k^0, in such a way as to obtain a covariant result. For a definite baryonic state $|n\rangle$, the second pole ensuring covariance comes from the corresponding Z graph (Figure 4.5). If $|n\rangle$ is a low-lying state, the type of

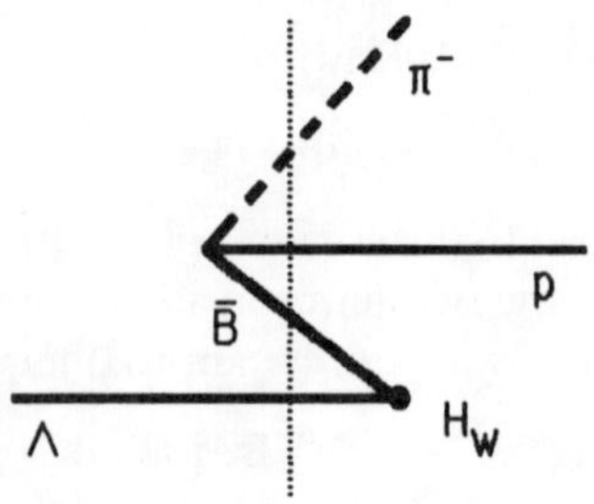

Figure 4.5 Z-graph contribution with $\Lambda\bar{B}p\pi^-$ intermediate state to the nonleptonic decay $\Lambda \to p\pi^-$.

state we are interested in, then the Z-graph pole is far away and gives only a small contribution, since it corresponds to a $B\bar{B}B\pi$ intermediate state. It is therefore almost equivalent to evaluating a single-particle intermediate-state contribution either with one term $|n\rangle$ in the expressions (4.232) and (4.233) or by a Feynman diagram with a $|n\rangle$ particle internal line.

Let us now summarize. We have

$$\langle B_2\pi^i|\mathscr{H}_W(0)|B_1\rangle \approx \frac{1}{(2\pi)^{3/2}}\frac{1}{(2\omega_\pi)^{1/2}}\frac{1}{f_\pi}$$
$$\times\Bigg\{\langle B_2|[Q_5^i,\mathscr{H}_W(0)]|B_1\rangle$$
$$-k_\mu(2\pi)^3\sum_{n_s}\frac{\langle B_2|j_5^{i\mu}(0)|n_s\rangle\langle n_s|\mathscr{H}_W(0)|B_1\rangle}{k^0+p_2^0-p_{n_s}^0+i\varepsilon}\Bigg|_{\boldsymbol{p}_{n_s}\approx\boldsymbol{p}_1}$$
$$+k_\mu(2\pi)^3\sum_{n_u}\frac{\langle B_2|\mathscr{H}_W(0)|n_u\rangle\langle n_u|j_5^{i\mu}(0)|B_1\rangle}{k^0-p_1^0+p_{n_u}^0-i\varepsilon}\Bigg|_{\boldsymbol{p}_{n_u}\approx\boldsymbol{p}_2}\Bigg\}, \tag{4.234}$$

where n_s and n_u are labels for the low-lying (three-quark) baryon states in the s and u channels.

Current algebra allows the first term to be simplified. It can easily be shown that

$$[Q_5^i,\mathscr{H}_W(0)] = -[Q^i,\mathscr{H}_W(0)], \tag{4.235}$$

where Q^i is the isotopic charge:

$$Q^i \equiv \int d\boldsymbol{x}\, j^{i0}(\boldsymbol{x}). \tag{4.236}$$

The relation (4.235) is an immediate consequence of

$$[Q_R^i,\mathscr{H}_W(0)] = 0, \tag{4.237}$$

where the right-handed charge is defined by

$$Q_R^i = Q^i + Q_5^i. \tag{4.238}$$

It is obvious that the right-handed charge commutes with $\mathscr{H}_W(0)$ if it does not contain right-handed quark fields. This is the case for the standard $(V-A)(V-A)$ interaction. On the other hand, as we shall see later, there are certain weak-interaction terms induced by QCD radiative corrections that do contain right-handed fields.

If (4.235) holds, we then have

$$\begin{aligned}&\langle B_2|[Q_5^i,\mathscr{H}_W(0)]|B_1\rangle\\ &= \langle B_2|\mathscr{H}_W(0)|Q^iB_1\rangle - \langle Q^iB_2|\mathscr{H}_W(0)|B_1\rangle.\end{aligned} \tag{4.239}$$

Sine Q^iB_1 and Q^iB_2 still belong to the $\frac{1}{2}^+$ baryon octet, we observe that the weak Hamiltonian is present in (4.234) combined with (4.239) only through *baryon-to-baryon matrix elements of low-lying states* $\langle B'|\mathscr{H}_W(0)|B\rangle$, where B and B′ are either the ground state or nearby excitations. Instead of calculating $\langle B_2\pi^i|\mathscr{H}_W(0)|B_1\rangle$, we only have to calculate $\langle B'|\mathscr{H}_W(0)|B\rangle$. If we apply the standard lowest-order nonrelativistic approximation of the quark model to calculate this matrix element, since $\mathscr{H}_W(0)$ is a four-fermion operator, the only way it can act within a baryon is by interaction of two quarks (Figure 4.6).

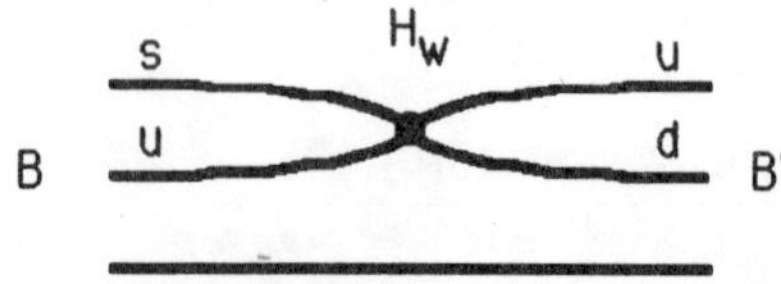

Figure 4.6 The matrix element $\langle B'|\ H_w(0)\ |B\rangle$.

We note that we assume that $\mathscr{H}_W(0)$ is normal-ordered in order to avoid self-energy contributions, which do not contribute to the decay. We have learned how to calculate such matrix elements in Section 2.1.4. The interesting point now is to understand the connection of this method, summarized in (4.234), with the direct-emission model above (Figure 4.2). We can understand the relation in terms of old-fashioned perturbation theory, first order in the weak interaction, and expansion into powers of the strong quark-pair creation interaction. The lowest order of this expansion will be the direct-emission graph, where the pion is directly created by the weak interaction and no further pair creation by the strong interaction is necessary. This contribution is much too small, and we assume that the main contribution comes from the two-step process of internal baryon conversion. The new calculation through (4.234) is a method of evaluating this two-step process. There is a distinction between two types of intermediate states in terms of old-fashioned perturbation theory. First there is a contribution of low-lying states that give a rapidly varying pole behaviour. This is the *pole contribution.* Secondly, there is a contribution of remote states, which is slowly varying but which may nevertheless be very large once it is summed, as shown in the actual calculation. This is the *commutator contribution* in (4.234).

It can be seen that in the simplest approximation (exact SU(3), $m_\pi=0$ and static limit) this commutator contribution corresponds to *internal* Z-graph intermediate states (Figure 4.4). These intermediate states have an energy corresponding to a five-quark (qqq$\bar{q}$q) intermediate state, but nevertheless dominate the S wave, because they give smaller powers of v/c at the interaction vertices than the three-quark states.

With (4.234), which expresses the two-step process (Figure 4.3(b)), we now have a way of calculating nonleptonic amplitudes, which is found to be successful, as we shall see in the next section.

However, this is not all. From the beginning, we have assumed that the weak interactions are described by the usual (V−A)(V−A) weak Hamiltonian. However, it is well known by now that the fundamental interaction that generates the weak effect is a trilinear coupling between fermions and gauge-vector bosons. Owing to the large mass of the W boson, the effect of this fundamental interaction can nevertheless be reduced to an effective four-quark (normal-ordered) operator. This is an effective Hamiltonian, to be used only at the lowest order in the Fermi coupling G_F. Since the W is heavy, QCD radiative corrections of the type of Figure 4.7 will involve $\alpha_s(Q^2)$ for large Q^2, or equivalently at short distances. The techniques of the renormalization group allow summation of the series of leading logarithmic corrections, of the form $[\alpha_s(\mu^2) \log (M_W^2/\mu^2)]^n$. This leads to an effective four-quark local interaction that includes these strong short-distance corrections. These short-distance corrections were not included in the quark-model methods for evaluating the nonleptonic matrix elements, described above.

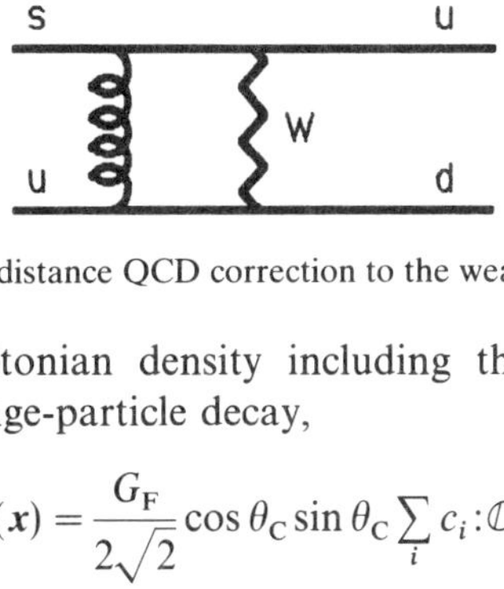

Figure 4.7 A typical short-distance QCD correction to the weak interaction.

The effective Hamiltonian density including the short-distance QCD corrections is, for strange-particle decay,

$$\mathscr{H}_W(x) = \frac{G_F}{2\sqrt{2}} \cos\theta_C \sin\theta_C \sum_i c_i :\mathcal{O}_i(x):, \tag{4.240}$$

where the operators $\mathcal{O}_i(x)$ are Lorentz scalar combinations of four-quark fields, and the coefficients c_i depend on the strong interactions. The important point to emphasize is that (4.240) is not the same as the usual standard (V−A)(V−A) interaction. The latter would correspond to a particular set of coefficients c_i. The actual c_i differ from their Born-approximation values by α_s corrections. Moreover, there are operators in (4.240) that are not of the (V−A)(V−A) type. These are the famous "penguin" operators discovered by Vainshtein, Zakharov and Shifman (1977). However, as we shall argue in Section 4.6.3, the *QCD short-distance corrections do not qualitatively change the picture of nonleptonic hyperon decays.*

The reader may have concluded from this short presentation that nonleptonic decays is a complicated topic. It is no wonder that there is not yet any consensus about this subject, after many years of work by many people. To avoid specialized discussions, we think it better to express mainly our personal point of view, with no intent of a polemic with other opinions. We also think it better to concentrate on hyperon decays, since, at present there is no convincing description of kaon decays.

The complexity of the topic is the main justification for the extensive development we give. But this is not without fruit, because this development results in an extensive and detailed test of the quark-model methods. Other similar decays, for instance the weak radiative decays like $\Sigma^+ \to p\gamma$, $\Lambda \to n\gamma$ can be successfully treated by the same methods (Gavela *et al.*, 1981b).

As to kaon decays, the fact that they are not so satisfactorily described can in fact be understood. In $K \to 2\pi$, we are reaching the limits of the possibilities of quark-model methods, because we have to treat three quasi-Goldstone bosons. It is not reasonable to use the nonrelativistic quark model in such a situation. It is probably not even reasonable to expect that the simple relativistic methods explained in Chapter 3 are able to handle the subtle problems of chiral symmetry and its associated Goldstone bosons. It could be argued that the pion is sometimes treated with success by the nonrelativistic quark model, as in the QPC model calculations of $N^* \to N\pi$. However, it must be observed that chiral symmetry is essential in $K \to 2\pi$, and therefore the Goldstone-boson properties of the π and K must be fully taken into account as in the cases of $\pi \to \mu\nu$ and $K \to \mu\nu$ (see Section 4.5.2).

4.6.2 The Most Naive Model of Hyperon Decays

As usual, we begin with the simplest assumptions and consider later the corrections to the simplest model. First, we take for the Hamiltonian density $\mathscr{H}_W$ the $(V-A)(V-A)$ interaction, neglecting for the moment the short-distance QCD corrections. Secondly, for the *S waves*, we consider only the *commutator term*, which is the matrix element of $\mathscr{H}_W^{pc}$ between two $\frac{1}{2}^+$ baryon states. For the P waves the commutator term is the matrix element of $\mathscr{H}_W^{pv}$ between $\frac{1}{2}^+$ baryons, but it can be demonstrated that (Lee–Swift (1964) Theorem)

$$\langle 8, \tfrac{1}{2}^+ | \mathscr{H}_W^{pv} | 8, \tfrac{1}{2}^+ \rangle = 0 \qquad (4.241)$$

in the SU(3) limit. We describe *P waves by the* $\frac{1}{2}^+$ *pole contribution* to the k_μ term; this contribution involves the weak interaction only through the matrix elements of $\mathscr{H}_W^{pc}$ between $\frac{1}{2}^+$ baryons. Then, for both S and P waves, we need only calculate the matrix elements

$$\langle \tfrac{1}{2}^+ | \mathscr{H}_W^{pc} | \tfrac{1}{2}^+ \rangle \qquad (4.242)$$

Thirdly, we calculate these matrix elements in the lowest-order nonrelativistic approximation, also setting the states to rest.

We shall consider later the corrections to this scheme: (i) resonance pole contributions in the k_μ term; (ii) relativistic corrections to the matrix elements of $\mathscr{H}_W(0)$; (iii) short-distance QCD corrections.

Let us now calculate the matrix elements (4.242) in the nonrelativistic limit. The formal expression has been given in Section 2.1.4 for $\Delta S = +1$:

$$\langle B'|H_W^{pc}|B\rangle = \delta(\boldsymbol{P}' - \boldsymbol{P}) \frac{G_F}{\sqrt{2}} \cos\theta_C \sin\theta_C \times 6$$

$$\times \langle \psi_{B'}|\, \tau^{(-)}(1) v^{(+)}(2)(1 - \boldsymbol{\sigma}(1)\cdot\boldsymbol{\sigma}(2))\, \delta(\boldsymbol{r}_1 - \boldsymbol{r}_2)\, |\psi_B\rangle, \tag{4.243}$$

where ψ_B and $\psi_{B'}$ are the internal wave functions. The calculation of spin–SU(3) matrix elements can be factorized by writing

$$\psi = \Phi \psi^s, \tag{4.244}$$

where ψ^s is the spatial symmetric wave function. We then have,

$$\begin{aligned}&\langle \psi_{B'}|\, \tau^{(-)}(1) v^{(+)}(2)(1 - \boldsymbol{\sigma}(1)\cdot\boldsymbol{\sigma}(2))\, \delta(\boldsymbol{r}_1 - \boldsymbol{r}_2)\, |\psi_B\rangle \\ &= \langle \Phi_{B'}|\, \tau^{(-)}(1) v^{(+)}(2)(1 - \boldsymbol{\sigma}(1)\cdot\boldsymbol{\sigma}(2))\, |\Phi_B\rangle \langle \psi^s_{B'}|\, \delta(\boldsymbol{r}_1 - \boldsymbol{r}_2)\, |\psi^s_B\rangle.\end{aligned} \tag{4.245}$$

Before discussing the important quantity, the baryon wave function at the center,

$$|\psi(0)|^2 \equiv \langle \psi^s|\, \delta(\boldsymbol{r}_1 - \boldsymbol{r}_2)\, |\psi^s\rangle, \tag{4.246}$$

which will give the absolute magnitude of the rate, let us concentrate on the calculation of the spin–SU(3) factor. In terms of spin–SU(3) wave functions of the baryon octet, we have

$$\begin{aligned}&\langle \Phi_{B'}|\, \tau^{(-)}(1) v^{(+)}(2)(1 - \boldsymbol{\sigma}(1)\cdot\boldsymbol{\sigma}(2))|\Phi_B\rangle \\ &= \tfrac{1}{2}\langle \varphi'_{B'}|\, \tau^{(-)}(1) v^{(+)}(2)\, |\varphi'_B\rangle \langle \chi'_{B'}|\, 1 - \boldsymbol{\sigma}(1)\cdot\boldsymbol{\sigma}(2)\, |\chi'_B\rangle \\ &\quad + \tfrac{1}{2}\langle \varphi''_{B'}|\, \tau^{(-)}(1) v^{(+)}(2)\, |\varphi''_B\rangle \langle \chi''_{B'}|\, 1 - \boldsymbol{\sigma}(1)\cdot\boldsymbol{\sigma}(2)\, |\chi''_B\rangle.\end{aligned} \tag{4.247}$$

Matrix elements between wave functions of different symmetry type have been eliminated since $1 - \boldsymbol{\sigma}(1)\cdot\boldsymbol{\sigma}(2)$ is symmetric for the exchange $1 \leftrightarrow 2$. Because of rotational invariance, the spins of B and B′ are necessarily the same, and they will be also the same as for the initial and final baryons B_1 and B_2 (let this spin be $+\frac{1}{2}$ for definiteness). We then have,

$$\langle \chi''|1 - \boldsymbol{\sigma}(1)\cdot\boldsymbol{\sigma}(2)\, |\chi''\rangle = 0, \tag{4.248}$$

$$\langle \chi'|1 - \boldsymbol{\sigma}(1)\cdot\boldsymbol{\sigma}(2)\, |\chi'\rangle = 4, \tag{4.249}$$

The first equation implies the $\Delta I = \frac{1}{2}$ rule, because there remains only the φ' part of the flavour wave function. Then, by direct inspection, it can be seen that, φ' being antisymmetric in the quarks 1 and 2, these are respectively in $I=\frac{1}{2}$ and $I=0$ states for B and B'.

The fact that the internal-baryon-conversion matrix elements (Figure 4.6) imply the $\Delta I=\frac{1}{2}$ *rule for baryons* has been noticed, in more or less general forms, by a number of people, starting with Miura and Minamikawa (1967; see also Körner, 1970; Pati and Woo, 1971). It is important to note that this fact is connected with color and the antisymmetry of baryon color wave functions. We should emphasize that this result ($\Delta I=\frac{1}{2}$ rule for a whole class of contributions to baryon nonleptonic decays) is specific to baryons and does not hold for $\mathrm{K}-\pi$ matrix elements. If $|\psi(0)|^2$ is large enough, we have for baryons a *rationale* for the $\Delta I=\frac{1}{2}$ rule that is lacking for $\mathrm{K}\rightarrow 2\pi$ decays.

We end with expressions of the form

$$(2\pi)^3\langle \mathrm{n}|\,\mathscr{H}_{\mathrm{W}}(0)\,|\Lambda\rangle = -\sqrt{3}\,G_{\mathrm{F}}\cos\theta_{\mathrm{C}}\sin\theta_{\mathrm{C}}\,|\psi(0)|^2, \tag{4.250}$$

where $|\psi(0)|^2$ is the matrix element (4.246). We get for the commutator term

$$\begin{aligned}&(2\pi)^3\langle \mathrm{p}\pi^-|\,\mathscr{H}_{\mathrm{W}}(0)\,|\Lambda\rangle\\ &= -\frac{1}{(2\pi)^{3/2}}\frac{1}{(2\omega_\pi)^{1/2}}\frac{1}{f_\pi}\sqrt{3}\,G_{\mathrm{F}}\cos\theta_{\mathrm{C}}\sin\theta_{\mathrm{C}}\,|\psi(0)|^2.\end{aligned} \tag{4.251}$$

In the last step, we have used the relations

$$Q(\pi^-)|\Lambda\rangle = 0, \tag{4.252}$$

$$Q(\pi^-)^\dagger|\mathrm{p}\rangle = -|\mathrm{n}\rangle. \tag{4.253}$$

We finally get the rate for the parity-violating amplitude:

$$\Gamma^{\mathrm{pv}} = \frac{3}{2\pi}\frac{E_{\mathrm{p}}}{m_\Lambda}k\frac{1}{f_\pi^2}(G_{\mathrm{F}}\cos\theta_{\mathrm{C}}\sin\theta_{\mathrm{C}})^2\,|\psi(0)|^4. \tag{4.254}$$

Let us now concentrate on the evaluation of the spatial matrix element (4.246), the baryon wave function at the center. The harmonic oscillator potential gives good results for peripheral processes, but, as already seen for meson annihilation, it is not realistic for a central quantity like $|\psi(0)|^2$. Let us quote for reference the value we obtain with the harmonic oscillator:

$$|\psi(0)|^2_{\mathrm{ho}} = (2\pi R^2)^{-3/2}. \tag{4.255}$$

Numerically, with our choice of parameters ($R^2=6\,\mathrm{GeV}^{-2}$),

$$|\psi(0)|^2_{\mathrm{ho}} \approx 4\times 10^{-3}\,\mathrm{GeV}^3. \tag{4.256}$$

On the other hand, a value deriving from more realistic potentials is

$$|\psi(0)|^2 \approx 10^{-2}\,\mathrm{GeV}^3. \tag{4.257}$$

The calculation of $|\psi(0)|^2$ for baryons, as opposed to mesons, has not yet been discussed much, and there are still prejudices about its evaluation, mainly because of difficulties in treating the three-quark system. A systematic method has been developed by Richard and collaborators (Richard and Taxil, 1983) but the explicit wave functions are complicated. The value (4.257) has also been obtained by using a superposition of Gaussians with different radii (Ono and Schöberl, 1982). This method is reliable for an approximate estimate of $|\psi(0)|^2$, and has the advantage that the wave functions are easy to handle. We adopt it, and it now remains to fix the potential. We could of course use the same potential as for mesons, with the relation, coming from assumption of confining color-octet forces,

$$V_{qq} = \tfrac{1}{2} V_{q\bar{q}}. \tag{4.258}$$

However, this turns out to be too strong an assumption when confronted with the data. We can at least follow the same line as for mesons. We determine the short- and large-distance behavior from QCD, and we determine the potential from the spectrum. We have to reproduce the ΔE_1 excitation energy (between the first orbitally excited states and the ground state), with a light-quark mass $m = 0.33$ GeV. The potential calculation of Ono and Schöberl (1982) which has the drawback of adopting a pure Coulomb behavior instead of the $1/r \log r$ of QCD, gives, with three Gaussians,

$$|\psi(0)|^2 \approx 8 \times 10^{-3}\,\text{GeV}^3, \tag{4.259}$$

whence the order of magnitude (4.257). It must be emphasized that the realistic value must be *above* the harmonic-oscillator value if we insist on fitting the same excitation energy ΔE_1 with the same quark mass for the two potentials; this may be verified in the calculations of Richard and Taxil (1983). The estimate (4.257) is confirmed by an analysis of electromagnetic hyperfine splittings (Le Yaouanc *et al.*, 1977b).

An important point is that this value is much higher than those currently considered in the literature. In particular, the value advocated by Isgur and Wise (1982) is

$$|\psi(0)|^2 \approx 2 \times 10^{-3}\,\text{GeV}^3 \tag{4.260}$$

Such a small value is obtained by assuming a harmonic-oscillator wave function with $R^2 \approx 10$ GeV^{-2}. Apart from questioning the use of the harmonic oscillator, we think that the main objection to this estimate is that R^2 has been chosen to describe the fine structure of the **56** ground-state multiplet (Isgur and Karl, 1979). We think that, for the determination of the potential, it is more reliable to fit the excitation energy ΔE_1, i.e. the main structure of the spectrum (as opposed to the hyperfine splitting Δ–N).

A comparison should also be made with the MIT bag model, as the MIT school has done much work in this area. The comparison cannot be straightforward, because the bag calculation is a semirelativistic one, for which (4.243) does not hold. Let us, however, define an *effective* $|\psi(0)|^2_{\text{bag}}$ by the formula

$$\langle \mathrm{n}|\mathscr{H}^{\mathrm{pc}}_{\mathrm{W}}|\Lambda\rangle_{\text{bag}} = \frac{1}{(2\pi)^3}\frac{G_{\mathrm{F}}}{\sqrt{2}}\cos\theta_{\mathrm{C}}\sin\theta_{\mathrm{C}}$$

$$\times\, 6 \times \langle \Phi_{\mathrm{n}}|\,\tau^{(-)}(1)v^{(+)}(2)(1-\boldsymbol{\sigma}(1)\cdot\boldsymbol{\sigma}(2))\,|\Phi_{\Lambda}\rangle\,|\psi(0)|^2_{\text{bag}}. \tag{4.261}$$

We find, on comparing with the results from the bag model,

$$|\psi(0)|^2_{\text{bag}} \sim 2\times 10^{-3}\,\mathrm{GeV}^3, \tag{4.262}$$

which is again a value much smaller than the one, (4.257), that we advocate. We believe there are several reasons why the bag estimate is so low. First, the use of the Dirac equation corresponds to taking account of some relativistic corrections, which tend to reduce the nonrelativistic estimate by a factor of about 2. Secondly, the bag model, with the standard parameters, gives a satisfactory description of the $(\mathbf{56}, 0^+)$ ground state, but would give much too low orbital excitations (de Grand, 1976). If the orbital excitations are fitted then much higher values for $|\psi(0)|^2{}_{\text{bag}}$ are obtained. Thirdly, the square-well potential tends to give smaller values than a more realistic (linear plus Coulomb) potential (Abe *et al.*, 80).

We therefore conclude that for a realistic potential that fits the orbital-excitation spectrum, and has the long- and short-distance behavior required by QCD intuition, our large value is reasonable, and much smaller values of $|\psi(0)|^2$ should be discarded. Let us now compute the absolute magnitude of the rate from the value for $|\psi(0)|^2$ obtained from a realistic potential for the spectrum, (4.257). We obtain numerically

$$\Gamma^{\mathrm{pv}}(\Lambda\to\mathrm{p}\pi^-) \approx 1.82\times 10^{-15}\,\mathrm{GeV}. \tag{4.263}$$

This time, in contrast with the direct-emission process, the agreement with the experimental value $\Gamma^{\mathrm{pv}} = 1.4\times 10^{-15}$ GeV is reasonable. We can conclude with Schmid (1977) that the quark model solves the old enhancement problem both of the absolute magnitude and the $\Delta I = \frac{1}{2}$ rule, at least for *hyperon* decays. This point must be particularly emphasized, because it is often hidden by the discussion of finer details and by the failure to explain the $\mathrm{K}\to 2\pi$ decays. The quark model roughly explains the relative order of magnitude of nonleptonic as compared with semileptonic *hyperon* decays, which has long been a mystery.

If we want to describe the whole set of parity-conserving and parity-violating hyperon decay amplitudes with their relative sign, we must pay attention to the conventions. First, for the relative sign of the S and P waves, the Particle Data Group tables use the convention, accepted for many years in current algebra, $\gamma_5 = i\gamma^0\gamma^1\gamma^2\gamma^3$, i.e. opposite in sign to the now-standard Bjorken and Drell convention. We therefore have to change our P-wave signs. Since the S wave (pv) of the transition Λ^0_- ($\Lambda \to p\pi^-$; the lower index stands for the pion charge) is conventionally chosen positive, in contradiction with our finding, we must also multiply both waves by a minus sign. As a result, we just have to change the signs of the S waves. As far as the signs relative to different SU(3) states are concerned, we are in agreement with the phase conventions of the Particle Data Group.

As to absolute magnitudes, we have to remark that the Particle Data Group factorizes a dimensionless factor $G_F\mu_c^2 = 2.27 \times 10^{-7}$. In addition, the invariant amplitudes used by them differ from our amplitudes M by normalization factors. (In the spirit of our approximations, we make $E_2 \approx m_2$. Remember that M is calculated in the rest frame of B_1). In summary, we have

$$M^{\mathrm{pv}} = -\frac{1}{(2\pi)^{3/2}}\frac{1}{(2\omega_\pi)^{1/2}} G_F\mu_c^2 A, \tag{4.264}$$

$$M^{\mathrm{pc}} = \frac{1}{(2\pi)^{3/2}}\frac{1}{(2\omega_\pi)^{1/2}} G_F\mu_c^2 \frac{Bk}{2m_2}. \tag{4.265}$$

In Table 4.8 we compare theoretical values of A and B with the experimental ones. We have followed Bonvin (1984) in extracting the $\Delta I = \frac{1}{2}$ part from the experimental data. This is in agreement with the fact that we do not claim to predict the small $\Delta I = \frac{3}{2}$ contribution that will come from the direct-emission process alone. Our results are $\Delta I = \frac{1}{2}$ as we retain the dominant pure $\Delta I = \frac{1}{2}$ internal-conversion baryon processes. We therefore compare directly with the experimental $\Delta I = \frac{1}{2}$ part. We now discuss the theoretical results. The discussion for S waves is easy, while P waves deserve detailed comments.

Table 4.8 Baryon nonleptonic decay amplitudes. Definition according to Particle Data Group (1982). Upper index stands for the hyperon charge, the lower one for the emitted pion charge. The data concern only the $\Delta I = \frac{1}{2}$ part as calculated by Bonvin (1984).

	A (parity-violating)		B (parity-conserving)	
	Theory	Experiment	Theory	Experiment
Λ^0_-	1.5	1.48 ± 0.01	4.8	10.02 ± 0.42
Σ^+_+	0	0.06 ± 0.01	16	19.07 ± 0.07
Σ^+_0	2.6	1.37 ± 0.02	-7.0	-13.31 ± 0.26
Σ^-_-	3.7	2.00 ± 0.03	5.6	0.25 ± 0.34
Ξ^-_-	3	2.09 ± 0.02	-10	-7.61 ± 0.34

As already observed, the S wave for $\Lambda \to p\pi^-$ is rather well predicted. Also, the algebraic prediction $S(\Sigma_+^+)=0$ is remarkably verified. For the other S waves (Σ_-^- or Σ_-^+, Ξ_-^-), although the order of magnitude is still good, the amplitude is too large by a factor of approximately two.

In this very simple model, *S waves* are given directly by the baryon matrix element through the *commutator term.* P waves are more involved, because to compute the baryon pole contribution it is necessary to estimate in addition axial-coupling matrix elements. But also, and what is more worrying, the calculation is ambiguous. It happens that most of the observed waves are obtained by a strong cancellation between large numbers due to the small energy denominators, of the order of the SU(3) mass breaking. The final result is very sensitive and unstable to the details of the calculation.

Let us consider for instance the *P-wave for* $\Lambda \to p\pi^-$. The amplitude is given by

$$M(\Lambda_-)=\frac{1}{(2\pi)^{3/2}}\frac{1}{(2\omega_\pi)^{1/2}}\frac{1}{f_\pi}k(2\pi)^3$$
$$\times\left\{\frac{\langle \mathrm{p}|j_5^z(\pi^-)|\mathrm{n}\rangle(2\pi)^3\langle \mathrm{n}|\mathscr{H}_{\mathrm{W}}^{\mathrm{pc}}|\Lambda\rangle}{m_\Lambda-m_\mathrm{N}}\right.$$
$$\left.-\frac{\langle \mathrm{p}|\mathscr{H}_{\mathrm{W}}^{\mathrm{pc}}|\Sigma^+\rangle(2\pi)^3\langle\Sigma^+|j_5^z(\pi^-)|\Lambda\rangle}{m_\Sigma-m_\mathrm{N}}\right\}. \tag{4.266}$$

The weak matrix elements of $\mathscr{H}_{\mathrm{W}}^{\mathrm{pc}}$ have already been calculated. It remains to evaluate the axial-current-matrix elements. But this has essentially been done when discussing baryon semileptonic decays, except for a difference in quantum numbers. We need here $\Delta S=0$, $\Delta Q=0$, ± 1 axial-current-matrix elements. We get finally

$$M=-\frac{1}{(2\pi)^{3/2}}\frac{1}{(2\omega_\pi)^{1/2}}\frac{1}{f_\pi}G_\mathrm{F}\cos\theta_\mathrm{C}\sin\theta_\mathrm{C}|\psi(0)|^2$$
$$\times 2\sqrt{3}\left(-\frac{5}{6}\frac{1}{m_\Lambda-m_\mathrm{N}}+\frac{1}{m_\Sigma-m_\mathrm{N}}\right) \tag{4.267}$$

There is a strong cancellation between the two terms in the parentheses because the mass differences are close and the algebraic coefficients are of the same order. In fact, on taking the limit $m_\Lambda-m_\mathrm{N}=m_\Sigma-m_\mathrm{N}$ the sign of (4.267) is inverted relatively to the empirical values for m_Σ and m_Λ. It might be wondered whether it is significant to take into account the Σ–Λ mass difference while so crude approximations are made in the v/c expansion, SU(3) breaking, etc. in the calculation of the matrix elements. Theoretically, the mass difference Σ–Λ is an effect combining SU(3) breaking and the spin–spin forces of the short-distance potential, which are second-order relativistic corrections (see e.g. Le Yaouanc *et al.*, 1978b). Since the residues of the poles are estimated at the lowest order, it seems that Σ–Λ should be neglected. The main conclusion is that, although the parity-conserving amplitude $P(\Lambda^0_-)$ is calculable in the above scheme of approximation, the result is not very significant, because it would be considerably altered by several different corrections.

The situation is similar for $P(\Sigma^-_-, \Sigma^+_+, \Sigma^+_0)$, as there are also strong cancellations in that case. Again, neglecting $m_\Sigma - m_\Lambda$ considerably alters the magnitude of Σ^-_-. The amplitudes Σ^+_+ and Σ^+_0 seem slightly more reliable. The great advantage of Ξ^-_- is that there is a unique pole contribution in the u channel, and therefore no cancellation at all.

A discussion of the current literature is now useful. It has been standard for a long time to calculate the $\frac{1}{2}^+$ baryon pole contribution by Yukawa pseudoscalar couplings instead of axial couplings for the pion strong emission. In principle, as observed in Sections 2.2.1 and 3.2.6, both procedures should be equivalent. If, however, the question is treated in a phenomenological way, without the quark model, different results may be obtained. Since the couplings are not all known from experiment, it is necessary to appeal to SU(3) symmetry, and applying this symmetry to axial couplings is *not equivalent* to applying it to pseudoscalar Yukawa couplings, because the relation between the two involves the baryon masses

$$g^{PS}_{BB'M} = (m_B + m_{B'}) \frac{1}{f_\pi} g^{PV}_{BB'M} \tag{4.268}$$

where M is some meson, and PS and PV denote the pseudoscalar (Yukawa) and pseudovector (axial) couplings. This equation is the Goldberger–Treiman relation obtained in the current-algebra framework. Since the actual baryon masses break the SU(3) symmetry, SU(3) cannot be exact simultaneously for g^{PS} and g^{PV}. Moreover, an f/d ratio has to be chosen, because there are two independent couplings between three SU(3) octets. The differences are not crucial in general; they are crucial in P-wave weak decays because of the above cancellations. A mass factor $m_B + m_{B'}$, which is different for two cancelling poles, or a small change in f/d from its SU(6) value $\frac{2}{3}$, can cause very considerable differences, especially for the transition Λ^0_-. For instance, taking SU(3) symmetry for Yukawa couplings with $f/d = \frac{2}{3}$, or SU(3) symmetry for axial couplings with $f/d = 0.54$, gives a difference of one *order of magnitude* for $P(\Lambda^0_-)$.

The first prescription has generally been used, and yields a very good fit to P waves (Gronau, 1972). However, as observed by Bonvin (1984), the SU(3) symmetry is experimentally much better for axial couplings. There are two pieces of evidence: first, the direct consideration of ΣΛπ and ΣΣπ strong couplings (Nagels *et al.*, 1979), and, secondly, the Cabibbo analysis of semileptonic decays. Including the recent analysis of Bourquin *et al.* (1983), they both converge towards SU(3) symmetry of axial couplings and $f/d = \frac{2}{3}$ ($f/d = 0.54$ was actually chosen by Bonvin). This is exactly the result of the quark model, which, disappointingly, gives a rather bad fit for the ratio of the different P waves (Table 4.8).

Finally, we retain our quark-model calculation, because it is more logical, and we make only a slight modification. We know that axial couplings are correctly described by the quark model in its nonrelativistic version, except for a discrepancy between the naive expectation $-\frac{5}{3}$ and the experimental value -1.22 for g_1/f_1 (see Section 4.2.1). We shall thus just include an overall corrective factor of approximately 0.7.

The conclusions to be drawn for P waves from comparison with experiment are as follows: (i) considering the transition Ξ^-_-, which is the most reliable, the general magnitude of P waves is correct; (ii) there are appreciable discrepancies for Σ^+_+, Σ^+_0 and Λ^0. Σ^-_- is very bad. This is not unexpected from the preceding discussion.

We also observe from Table 4.8 that the signs are good for all the waves, except possibly $P(\Sigma^-_-)$. On the whole, the most naive quark model once more meets with encouraging success. However, corrections are needed: they will be discussed in the following sections.

4.6.3 Short-Distance QCD Corrections

We have considered up to now the old local four-fermion (V−A)(V−A) interaction. But in fact the weak transition results from the exchange of W bosons, and what we really get is not a four-quark local Hamiltonian, but rather an amplitude proportional to

$$\int \mathrm{d}x\,\mathrm{d}y\, D(m_W; x-y)\,\langle \mathrm{B}_2\pi^i|\,\mathrm{T}\,\{j^-_\mu(x), j^{+\mu}(y)\}|\mathrm{B}_1\rangle, \tag{4.269}$$

where $D(m_W; x-y)$ is the W propagator. Because m_W is very large with respect to the hadronic scales, the transition operator can be expressed in terms of local four-fermion operators according to the Wilson short-distance expansion of the product of two currents (Wilson, 1969; Altarelli and Maiani, 1974; Gaillard and Lee, 1974a). There is a set of operators $\mathcal{O}_i(x)$ such that

$$\mathrm{T}\{j^-_\mu(x), j^{+\mu}(y)\} \underset{y\to x}{\simeq} \sum_i c_i(x-y)\,\mathcal{O}_i(x). \tag{4.270}$$

The $\mathcal{O}_i$ are local operators that can be expressed as local normal products of the basic fields, quarks and gluons, at the point x. It can be shown that for weak transition matrix elements of order $1/m_W^2$, which correspond to the usual first-order weak interactions, the only relevant local operators are *normal products of four quark fields*. Denoting these operators by $\mathcal{O}_i$, we can write the matrix element under consideration in the form

$$M \equiv (2\pi)^3 \langle B_2 \pi^i | \mathcal{H}_{\text{eff}} | B_1 \rangle$$
$$= (2\pi)^3 \frac{G_F}{2\sqrt{2}} \cos\theta_C \sin\theta_C \langle B_2 \pi^i | \sum_i c_i \mathcal{O}_i | B_1 \rangle. \tag{4.271}$$

The c_i are numbers coming from the integration over the relative position $r = x - y$:

$$\int d^4 r\, D(m_W; r)\, c_i(r) = \frac{c_i}{m_W^2}. \tag{4.272}$$

The coefficients $c_i(r)$ depend only on the strong-interaction QCD Lagrangian, while the c_i on the right-hand side also depend logarithmically on the mass m_W. They can be calculated by taking matrix elements between suitable states, for instance quark states. In this latter case, owing to the QCD asymptotic freedom, the matrix elements can be calculated by perturbative Feynman diagrams. Since QCD contains ultraviolet divergences, renormalization is necessary and a dependence is introduced on some renormalization mass μ. However, it is important to emphasize that the physical transition amplitude should be independent of this renormalization point μ. Therefore the coefficient $c_i(r)$ and the matrix elements of $\mathcal{O}_i(x)$ should depend on μ in such a way as to eventually have (4.271) independent of μ. Finally, after the integration (4.272), coefficients c_i are obtained that depend on g^2, μ, m_W and current-quark masses. The power m_W^{-2} having been extracted, the dependence of c_i on m_W is only logarithmic. We do not present the actual calculation of the c_i Wilson coefficients, but only the results. We can recover, of course, the previous weak Hamiltonian in the absence of QCD effects at short distance, in the limit $\alpha_s \to 0$.

For a $\Delta S = +1$ transition, a standard set of operators is the following. We introduce (Donoghue *et al.*, 1980) the useful auxiliary four-fermion operators (conventionally taken at zero space–time coordinates)

$$\left.\begin{aligned}
\mathcal{H}_A &= \bar{d}\gamma^\mu(1-\gamma_5)u\, \bar{u}\gamma_\mu(1-\gamma_5)s,\\
\mathcal{H}_B &= \bar{u}\gamma^\mu(1-\gamma_5)u\, \bar{d}\gamma_\mu(1-\gamma_5)s,\\
\mathcal{H}_C &= \bar{d}\gamma^\mu(1-\gamma_5)d\, \bar{d}\gamma_\mu(1-\gamma_5)s,\\
\mathcal{H}_D &= \bar{s}\gamma^\mu(1-\gamma_5)s\, \bar{d}\gamma_\mu(1-\gamma_5)s,
\end{aligned}\right\} \tag{4.273}$$

where the color indices are saturated within each bilinear (e.g. $\bar{u}_\alpha \ldots u_\alpha\, \bar{u}_\beta \ldots s_\beta$). Then we set

$$\left.\begin{aligned} \mathcal{O}_1 &= \mathcal{H}_A - \mathcal{H}_B, \\ \mathcal{O}_2 &= \mathcal{H}_A + \mathcal{H}_B + 2\mathcal{H}_C + 2\mathcal{H}_D, \\ \mathcal{O}_3 &= \mathcal{H}_A + \mathcal{H}_B + 2\mathcal{H}_C - 2\mathcal{H}_D, \\ \mathcal{O}_4 &= \mathcal{H}_A + \mathcal{H}_B - \mathcal{H}_C. \end{aligned}\right\} \tag{4.274}$$

These combinations are selected for reasons of symmetry. In particular, $\mathcal{O}_1$ is antisymmetric under $s \leftrightarrow u$ exchange, while $\mathcal{O}_{2,3,4}$ are symmetric. $\mathcal{O}_{1,2}$ are SU(3) octets, while $\mathcal{O}_{3,4}$ behave as **27**. Finally, $\mathcal{O}_{1,2,3}$ are $\Delta I = \frac{1}{2}$, while $\mathcal{O}_4$ is $\Delta I = \frac{3}{2}$.

The operator $\mathcal{H}_A$ is in fact the one present in the standard (V−A)(V−A) interaction considered up to now. It can be expressed in terms of the operators $\mathcal{O}_i$ according to

$$\mathcal{H}_A = \tfrac{1}{2}\mathcal{O}_1 + \tfrac{1}{10}\mathcal{O}_2 + \tfrac{1}{15}\mathcal{O}_3 + \tfrac{1}{3}\mathcal{O}_4. \tag{4.275}$$

Therefore $c_1 = 1$, $c_2 = \frac{1}{5}$, $c_3 = \frac{2}{15}$ and $c_4 = \frac{2}{3}$, and no further operator in the set $\{\mathcal{O}_i\}$ would give exactly the standard old (V−A)(V−A) model. However, owing to the strong interactions, c_i deviate from these values by quantities proportional to α_s times logarithms of m_W/μ: for example,

$$c_1 = 1 + \frac{\alpha_s}{\pi} \ln \frac{m_W^2}{\mu^2} + O(\alpha_s^2). \tag{4.276}$$

In addition, there are two new operators $\mathcal{O}_5$ and $\mathcal{O}_6$, not present in the expression for $\mathcal{H}_A$ because they are generated only at first order in α_s. They are found by making the short-distance expansion of the diagram of Figure 4.8, the so-called penguin diagram (Vainshtein *et al.*, 1977).

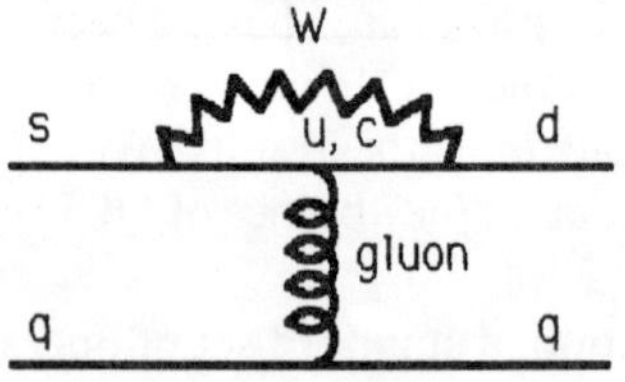

Figure 4.8 The pure $\Delta I = \frac{1}{2}$ penguin diagram.

These new contributions are pure $\Delta I = \frac{1}{2}$, since they have an $s \to d$ transition, that is $\Delta I = \frac{1}{2}$, and the qqg vertex is flavor-singlet. The new operators $\mathcal{O}_5$ and $\mathcal{O}_6$ can be written as

$$\mathcal{O}_6 = [\bar{u}\gamma^\mu(1+\gamma_5)u + \bar{d}\gamma^\mu(1+\gamma_5)d + \bar{s}\gamma^\mu(1+\gamma_5)s] \times \bar{d}\gamma_\mu(1-\gamma_5)s \tag{4.277}$$

$$\begin{aligned} \mathcal{O}_5 = {} & [\bar{u}_\alpha\gamma^\mu(1+\gamma_5)u_\beta + \bar{d}_\alpha\gamma^\mu(1+\gamma_5)d_\beta + \bar{s}_\alpha\gamma^\mu(1+\gamma_5)s_\beta] \\ & \times \bar{d}_\beta\gamma_\mu(1-\gamma_5)s_\alpha - \tfrac{2}{3}\mathcal{O}_6. \end{aligned} \tag{4.278}$$

Note that in $\mathcal{O}_5$ color indices of the quarks have been given explicitly (as Greek indices) because the bilinears are not color singlets. The fact that $\mathcal{O}_5$ and $\mathcal{O}_6$ are only induced by graphs involving gluons is manifested by two facts: (i) the presence of right-handed currents, which are induced by the vector gluon–quark coupling; (ii) the presence of neutral currents that are not color singlets — also generated by the gluon–quark coupling.

Expanding in α_s gives

$$c_6 = O(\alpha_s^2), \tag{4.279}$$

$$c_5 = -\frac{1}{3}\frac{\alpha_s}{4\pi}\ln\frac{m_c^2}{\mu^2} + O(\alpha_s^2). \tag{4.280}$$

At higher orders there are logarithms of m_W/μ, but at lowest order there is only the logarithm of m_c/μ, which is much smaller. Very heavy flavours (like the t quark, which can enter in the fermion loop of Figure 4.8) can change this estimate.

By a generalization of the arguments proving the $\Delta I = \frac{1}{2}$ rule for baryons in the old (V − A)(V − A) model, it can be proved that

$$\langle B'|\mathcal{O}_2|B\rangle = \langle B'|\mathcal{O}_3|B\rangle = \langle B'|\mathcal{O}_4|B\rangle = 0 \tag{4.281}$$

This is a generalization of the $\Delta I = \frac{1}{2}$ rule, since $\mathcal{O}_4$ is the only $\Delta I = \frac{3}{2}$ operator. This implies, leaving apart the penguin contributions,

$$\langle B'|\mathcal{H}_W^{\text{eff}}|B\rangle = \frac{G_F}{2\sqrt{2}}\cos\theta_C\sin\theta_C\, c_1 \langle B'|\mathcal{O}_1|B\rangle$$

$$= \frac{G_F}{\sqrt{2}}\cos\theta_C\sin\theta_C\, c_1 \langle B'|\mathcal{H}_A|B\rangle, \tag{4.282}$$

the last expression following from (4.275) and (4.281).

The interesting conclusion is that, apart from the penguin contributions (operators $\mathcal{O}_5$ and $\mathcal{O}_6$), the difference from the previous naive calculation consists only *in a general factor c_1 independent of the hadron states.* It is therefore important to see whether the penguin contribution changes this conclusion.

Let us consider the contribution of $\mathcal{O}_5$ and $\mathcal{O}_6$ (Vainshtein *et al.*, 1977). Here something qualitatively new happens (Donoghue *et al.*, 1980). If we apply the analysis of Section 4.6.2, we can again write the amplitude as a commutator plus pole contributions. The weak side of the pole contribution is once more given by baryon-to-baryon matrix elements $\langle B'|\mathcal{O}_{5,6}|B\rangle$. However, as explained below, in contrast with the standard (V − A)(V − A) interaction with only the operators $\mathcal{O}_i (i = 1, \ldots, 4)$, the commutator contribution is not given now by such baryon matrix elements. As we have seen in Section 4.6.1, the commutator can be reduced to a baryon matrix element of $\mathscr{H}_W(0)$ if the relation

$$[Q_5^j, \mathscr{H}_W(0)] = -[Q^j, \mathscr{H}_W(0)] \tag{4.283}$$

holds.

However, since $\mathcal{O}_{5,6}$ contain right-handed currents, the demonstration leading to (4.283) no longer holds. It is still true, however, that

$$[Q_5^j, \tilde{\mathcal{O}}_{5,6}] = -[Q^j, \tilde{\mathcal{O}}_{5,6}], \tag{4.284}$$

where $\tilde{\mathcal{O}}_{5,6}$ denote the operators $\mathcal{O}_{5,6}$ *without normal ordering*. This comes from the fact that $\tilde{\mathcal{O}}_{5,6}$ contain only flavor-*singlet* right-handed currents, and, since Q_R^j bears only isospin $I=1$ (j labelling the pion state),

$$[Q_R^j, \bar{\psi}_R \gamma_\mu \psi_R] = 0, \tag{2.285}$$

it follows that

$$[Q_R^j, \tilde{\mathcal{O}}_{5,6}] = 0, \tag{4.286}$$

since $\tilde{\mathcal{O}}_{5,6}$ are products of currents. On the other hand, the normal-ordered $\mathcal{O}_{5,6}$, which govern the physical transition matrix element, are *not* products of currents, and therefore (4.284) is not *a priori* valid for $\mathcal{O}_{5,6}$. In fact, applying Wick's theorem,

$$\mathcal{O}_5 = \tilde{\mathcal{O}}_5 + \frac{32}{9}\langle 0|\bar{d}d|0\rangle \bar{d}(1-\gamma_5)s + \frac{32}{9}\langle 0|\bar{s}s|0\rangle \bar{d}(1+\gamma_5)s. \tag{4.287}$$

The additional terms come from the contractions of the quark fields. Such contractions are also present for the operators $\mathcal{O}_i\,(i=1,\ldots,4)$. What is specific to $\mathcal{O}_{5,6}$ is that they contain both left- and right-handed currents, leading to vacuum expectation values of scalar and pseudoscalar densities, as coefficients of scalar (S) and pseudoscalar (P) bilinear operators: $\langle 0|\bar{d}d|0\rangle = \langle 0|\bar{d}_L d_R + \bar{d}_R d_L|0\rangle$, and obviously such condensates cannot contribute to the contractions of $\mathcal{O}_i(i=1,\ldots,4)$, which contain only left-handed fields. It is clear that such combinations of bilinear P and S operators do not commute with Q_R^j in general. We have

$$[Q_R^j, \bar{d}(1+\gamma_5)s] = 2[Q_R^j, \bar{d}_L s_R] = 0, \tag{4.288}$$

but, in contrast,

$$[Q_R^j, \bar{d}(1-\gamma_5)s] = 2[Q_R^j, \bar{d}_R s_L] \neq 0, \tag{4.289}$$

since Q_R^j may contain d_R. For the particular case $\pi^- = -d\bar{u}$

$$[Q_R(\pi^-), \bar{d}(1-\gamma_5)s] = -\bar{u}(1-\gamma_5)s, \tag{4.290}$$

and finally

$$[Q_5(\pi^-), \mathcal{O}_5] = -[Q(\pi^-), \mathcal{O}_5] - \frac{64}{9}\langle 0|\,\bar{d}d\,|0\rangle\, \bar{u}(1-\gamma_5)s, \qquad (4.291)$$

The second term is present because the vacuum expectation value $\langle 0|\,\bar{d}d\,|0\rangle \neq 0$, inducing the spontaneous breaking of chiral symmetry (1.40). This phenomenon is at the base of the quasi-Goldstone character of the pion and of PCAC. Let us summarize the reasons why $\mathcal{O}_5$ and $\mathcal{O}_6$ do not satisfy the commutation relations (4.283) that are satisfied by $\mathcal{O}_i$ $(i=1,\ldots,4)$. First, we are considering *normal-ordered* four-fermion interactions, i.e. the part that is actually responsible for the decay and not the parts of the interaction that give only quark self-energies. Secondly, the four-fermion interaction contains left- and right-handed currents, which are induced through QCD corrections. Thirdly, there is spontaneous breaking of chiral symmetry. If one of these conditions is dropped, we have the usual commutation relations. It is therefore clear that the *anomalous* term (the last term in the right-hand side of 4.291) is something very specific to QCD.

As a simple calculation shows, this last term corresponds to the direct-emission contribution generated by $\mathcal{O}_5$. In contrast with that generated by $\mathcal{O}_i$ $(i=1,\ldots,4)$, it does not vanish with k_μ, and it is included in the commutator term. Although a number of authors calculate this contribution as direct emission, we prefer to stick to our method, based on the reduction to the commutator.

The first term on the right-hand side of (4.291) (S wave) and the $\frac{1}{2}^+$ pole contribution to the P wave (the $\mathcal{O}_5$ contribution to (4.266)), are given by the one-baryon matrix elements $\langle B'|\,\mathcal{O}_5|B\rangle$, which give in the nonrelativistic approximation an expression similar to that for $\mathcal{O}_1$, proportional to $|\psi(0)|^2$. The magnitude is also of the same order as for $\mathcal{O}_1$, but, as it is multiplied by a much smaller c_5 coefficient, the final contribution to the amplitude is negligible compared with that of $c_1\mathcal{O}_1$. A possible exception could be the P waves, when strong cancellations occur in the $\mathcal{O}_1$ contribution, as for Λ^0_-.

Let us now consider the second term in the expression (4.291) for the commutator. It contributes, for example to the amplitude for $\Lambda \to p\pi^-$, through a term which we denote $M(5)$,

$$M(5) = -\frac{1}{(2\pi)^{3/2}}\frac{1}{(2\omega_\pi)^{1/2}}\frac{1}{f_\pi}\frac{G_F}{2\sqrt{2}}\cos\theta_C \sin\theta_C$$

$$\times\, c_5(2\pi)^3\,\frac{64}{9}\,\langle 0|\,\bar{d}d\,|0\rangle\langle p|\,\bar{u}(1-\gamma_5)s\,|\Lambda\rangle. \qquad (4.292)$$

We note that there is a contribution to both S and P waves. What we have to calculate first is a matrix element of a bilinear operator, and then we need an evaluation of the vacuum expectation value $\langle 0|\bar{d}d|0\rangle$.

For the *bilinear matrix element*, we could use the quark model as in the elementary-pion-emission model (Section 2.2.1). However, it is also possible, as suggested by Vainshtein *et al.* (1977), to use the fact that $\bar{u}(1-\gamma_5)s$ is a divergence of currents. We then have

$$\bar{u}s = \mathrm{i}\partial_\mu(\bar{u}\gamma^\mu s)/(\hat{m}_\mathrm{s}-\hat{m}_\mathrm{u}), \tag{4.293}$$

$$\bar{u}\gamma_5 s = -\mathrm{i}\partial_\mu(\bar{u}\gamma^\mu\gamma_5 s)/(\hat{m}_\mathrm{s}+\hat{m}_\mathrm{u}), \tag{4.294}$$

where $\hat{m}_\mathrm{s}$ and $\hat{m}_\mathrm{u}$ are the *current* masses of the quarks, not to be confused with the constituent-quark masses used in the quark-model calculations of the spectrum and transitions. The current masses $\hat{m}$ are small, depend on the renormalization scale, and express, in the QCD Lagrangian, the small departure of QCD from exact chiral symmetry (whence the *quasi*-Goldstone character of the pion); see Section 1.2.4. As an order of magnitude, we propose (Gasser and Leutwyler, 1982).

$$\hat{m} = 7\,\mathrm{MeV}, \qquad \hat{m}_\mathrm{s} = 180\,\mathrm{MeV}, \tag{4.295}$$

where $\hat{m}$ is the average of $\hat{m}_\mathrm{u}$ and $\hat{m}_\mathrm{d}$.

Using the baryon SU(6) wave functions (for the axial current we can neglect the pseudoscalar induced term from the K pole, since $q^2 = m_\pi^2 \approx 0$), we can write the matrix elements of the current as

$$(2\pi)^3\langle \mathrm{p}|\bar{u}\gamma^\mu s|\Lambda\rangle \approx \sqrt{\tfrac{3}{2}}\,\bar{p}\gamma^\mu\Lambda \tag{4.296}$$

$$(2\pi)^3\langle \mathrm{p}|\bar{u}\gamma^\mu\gamma_5 s|\Lambda\rangle \approx -\tfrac{3}{5}\sqrt{\tfrac{3}{2}}\,g_\mathrm{A}\bar{p}\gamma^\mu\gamma_5\Lambda = -0.87\bar{p}\gamma^\mu\gamma_5\Lambda, \tag{4.297}$$

with the $\mathrm{n}\rightarrow\mathrm{p}$ axial coupling $g_\mathrm{A} = 1.25$. We finally obtain

$$(2\pi)^3\langle \mathrm{p}|\bar{u}s|\Lambda\rangle \approx \sqrt{\tfrac{3}{2}} \tag{4.298}$$

$$(2\pi)^3\langle \mathrm{p}|\bar{u}\gamma_5 s|\Lambda\rangle = -0.87\frac{m_\mathrm{p}+m_\Lambda}{\hat{m}+\hat{m}_\mathrm{s}}\,\bar{p}\gamma_5\Lambda \approx -10\,\bar{p}\gamma_5\Lambda. \tag{4.299}$$

The result for the scalar density is in fact the same as in the nonrelativistic quark model. In contrast the result for the pseudoscalar density is very different from the nonrelativistic calculation, because of the mass factor $(m_\mathrm{p}+m_\Lambda)/(\hat{m}+\hat{m}_\mathrm{s}) \gg 1$.

The *vacuum expectation value* $\langle 0|\bar{d}d|0\rangle$ has a standard evaluation connected with the above estimate of quark masses (4.295):

$$\langle 0|\bar{d}d|0\rangle \approx -(0.230\,\mathrm{GeV})^3, \qquad \mu = 1\,\mathrm{GeV}. \tag{4.300}$$

To find a scale for the penguin contribution, we shall compare it to the naive estimate of the internal-conversion process (4.251). We have, for the S wave of $\Lambda \to p\pi^-$,

$$\frac{M(5)^{\mathrm{pv}}}{M(\mathrm{naive})^{\mathrm{pv}}} = \frac{16}{9} c_5 \frac{\langle 0|\bar{d}d|0\rangle}{|\psi(0)|^2} \approx -2c_5. \tag{4.301}$$

We obtain $c_5 = -0.006$ with $\Lambda = 0.2\,\mathrm{GeV}$, $\mu = 1\,\mathrm{GeV}$ and a charmed quark mass $m_c(\mu) = 1.3\,\mathrm{GeV}$. However, for such a low value of m_c, the calculation of c_5 is questionable. In principle, for the leading-log approximation to make sense, we require $m_c \gg \mu$. On the other hand, a meaningful estimate of the different parameters at the scale μ, $\alpha_s(\mu)$, $m_c(\mu)$ and $\langle\bar{\psi}\psi\rangle(\mu)$ requires $\mu \gg \Lambda$. Therefore we may choose something like $\mu = (\Lambda m_c)^{1/2}$, which gives $\mu \approx 0.6\,\mathrm{GeV}$. This leads to a somewhat larger result, $c_5 = -\frac{1}{27}$, and we obtain

$$\frac{M(5)^{\mathrm{pv}}}{M(\mathrm{naive})^{\mathrm{pv}}} \approx 0.045. \tag{4.302}$$

Anyway, it can be safely concluded that *the penguin contribution to hyperon decays is negligible*, at least in perturbative QCD. The same conclusion is found for P waves. We have, for example for the transition $\Lambda \to p\pi^-$,

$$\frac{M(5)^{\mathrm{pc}}}{M(\mathrm{naive})^{\mathrm{pc}}} \approx -\frac{1}{2} c_5, \qquad \mu = 1\,\mathrm{GeV}. \tag{4.303}$$

Our final conclusion on short-distance corrections is therefore that only the corrections to $\mathcal{O}_i\,(i = 1, \ldots, 4)$ are significant. The effect is then only to multiply the naive result by $c_1 \sim 2$, and this would lead to systematically too large decay amplitudes, as the naive estimate of Section 4.6.2, equivalent to taking $c_1 = 1$, was already of the right order of magnitude.

4.6.4 Relativistic Corrections

As already emphasized, the quantity to be determined for nonleptonic hyperon decays, as calculated through PCAC, is the baryon-to-baryon matrix element $\langle B'|\mathcal{H}_W|B\rangle$. We have so far used the standard nonrelativistic quark-model methods. But we shall now consider higher-order corrections, as we have done for other processes. The problem is analogous to that of meson annihilation in e^+e^-, where the amplitude is also the wave function at zero interquark distance. The main corrections are the short-distance QCD corrections of the form $[\alpha_s(\mu^2)\ln(m_W{}^2/\mu^2)]^n$, which are calculable (Section 4.6.3) owing to the fact that QCD is asymptotically free.

There is, however, another type of corrections, usually called relativistic corrections, which are of order v^2/c^2, and which come from soft momentum regions, described by the relativistic potential methods discussed in Chapter 3. Although they do not seem to be essential, we shall discuss them briefly, since they have been considered by many authors.

For the sake of simplicity, we start from the four-fermion interaction. Furthermore, we assume that the baryons are at rest. If we first assume that the spatial wave function, expressing the state of the baryon in terms of creation operators of quarks, is just the nonrelativistic one, then we simply have to consider the relativistic quark spinors instead of two-component Pauli spinors. This amounts to the substitution of spinors corresponding to annihilation operators,

$$\chi_s \to \left(\frac{m}{E}\right)^{1/2} u_s(\boldsymbol{p}), \tag{4.304}$$

and those to creation operators,

$$\chi_s^\dagger \to \left(\frac{m}{E}\right)^{1/2} \bar{u}_s(\boldsymbol{p}), \tag{4.305}$$

and then to use the full expression for $\mathscr{H}_W$, as, for example, for the parity-conserving part,

$$\mathscr{H}_W^{pc} = \frac{G_F}{\sqrt{2}} \cos\theta_C \sin\theta_C \, [\bar{d}\gamma^\mu u \, \bar{u}\gamma_\mu s + \bar{d}\gamma^\mu\gamma_5 u \, \bar{u}\gamma_\mu\gamma_5 s], \tag{4.306}$$

instead of its nonrelativistic reduction. The calculation is straightforward. Performing an expansion to obtain the v^2/c^2 corrections, we find that we need only make the following substitution in the nonrelativistic expression:

$$\begin{aligned} |\psi(0)|^2 &= \frac{1}{(2\pi)^3} \int d\boldsymbol{q} \int \prod_i d\boldsymbol{p}_i \, \delta(\Sigma \boldsymbol{p}_i) \, \psi^\dagger(\boldsymbol{p}_1 + \boldsymbol{q}, \boldsymbol{p}_2 - \boldsymbol{q}, \boldsymbol{p}_3) \, \psi(\boldsymbol{p}_1, \boldsymbol{p}_2, \boldsymbol{p}_3) \\ &\to \frac{1}{(2\pi)^3} \int d\boldsymbol{q} \int \prod_i d\boldsymbol{p}_i \, \delta(\Sigma \boldsymbol{p}_i) \, \psi^\dagger(\boldsymbol{p}_1 + \boldsymbol{q}, \boldsymbol{p}_2 - \boldsymbol{q}, \boldsymbol{p}_3) \\ &\times \left[1 - \frac{\boldsymbol{p}_1^2 + \boldsymbol{p}_2^2 + (\boldsymbol{p}_1 + \boldsymbol{q})^2 + (\boldsymbol{p}_2 - \boldsymbol{q})^2 + 4\boldsymbol{p}_1 \cdot \boldsymbol{p}_2}{8m^2}\right] \psi(\boldsymbol{p}_1, \boldsymbol{p}_2, \boldsymbol{p}_3). \end{aligned} \tag{4.307}$$

For the harmonic-oscillator wave functions, we find that this amounts to the simple substitution

$$|\psi(0)|^2 \to |\psi(0)|^2 \left(1 - \frac{1}{4m^2 R^2}\right) \tag{4.308}$$

This is the result found by Bonvin (1984). The correction factor is 0.61 for our set of parameters. A more refined calculation (Cortés *et al.*, 1978) gives a smaller correction ~ 0.87. In fact, this correction does not appear to be essential, compared with the short-distance factor c_1. Moreover, in contrast with the initial assumption of simply replacing Pauli spinors by free Dirac spinors, the two-component wave function (denoted by $\langle \boldsymbol{p}_i, \boldsymbol{r}_i | \Psi \rangle$ in (3.86)) is also submitted to relativistic corrections as compared with the Schrödinger wave function (see Section 3.3). This is not taken into account in the above references. According to a study of mesons made in a Bethe–Salpeter framework (Durand, 1982), the relativistic wave function at the center, $|\psi(0)|^2$, is larger than the nonrelativistic one, with the mass spectrum being constrained to be the same. It is interesting to note that a study of the hydrogen atom leads to the same conclusion for a Coulomb potential. The wave function does indeed have a singularity; and the calculation of the weak matrix element for the Dirac equation in a Coulomb potential gives finally an increase of the effective $|\psi(0)|^2$ that goes to infinity when $\alpha \to 1$ (Abe *et al.*, 1980). To summarize: we suspect that the reduction effect of the nonleptonic amplitude due to the relativistic corrections, if any, is not very large. But nothing very convincing can be said before there have been improved calculations of the corrected wave functions, which may be difficult to carry out.

4.6.5 Corrections Due to Excited Baryon Intermediate States

One of the important assumptions of the preceding Section concerns the evaluation of the k_μ term in (4.229). In the transformed expressions (4.232) and (4.233), based on the insertion of a complete set of intermediate states, we have retained only the ground-state $\frac{1}{2}^+$ baryons. In principle, all the states with $J^P = \frac{1}{2}^+$ or $\frac{1}{2}^-$ contribute. In our formulation, the baryons B and B′ are assumed to be at rest, and $J \neq \frac{1}{2}$ is forbidden by rotational invariance. Moreover, the states $\frac{1}{2}^-$ contribute to the S waves, and $\frac{1}{2}^+$ to the P waves, as we shall see below. The philosophy of *nearby singularities* would select those baryons with lowest masses; namely, as observed many years ago by Okubo (1968), the principal candidates in addition to the ground-state baryons, would be the $\Lambda(1405)$ and the $P_{11}(1440)$. However, this point of view only takes the masses into account. Another point of view is that of the quark model in the limit of SU(6) symmetry; in this situation, a whole multiplet, like the $(\mathbf{70}, 1^-)$ or the $(\mathbf{56}, 0^+)$ radially excited, should be considered, because all the members have the same mass in this SU(6) limit, and the *couplings* are therefore the essential criterion discriminating the various resonance contributions. It is true that mass splittings breaking the SU(6) symmetry will change this conclusion, but there is no reason for the effect of couplings to disappear. Moreover, there is no general reason for the lowest-mass member of the multiplet to have the strongest couplings, nor to dominate over the sum of the contributions of the other members of the multiplet.

A striking example concerns the S-wave Σ^+_+. Theoretically $A(\Sigma^+_+)=0$ from the commutator alone, as is indeed found experimentally. This prediction would be destroyed by an isolated $\Lambda(1405)$ contribution. But in the quark model it is found that, in the limit of SU(6) symmetry, all the $\frac{1}{2}^-$ resonance contributions from $(\mathbf{70}, 1^-)$ cancel each other, thus maintaining the good result $A(\Sigma^+_+)=0$ (Le Yaouanc *et al.*, 1979). This clearly reveals the failure of considerations based only on the pole position, and neglecting the coupling.

This quark-model point of view also contrasts with earlier attempts to take into account final-state interactions (Reid and Trofimenkoff, 1972). In such attempts, only the *s*-channel singularities (πN scattering for Σ and Λ decay) are considered, while in the SU(6)-symmetric quark model the strange resonances that appear in the *u* channel ($\pi\Lambda$, $\pi\Sigma$ scattering) are to be considered on exactly the same footing. Moreover, in the case of the $\frac{1}{2}^-$ resonance contribution to S waves, nonstrange resonances (*s* channel) *exactly cancel each other* in the SU(6) limit (Le Yaouanc *et al.*, 1979). We are therefore completely at odds with the calculation of Reid and Trofimenkoff (1972).

Let us proceed to actual calculations according to the expressions (4.232) and (4.233). We have to calculate the matrix elements $\langle B'|\mathscr{H}_W|B\rangle$, where B or B$'$ is necessarily the initial or the final ground-state $\frac{1}{2}^+$ baryon, and the second baryonic state is a resonance of the *s* or *u* channels. Since B_1 is at rest, according to our expression (4.232), the *s*-channel resonance must also be taken at rest. The same conclusion does not hold for the *u*-channel resonance, according to (4.233), as the momentum should be $\boldsymbol{p}_2=-\boldsymbol{k}$. For the sake of simplicity, we assume, just as in the calculation of the baryon pole contribution to the P wave in Section 4.5.2, that the baryons are all at rest. It seems reasonable to neglect $\boldsymbol{k}$, since $|\boldsymbol{k}|$ is small, except in the overall factor k_μ in front of the quantity that is to be calculated. Since $\mathscr{H}_W$ is a scalar, it is then necessary for the resonance to have the spin of the initial or final baryon, and the only possible states are then $J=\frac{1}{2}$, as mentioned above. Moreover $\mathscr{H}_W^{pc}$ then induces transitions from a ground state only to $J^P=\frac{1}{2}^+$ resonances, and $\mathscr{H}_W^{pv}$ to $J^P=\frac{1}{2}^-$ resonances. Therefore $\frac{1}{2}^+$ resonances contribute only to P waves, and $\frac{1}{2}^-$ resonances only to S waves. We also have to calculate axial-current-matrix elements $\langle B'|j_5^{i\mu}|B\rangle$. Similarly, one of the baryons is a ground-state $\frac{1}{2}^+$ baryon (B_1 or B_2), and the other a resonance with $J^P=\frac{1}{2}^\pm$. For similar reasons, we assume both to be at rest, and neglect the transfer $\boldsymbol{k}$ between them.

The $(\mathbf{70}, 1^-)$ $\frac{1}{2}^-$ intermediate-state contribution (S wave) can be reliably calculated (Le Yaouanc *et al.*, 1979) since it appears in the lowest approximation to the matrix elements of $\mathscr{H}_W^{pv}$ and $j_5^{i\mu}$. For $\langle B'|\mathscr{H}_W^{pv}|B\rangle$, the non-relativistic expression has been given in Section 2.1.4:

$$\langle \mathrm{B}'|H_{\mathrm{W}}^{\mathrm{pv}}|\mathrm{B}\rangle = \delta(\boldsymbol{P}-\boldsymbol{P}')\frac{G_{\mathrm{F}}}{\sqrt{2}}\cos\theta_{\mathrm{C}}\sin\theta_{\mathrm{C}}$$

$$\times 6\langle\psi_{\mathrm{B}'}| -\frac{1}{m}\tau^{(-)}(1)v^{(+)}(2)\{(\boldsymbol{\sigma}(1)-\boldsymbol{\sigma}(2))\,\delta(\boldsymbol{r}_1-\boldsymbol{r}_2)(\boldsymbol{p}_1-\boldsymbol{p}_2)$$
$$+[\boldsymbol{\sigma}(1)-\boldsymbol{\sigma}(2)+\mathrm{i}(\boldsymbol{\sigma}(1)\times\boldsymbol{\sigma}(2))][-\mathrm{i}\boldsymbol{\nabla}_{\boldsymbol{r}_1}\delta(\boldsymbol{r}_1-\boldsymbol{r}_2)]\}\,|\psi_{\mathrm{B}}\rangle. \quad (4.309)$$

In the calculation of $\langle \mathrm{B}'|j_5^{i\mu}|\mathrm{B}\rangle$ we can neglect the space component, because $\boldsymbol{j}_5$ is momentum-independent (2.51) and therefore $\langle \mathrm{B}'|\boldsymbol{j}_5|\mathrm{B}\rangle$ is proportional to the transition form factor $0^+\to 1^-$, which contains a factor $|\boldsymbol{k}|$ (due to the orthogonality of the spatial wave functions). In contrast $\langle \mathrm{B}'|j_5^{i0}|\mathrm{B}\rangle$ is nonzero, even at $\boldsymbol{k}=0$ (remember that the $\frac{1}{2}^-$ baryon wave function has as a factor the internal quark momentum $\boldsymbol{p}$), and in fact we set $\boldsymbol{k}=0$. Then, from the considerations of Section 2.2.1,

$$\langle \mathrm{B}'|j_5^{i0}(0)|\mathrm{B}\rangle = \frac{1}{(2\pi)^3}3\langle\psi_{\mathrm{B}'}| -\tau^i(3)2\mathrm{i}\,\boldsymbol{\sigma}(3)\cdot\boldsymbol{\nabla}_{\boldsymbol{r}_3}|\psi_{\mathrm{B}}\rangle, \quad (4.310)$$

where $\tau^i(3)$ is the flavor operator corresponding to j_5^{i0} acting over the third quark.

For the next level, corresponding to two harmonic-oscillator excitations, which contains the SU(6) multiplets $(\mathbf{56},0^+)$, $(\mathbf{70},0^+)$, $(\mathbf{56},2^+)$, $(\mathbf{70},2^+)$, $(\mathbf{20},1^+)$, the $\frac{1}{2}^+$ states will contribute to the P waves (parity-conserving). For the $\frac{1}{2}^+$ radial excitation $(\mathbf{56},0^+)$ the lowest v/c approximation for $\mathscr{H}_{\mathrm{W}}^{\mathrm{pc}}$ gives a nonzero result. On the other hand, the lowest approximation (4.310) fails,

$$\langle \mathrm{B}'|j_5^{i0}|\mathrm{B}\rangle = 0, \quad (4.311)$$

because B and B′ are at rest and have the same parity, while j_5^{i0} is odd under parity. We must then consider $\boldsymbol{j}_5^i$, which is of even parity, but with a better approximation in v/c. The next order in v/c of $\boldsymbol{j}_5^i$ has already been discussed (Section 3.2.5, (3.60)) and it amounts to making the substitution

$$\sigma_z\to\sigma_z\left(1-\frac{1}{2m^2}\boldsymbol{p}_{\mathrm{T}}^2\right), \quad (4.312)$$

and we therefore essentially have to calculate spatial matrix elements $\langle\psi_{\mathrm{B}'}|\boldsymbol{p}_{\mathrm{T}}^2|\psi_{\mathrm{B}}\rangle$. The necessity of including baryon excitations of negative parity in S waves has been stressed by Le Yaouanc *et al.* (1979), who have computed the $(\mathbf{70},1^-)$ contribution using the harmonic-oscillator potential.

With the same potential, Bonvin (1984) has computed the contribution of the radially excited $(\mathbf{56},0^+)$. It must be remembered, however, that the harmonic potential is not a realistic one. Certain quantities should not be very sensitive to the use of simple Gaussian wave functions, but this is not the case for the weak matrix element, where small interquark distances are involved.

Moreover, for radial excitations, the precise position of the nodes may be crucial, and the harmonic-oscillator calculation probably cannot be trusted. It must be remembered that the ratio of the meson wave functions at the center of a ground state and a radially excited meson is incorrectly predicted by the harmonic-oscillator model.

We shall retain only some qualitative conclusions. (i) The resonance corrections are of order $\delta m/m$ with respect to the principal term. Here m is the quark mass and δm is the SU(3) breaking. This is easily seen, as there is a front factor k_0 or $|\boldsymbol{k}|$ of order δm ($k_0 = |\boldsymbol{k}|$), the product of the matrix elements of axial and weak operators is of order v^2/c^2, and the energy denominators are of order mv^2/c^2. Therefore these are essentially *linear corrections in the SU(3) breaking.* (ii) the sign of the $\frac{1}{2}^-$ intermediate-states correction to S waves is well defined. This contribution is *opposite in sign* to the commutator term. Therefore it depresses the S wave, and this is just what is needed to correct the naive model (if we disregard the factor c_1). In fact, most of the S waves were too large by a factor two, and the (**70**, 1^-) intermediate states give precisely a correction of -50% with the harmonic oscillator. The sign of the correction from the $\frac{1}{2}^+$ excited baryons to the P waves is perhaps less certain, since the other SU(6) multiplets of the same energy have not been taken into account. It is nevertheless interesting that it enhances the P wave, thus correcting most P waves of the naive model in the right direction.

In conclusion, it must be emphasized that the most naive model of hyperon nonleptonic decays works strikingly well, as magnitudes and signs come out correctly. Resonance corrections improve the agreement further, although here more realistic calculations are still needed.

Finally, concerning the QCD short-distance corrections, we must emphasize that the penguin contributions are small. They could show up more clearly in the decay $\Omega^- \to \Xi\pi$, where the main baryon-internal-conversion contribution is absent, and the penguin diagram contributes (Finjord, 1978). On the other hand, the c_1 coefficient seems to lead to a general overestimation of the decay amplitudes, at least if our evaluation of $|\psi(0)|^2$ is to be trusted.

Appendix

A.1 METRIC

We use the Lorentz metric

$$g_{\mu\nu} = \begin{pmatrix} 1 & 0 & 0 & 0 \\ 0 & -1 & 0 & 0 \\ 0 & 0 & -1 & 0 \\ 0 & 0 & 0 & -1 \end{pmatrix}. \tag{A.1}$$

A.2 DIRAC MATRICES AND SPINORS

For the Dirac matrices we use the following representation:

$$\left.\begin{aligned} \gamma^0 &= \begin{pmatrix} 1 & 0 \\ 0 & -1 \end{pmatrix}, \qquad \gamma^5 = \gamma_5 = \begin{pmatrix} 0 & 1 \\ 1 & 0 \end{pmatrix}, \\ \gamma^i &= \begin{pmatrix} 0 & \sigma^i \\ -\sigma^i & 0 \end{pmatrix}, \qquad \sigma_{\mu\nu} = \tfrac{i}{2}[\gamma_\mu, \gamma_\nu], \end{aligned}\right\} \tag{A.2}$$

where σ_i are the Pauli matrices:

$$\left.\begin{aligned} \sigma^i &= \sigma_i, \qquad \sigma^1 = \begin{pmatrix} 0 & 1 \\ 1 & 0 \end{pmatrix}, \\ \sigma^2 &= \begin{pmatrix} 0 & -\mathrm{i} \\ \mathrm{i} & 0 \end{pmatrix}, \qquad \sigma^3 = \begin{pmatrix} 1 & 0 \\ 0 & -1 \end{pmatrix}. \end{aligned}\right\} \tag{A.3}$$

$\boldsymbol{\alpha}$ and β are defined as

$$\beta = \gamma^0, \qquad \alpha^i = \beta\gamma^i. \tag{A.4}$$

Covariant spinors are normalized according to

$$\bar{u}_s(p)u_s(p) = 1, \qquad \bar{v}_s(p)v_s(p) = -1, \tag{A.5}$$

with

$$\bar{w} = w^\dagger \gamma^0$$

for any spinor w.

The spinors u and v correspond to positive- and negative-energy solutions of the Dirac equation. For free spinors,

$$\left.\begin{aligned} &(\not{p}-m)u_s(p) = 0, \qquad (\not{p}+m)v_s(p) = 0, \\ &u_s(p) = \left(\frac{E+m}{2m}\right)^{1/2} \begin{pmatrix} \chi_s \\ \dfrac{\boldsymbol{\sigma}\cdot\boldsymbol{p}}{E+m}\chi_s \end{pmatrix}, \\ &v_s(p) = \left(\frac{E+m}{2m}\right)^{1/2} \begin{pmatrix} \dfrac{\boldsymbol{\sigma}\cdot\boldsymbol{p}}{E+m}\chi_s \\ \chi_s \end{pmatrix}, \end{aligned}\right\} \tag{A.6}$$

where χ_s is a Pauli spinor.

In Section 3.3 we use noncovariant projectors on positive- and negative-energy spinors:

$$\left.\begin{aligned} \Lambda_+(\boldsymbol{p}) &= \sum_s u_s(\boldsymbol{p})u_s^\dagger(\boldsymbol{p}), \\ \Lambda_-(\boldsymbol{p}) &= \sum_s v_s(\boldsymbol{p})v_s^\dagger(\boldsymbol{p}), \end{aligned}\right\} \tag{A.7}$$

where the normalization is now

$$u_{s'}^\dagger(\boldsymbol{p})u_s(\boldsymbol{p}) = v_{s'}^\dagger(\boldsymbol{p})v_s(\boldsymbol{p}) = \delta_{ss'}. \tag{A.8}$$

For free Dirac spinors we have

$$\Lambda_\pm(\boldsymbol{p}) = \frac{1}{2}\left(1 \pm \frac{\boldsymbol{\alpha}\cdot\boldsymbol{p} + \beta m}{E}\right). \tag{A.9}$$

These must not be confused with the covariant projectors defined in the appendix of Bjorken and Drell (1964):

$$\Lambda_+(p) = \sum_s u_s(p)\bar{u}_s(p), \qquad \Lambda_-(p) = -\sum_s v_s(p)\bar{v}_s(p). \tag{A.10}$$

A.3 FIERZ TRANSFORMATION

For any four fermionic fields $q_i (i=1, \ldots, 4)$ the following identities hold (Fierz, 1937):

$$\bar{q}_1\gamma_\mu(1-\gamma_5)q_2\bar{q}_3\gamma_\mu(1-\gamma_5)q_4 = \bar{q}_3\gamma_\mu(1-\gamma_5)q_2\bar{q}_1\gamma_\mu(1-\gamma_5)q_4, \quad (A.11)$$

$$\bar{q}_1\gamma_\mu(1-\gamma_5)q_2\bar{q}_3\gamma_\mu(1+\gamma_5)q_4 = -2\bar{q}_3(1-\gamma_5)q_2\bar{q}_1(1+\gamma_5)q_4. \quad (A.12)$$

A.4 SU(3)

The Gell-Mann λ matrices are defined (Carruthers, 1966) as ($\lambda_i = \lambda^i$)

$$\left.\begin{aligned}
&\lambda_1 = \begin{pmatrix} 0 & 1 & 0 \\ 1 & 0 & 0 \\ 0 & 0 & 0 \end{pmatrix}, \qquad \lambda_2 = \begin{pmatrix} 0 & -\mathrm{i} & 0 \\ \mathrm{i} & 0 & 0 \\ 0 & 0 & 0 \end{pmatrix}, \\
&\lambda_3 = \begin{pmatrix} 1 & 0 & 0 \\ 0 & -1 & 0 \\ 0 & 0 & 0 \end{pmatrix}, \qquad \lambda_4 = \begin{pmatrix} 0 & 0 & 1 \\ 0 & 0 & 0 \\ 1 & 0 & 0 \end{pmatrix}, \\
&\lambda_5 = \begin{pmatrix} 0 & 0 & -\mathrm{i} \\ 0 & 0 & 0 \\ \mathrm{i} & 0 & 0 \end{pmatrix}, \qquad \lambda_6 = \begin{pmatrix} 0 & 0 & 0 \\ 0 & 0 & 1 \\ 0 & 1 & 0 \end{pmatrix}, \\
&\lambda_7 = \begin{pmatrix} 0 & 0 & 0 \\ 0 & 0 & -\mathrm{i} \\ 0 & \mathrm{i} & 0 \end{pmatrix}, \qquad \lambda_8 = \begin{pmatrix} \frac{1}{\sqrt{3}} & 0 & 0 \\ 0 & \frac{1}{\sqrt{3}} & 0 \\ 0 & 0 & -\frac{2}{\sqrt{3}} \end{pmatrix},
\end{aligned}\right\} \quad (A.13)$$

with the following normalization and commutation rules:

$$\left.\begin{aligned} \mathrm{Tr}(\lambda_a\lambda_b) &= 2\delta_{ab}, \\ [\lambda_a, \lambda_b] &= 2\mathrm{i}f^{abc}\lambda_c. \end{aligned}\right\} \quad (A.14)$$

f^{abc} are the structure constants of SU(3), they are antisymmetric for the exchange of any two indices, and their values are

$$\left.\begin{aligned} &f^{123} = 1, \qquad f^{147} = f^{246} = f^{257} = f^{345} = \tfrac{1}{2}, \\ &f^{156} = f^{367} = -\tfrac{1}{2}, \qquad f^{458} = f^{678} = \tfrac{1}{2}\sqrt{3}. \end{aligned}\right\} \quad (A.15)$$

In a given representation of the group, we call T_a the representative of the generators of the Lie algebra. T_a is normalized according to

$$\mathrm{Tr}(T_aT_b) = \tfrac{1}{2}\delta_{ab}, \qquad [T_a, T_b] = \mathrm{i}f^{abc}T_c. \quad (A.16)$$

In the fundamental representation,

$$T_a = \tfrac{1}{2}\lambda_a. \tag{A.17}$$

The Casimir operator C_2 is defined by

$$\sum_a T_a T_a = C(R)\,1, \tag{A.18}$$

where 1 is the identity matrix and $C(R)$ is a number depending on the representation R

$$\left.\begin{aligned} &C(3) = C(\bar{3}) = \tfrac{4}{3}, \\ &C(8) = 3, \\ &C(6) = C(\bar{6}) = \tfrac{10}{3}, \\ &C(10) = C(\overline{10}) = 6, \\ &C(27) = 8. \end{aligned}\right\} \tag{A.19}$$

The anticommutator of λ matrices is given by

$$\{\lambda_a, \lambda_b\} = \tfrac{4}{3}\delta_{ab} 1 + d_{abc}\lambda_c, \tag{A.20}$$

where d_{abc} are symmetric under the exchange of two indices, and are given in the literature (Carruthers, 1966).

The matrix element of an SU(3) octet operator between two SU(3) octet states has the form

$$\langle V_a|\, \mathcal{O}_b \,|V_c\rangle = f_{abc}F + d_{abc}D, \tag{A.21}$$

where the reduced matrix elements F and D depend on the physics. These are called the F and D couplings.

A.5 FIRST AND SECOND QUANTIZATION

A second-quantization operator is a polynomial in the fields. For example, consider a quadratic operator in the quark fields of the form:

$$\mathcal{O} = q^{\dagger}(\boldsymbol{y})\,\mathcal{O}q(\boldsymbol{x}), \tag{A.22}$$

where $\mathcal{O}$ is the matrix in Dirac, flavor and color space.

The matrix element of (A.22) in the one quark sector is obtained simply by replacing the fields q by the quark wave functions ψ:

$$\langle \mathrm{f}\,|\,\mathcal{O}\,|\mathrm{i}\rangle = \psi_{\mathrm{f}}(\boldsymbol{y})\,\mathcal{O}\psi_{\mathrm{i}}(\boldsymbol{x}), \tag{A.23}$$

where the wave functions are Dirac vectors:

$$\psi(\boldsymbol{x}) = (2\pi)^{-3/2}\sum_s \int \mathrm{d}\boldsymbol{p}\left(\frac{m}{p^0}\right)^{1/2} \mathrm{e}^{\mathrm{i}p\cdot x}u_s(p)\,\tilde{\psi}_s(\boldsymbol{p}), \tag{A.24}$$

with $p^0 = E = (\boldsymbol{p}^2 + m^2)^{1/2}$; $u_s(p)$ is defined in (A.6), and $\tilde{\psi}_s(\boldsymbol{p})$ is the quark wave function in momentum space:

$$|\mathrm{q}\rangle = \sum_s \int \mathrm{d}\boldsymbol{p}\, \tilde{\psi}_s(\boldsymbol{p}) b_s^\dagger(\boldsymbol{p}) |0\rangle, \tag{A.25}$$

$|0\rangle$ being the vacuum.

In the one antiquark sector $\mathcal{O}$ (A.22) may be expressed in terms of the charge-conjugate fields

$$q^{\mathrm{c}} = \mathrm{i}\gamma^2 q^*, \tag{A.26}$$

$$\mathcal{O} = -q^{\mathrm{c}\dagger}(\boldsymbol{x})(\mathrm{i}\gamma^2)\mathcal{O}^{\mathrm{t}}(\mathrm{i}\gamma^2)q^{\mathrm{c}}(\boldsymbol{y}), \tag{A.27}$$

$\mathcal{O}^{\mathrm{t}}$ being the transpose of $\mathcal{O}$.

The matrix element is then obtained by replacing the q^{c} fields by the antiquark wave functions defined analogously to (A.24) and (A.25) with $b^\dagger$ replaced by $d^\dagger$.

If the first-quantization operators are written in terms of a kernel (sometimes called an infinite matrix),

$$\langle \mathrm{f} | \mathcal{O} | \mathrm{i} \rangle = \int \mathrm{d}\boldsymbol{r}\, \mathrm{d}\boldsymbol{r}'\, \psi_{\mathrm{f}}^\dagger(\boldsymbol{r}') K(\boldsymbol{r}', \boldsymbol{r}) \psi_{\mathrm{i}}(\boldsymbol{r}), \tag{A.28}$$

then, from (A.23),

$$K(\boldsymbol{r}', \boldsymbol{r}) = \mathcal{O}\, \delta(\boldsymbol{x} - \boldsymbol{r}')\, \delta(\boldsymbol{y} - \boldsymbol{r}). \tag{A.29}$$

In the common case where the second-quantized operator is of the form

$$\mathcal{O} = q^\dagger(\boldsymbol{x}) \Gamma q(\boldsymbol{x}), \tag{A.30}$$

its first-quantized form is

$$\Gamma \delta(\boldsymbol{x} - \boldsymbol{r})\, \delta(\boldsymbol{x} - \boldsymbol{r}') = \Gamma \langle \boldsymbol{r}' |\, \delta(\boldsymbol{x} - \mathbf{r})\, | \boldsymbol{r} \rangle \tag{A.31}$$

where $\mathbf{r}$ is the position operator corresponding to the kernel:

$$\boldsymbol{r} \delta(\boldsymbol{r} - \boldsymbol{r}'). \tag{A.32}$$

A.6 PHASE CONVENTION FOR WAVE FUNCTIONS

To combine quark spins and isospins, we use the phase given by the Clebsch–Gordan coefficients. This is also the case for mesons, where we take the Clebsch–Gordan coefficients with the ordering quark first and antiquark second; the antiquark doublets in the standard representation are given by multiplying the adjoint representation by $\mathrm{i}\sigma^2$ and $\mathrm{i}\tau^2$ in spin and isospin spaces respectively.

For SU(3) multiplets we use de Swart's (1963) conventions: the signs are left unchanged for the operators $I_\pm$ and $V_\pm$ when applied to the states u, d, s and $-\bar{u}, \bar{d}, \bar{s}$. The relative sign between the meson octet and singlet has been chosen to have the mixing angles compatible with the values given by the Particle Data Group (1984).

To combine orbital angular momentum and spin, we use the Clebsch–Gordan phase convention with L taken before S.

The combination of the different symmetry representations to build up the total wave function is done according to the phase convention chosen in (1.73).

The spatial wave functions for the excitations are taken to be real in configuration space. For the harmonic oscillator, we take the orbital excitations in configuration space with the solid spherical harmonics multiplied by a positive factor. Note that this implies a phase $(-\mathrm{i})^L$ in momentum space.

A.7 PHYSICAL CONSTANTS AND CONVENTIONS

The charge e is related, in Heaviside–Lorentz units, to the fine-structure constant α by

$$\mathrm{e} = (4\pi\alpha)^{1/2}, \qquad \alpha = \tfrac{1}{137}. \tag{A.33}$$

Our notation for the pion decay constant (see (3.138) and (4.217)) is

$$f_\pi = \sqrt{2}\, F_\pi = 130\ \mathrm{MeV}. \tag{A.34}$$

Literature survey

We shall try here to complete the bibliography of the book by adding

(i) references to some historically fundamental or original papers;
(ii) a few books and review articles that could be of practical help to the reader; and
(iii) references to more specialized work that has not been treated but that could in principle have been part of the book.

GENERAL INTRODUCTION (CHAPTER 1)

The notion of quarks as the standard picture of hadron structure was established in a number of steps: the algebraic concept introduced by Gell-Mann (1964a,b) and Zweig (1964), further developed in SU(6) by Gürsey and Radicati (1964) and Gürsey, Pais and Radicati (1964); the rapid development of hadron spectroscopy and the study of transitions within the quark model (see e.g. Morpurgo, 1969; Becchi and Morpurgo, 1965, 1966; Dalitz, 1965, 1966, 1967a,b; Dalitz and Sutherland, 1966; Faiman and Hendry, 1968; Copley *et al.*, 1969); the introduction of the parton model to describe deep-inelastic electron–nucleon scattering (Bjorken and Paschos, 1969; Feynman, 1969) and the discovery, combining electron and neutrino data, that these partons had the quark quantum numbers (for useful reviews see Feynman, 1972; Close, 1979; Dalitz, 1976); the introduction of the concept of color (anticipated by Greenberg, 1964) by Gell-Mann (1972); the prediction of the charm quantum number (Glashow, Iliopoulos and Maiani, 1970), of its mass (Gaillard and Lee, 1974b), and its discovery in 1974.

Decisive developments have been the Standard Model of the electroweak theory and quantum chromodynamics (QCD), based respectively on the gauge groups $SU(2)\times U(1)$ and $SU(3)_c$.

A summary of weak-interaction phenomenology prior to the electroweak gauge theory can be found in Marshak *et al.* (1968). The electroweak unification was anticipated by Glashow (1961) and made into a consistent renormalizable theory of leptons by Weinberg (1967) and Salam (1968) by the introduction of the Higgs (1964) mechanism. The Glashow–Iliopoulos–Maiani mechanism (Glashow *et al.*, 1970), completing the Cabibbo (1963) theory, made possible the extension of electroweak theory to quarks, and hence to the hadron sector. The generalization to three quark families was made by Kobayashi and Maskawa (1972), and provides a natural mechanism for *CP* violation in the standard model.

Quantum chromodynamics, proposed by Fritzsch and Gell-Mann (1972), was proven to be asymptotically free (Gross and Wilczek, 1973; Politzer, 1973), and is reviewed by Politzer (1974), and Marciano and Pagels (1978). Applications of perturbative QCD are reviewed by Altarelli (1982, 1985). Further references on perturbative QCD applied to OZI-forbidden decays are given in Shifman (1981).

For an overall view of gauge-field theories see Abers and Lee (1973), Weinberg (1974), Iliopoulos (1976), Leite Lopes (1981) and the selected papers edited by Mohapatra and Lai (1981). There is a special chapter on the subject in the textbook on field theory by Itzykson and Zuber (1980). Faddeev and Slavnov (1980) give an account at an advanced level.

Elements of group theory and of flavor SU(3) can be found in Gell-Mann and Ne'eman (1964) and Carruthers (1966).

Current algebra (Gell-Mann, 1964a,b) and PCAC (see e.g. Gell-Mann and Lévy, 1960a,b) are extensively reviewed in Adler and Dashen (1968). The conceptual content of PCAC is clarified in Dashen and Weinstein (1969). Good reviews on explicit breaking of chiral symmetry can be found in Pagels (1975) and Gasser and Leutwyler (1982). The dynamical breaking of chiral symmetry (Nambu and Jona-Lasinio, 1961a,b; Gell-Mann *et al.*, 1968), giving a Nambu–Goldstone pion (Nambu and Jona-Lasinio, 1961a,b; Goldstone, 1961), has recently been studied by Finger *et al.* (1979), Amer *et al.* (1983), Le Yaouanc *et al.* (1984, 1985a,b), Adler and Davis (1984) and Bernard, Brockmann and Weise (1985) in potential models.

The axial anomaly — first discussed in two dimensions by Schwinger (1951) — was discovered in four dimensions by Adler (1969) and Bell and Jackiw (1969). A good analysis of the implications for $\pi^0 \to \gamma\gamma$ can be found in Adler (1970). The role of the axial anomaly in QCD was first stressed by 't Hooft (1976).

Concerning the nonperturbative aspects of QCD, lattice QCD (Wilson, 1974), the most ambitious approach, is reviewed by Creutz (1981), Kogut (1983) and Hasenfratz (1983). QCD duality sum rules were introduced by Shifman, Vainshtein and Zakharov (1978, 1979), and have been reviewed by Reinders, Rubinstein and Yazaki (1985).

Besides lattice QCD, other approaches have been proposed for explaining quark confinement, starting from the work of Nielsen and Olesen (1973), based on an analogy with the Meissner effect of super-conductivity ('t Hooft, 1979, 1980; Mandelstam, 1979; Baker *et al.*, 1985).

The bag model (Chodos *et al.*, 1974a,b; De Grand *et al.*, 1975), reviewed by Hasenfratz and Kuti (1978), can find a rationale in a soliton picture for the confining mechanism (Friedberg and Lee, 1978; Lee, 1979, 1981).

Saturation of colored forces for color-singlet bound states was emphasized in a simple way by Nambu (1966) and Lipkin (1973a,b). A good account of the spectrum of hadrons according to the nonrelativistic quark model is given by Dalitz (1967b). The baryon wave functions for excited states in the harmonic oscillator model were tabulated in Faiman (1968), and the systematic classification of baryon states was pursued with a classification of the types of SU(6) breaking forces (Greenberg and Resnikoff, 1967) by Dalitz and collaborators (Horgan and Dalitz, 1973; Dalitz, 1975).

A major breakthrough has been the study of spin-dependent forces in QCD (De Rújula *et al.*, 1975), which has lead to a rich phenomenology of hadrons using QCD-inspired potentials. For baryons made out of light quarks, this program is still being carried out (see e.g. Gromes and Stamatescu, 1976, 1979; Gromes, 1984b; Isgur and Karl, 1978, 1979; Isgur *et al.*, 1978). Charmonium spectroscopy was first developed systematically by the Cornell group (Eichten *et al.*, 1975, 1976, 1978) and Schnitzer and collaborators (Kang and Schnitzer, 1975). Useful reviews on charm spectroscopy are given by Jackson (1976), Gottfried (1977), Appelquist, Barnett and Lane (1978), Martin (1977) and Flamm and Schöberl (1982).

Other interesting reviews on quarkonia are given by Quigg and Rosner (1979) and by Ono (1983), Kühn (1986) (with emphasis on the bottonium $b\bar{b}$ states), Buchmüller (1984) and Martin (1984) (with predictions for the toponium $t\bar{t}$ system).

General books on the quark model are Kokkedee (1969), Close (1979) (who deals also with the parton model) and Flamm and Schöberl (1982).

NONRELATIVISTIC APPROXIMATION FOR TRANSITIONS (CHAPTER 2)

The nonrelativistic quark model mainly demonstrates its value by its practical achievements, and developments are mainly to be found in phenomenological papers. On the other hand, the methods are quite standard in other fields of quantum mechanics and are often to be found in classical treatises; therefore, we just quote original papers in the cases where the model is really specific to particle physics.

As far as the standard problems of elementary-quantum-emission (Sections 2.1.1, 2.1.2 and 2.2.1) are concerned, reference can be made to any good book on atomic or nuclear physics, such as Blatt and Weisskopf (1952). It is important to note, however, that in atomic and nuclear physics there are additional approximations as compared with particle physics. Pion emission has been treated in a nonrelativistic quark model by Becchi and Morpurgo (1966).

The first description of leptonic annihilation was due to Van Royen and Weisskopf (1967). The treatment of OZI-forbidden annihilation at lowest order is essentially the same as for positronium photonic annihilation (Landau and Lifshitz, 1972); the calculation of the various annihilation processes is systematically presented by Novikov *et al.* (1978); the specifically QCD aspect is due to Appelquist and Politzer (1975).

For the various pair-creation models of OZI-allowed strong decays, we refer to the original papers (Micu, 1969; Carlitz and Kislinger, 1970; Le Yaouanc *et al.*, 1973, 1975b; Eichten *et al.*, 1978; Alcock and Cottingham, 1984; Dosch and Gromes, 1986; Kokoski and Isgur, 1985). For rearrangement models see Mott and Massey (1949) and Le Yaouanc *et al.* (1978a).

For nonleptonic decay-matrix elements, the general background is found in the book of Marshak *et al.* (1969); the quark-model calculation is presented in Le Yaouanc *et al.* (1979).

For all the chapters of the present book, we recommend the general background of quantum mechanics and quantum field theory given by Messiah (1959) and Bjorken and Drell (1964, 1965). Particular attention must be paid to perturbation theory, both time-independent and time-dependent.

PHENOMENOLOGY OF THE NONRELATIVISTIC MODEL (CHAPTER 4)

Chapter 4 is rather long as it is, but it still does not cover all of the phenomenological aspects. We try here to make up for the omissions by providing relevant references; however, the coverage remains incomplete.

A synthesis of the nonrelativistic quark-model results for transitions in the ground state (L=0) is found in the lectures of Morpurgo (1969). These results are closely connected to the previous algebraic SU(6) approach of Gürsey and Radicati (1964) and Gürsey *et al.* (1964). Since then some effort has been devoted to improve the understanding of the light-quark ground-state static and transition properties; the question is closely connected to that of SU(3) and SU(6) breaking (the latter is in fact a relativistic effect, and, although this is not always obvious in the literature, it should be treated as such). Magnetic moments have been repeatedly discussed by Franklin (see e.g. Franklin, 1979); a good review is that by Rosner (1980). SU(6) configuration mixing was first studied by the present authors in connection with a large series of properties of the ground state (neutron charge radius, Δ-decay problems, Σ–Λ splitting, structure functions etc.) (Le Yaouanc *et al.*, 1977b; see further references below in the section of the survey dealing with relativistic aspects). It has been the subject of several papers by Isgur, Karl and collaborators (e.g. Isgur *et al.*, 1978).

The specific predictions of *explicit* quark models, as opposed to the initial SU(6)-symmetry scheme, have been especially displayed by the work on *baryon orbital excitations*, following the breakthrough in baryon spectroscopy (Greenberg, 1964; Dalitz, 1967b; Faiman and Hendry, 1968). The two pioneering works are those on strong and radiative decays by Faiman and Hendry (1968, 1969) and Copley *et al.* (1969). Photoproduction became one of the most convincing applications of the quark model following the appearance of the latter paper. The paper of Feynman *et al.* (1971), although dressed in covariant clothes, was actually one of the most extensive pieces of work in the nonrelativistic quark model, covering almost all the possible applications at the time.

In the 70s, *SU(6) analysis* was the most popular direction of work. In *strong decays*, the very important contribution of Rosner and collaborators (Colglazier and Rosner, 1971; Petersen and Rosner, 1972; Faiman and Plane, 1972; Faiman and Rosner, 1973) was to show complete failure of $SU(6)_W$ symmetry. It stimulated the work on *explicit* quark models beyond the old $\boldsymbol{\sigma}\cdot\boldsymbol{k}_\pi$ interaction, already initiated by Mitra and Ross (1967), with their introduction of the recoil term; this term is naturally included in the Feynman *et al.* (1971) model, extensively applied by Moorhouse and Parsons (1973), but a conceptually crucial step in the development of explicit quark models of strong decays was the formulation of the quark-pair-creation model (Micu, 1969; Carlitz and Kislinger 1970; Le Yaouanc *et al.*, 1973, 1975b), first applied to baryon orbital-excitation decays, but which had later many new applications (Le Yaouanc *et al.*, 1978a; Busetto and Oliver, 1983; Gavela *et al.*, 1980a; Heikkilä, Tornqvist and Ono, 1984; Tornqvist and Ono, 1984; Ader, Bonnier and Sood, 1980; Dover, Fishbane and Furui, 1986; for charmonium see below). An extensive study by the Bordeaux group (Ader *et al.*, 1982) has displayed the universality of the γ constant.

The SU(6) analysis has also been applied to photoproduction and to overall fits of the decays (Hey, Lichtfield and Cashmore, 1975). Finally, it must be noted that, for quite a while, the work on SU(6) analysis was connected with the SU(3)×SU(3) null-plane symmetry originating in Weinberg's (1969) analysis of pion couplings, through the Buccella-Melosh transformation (Buccella, 1970; Melosh, 1974). A good deal of discussion followed (Fuchs, 1975; Carlitz and Tung, 1976). An elucidation of the transformation in terms of explicit quark models is found in our own papers (Le Yaouanc *et al.*, 1973; Gavela *et al.*, 1982c) and in Buccella, Savoy and Sorba (1974).

In the first part of Close's 1979 book there is a very nice presentation of the discussions of this period (up to about 1975). It must be said that the algebraic approaches have now been superseded by the development of the explicit quark models. An intermediate approach is represented by the work of Koniuk and Isgur (1980), which uses the impressive achievements of Isgur and Karl in baryonic spectroscopy (in particular for the SU(6) mixing patterns), but fits the pion-decay reduced matrix elements.

The superiority of explicit quark models has been particularly illustrated since 1975 by *charmonium* phenomenology. We have here four main bodies of work: (i) radiative transitions; (ii) strong decays above the $D\bar{D}$ threshold; (iii) leptonic annihilation; (iv) OZI-forbidden annihilation under the $D\bar{D}$ threshold. The phenomenology of charmonium is marked by the general contribution of the Cornell group; Jackson (1976) and Gottfried (1977) have given useful reviews of this subject, and we have already quoted the two detailed papers by Eichten *et al.* (1978, 1980). Radiative transitions will be further referenced below when we cover relativistic aspects. As far as the typical strong decays above the ψ(3.770) are concerned, explicit calculations have been essentially done using the Orsay QPC model, although qualitative ideas can be found in the papers of the Cornell group. The use of the QPC model was initiated by Barbieri *et al.* (1975b) before the precise study of the decays of the ψ(4.028,4.414). The interpretation of these decays in the standard $q\bar{q}$ model, as opposed to the molecular interpretation (De Rújula *et al.*, 1976, 1977), was first due to the present authors (Le Yaouanc *et al.*, 1977a). Detailed calculations have been made by Ono (1981a) and extended to the Ys above the $B\bar{B}$ threshold. Chaichian and Kögerler (1980) have performed other calculations. Leptonic annihilation is closely connected with the discussion of the potential, on which there exists a great deal of literature. Regarding exclusive OZI-forbidden decays and perturbative QCD, we must mention the important work reported at the Bonn conference (Shifman, 1981). QCD radiative corrections will be mentioned below.

On the particular topic of two-photon annihilation, there are references in the report of Barnes (1986).

A specialized but very important domain of hadron phenomenology is constituted by *weak nonleptonic* interactions. It is a very old field, but is still not exhausted and is very active. The prototype of such processes is the famous (or infamous!) $K \to \pi\pi$ decay. We feel that it lies outside the scope of the methods covered in this book, and we shall not recommend particular papers in the immense literature devoted to the problem, because, first, it would necessarily be unfair, and, secondly, because there is still controversy and we feel that a decisive breakthrough has yet to be made. Hyperon decays are on much firmer ground, and the contribution of the quark model is here much more deserving of mention. One of the most striking achievements is the explanation of the $\Delta I=\frac{1}{2}$ rule, beginning with the work of Miura and Minamikawa (1967) (see the text). The work of Gronau (1972) has been important for the subsequent quark-model calculations. Schmid (1977) pointed out some essential ideas (see also Le Yaouanc *et al.* (1977b)), which are still overlooked by many people. The explanation of the P/S wave discrepancy can be found in Le Yaouanc *et al.* (1979). Pham (1984) has proposed an argument confirming the role of $\frac{1}{2}^-$ resonances. Finjord (1978) has pointed out the interest of Ω^- decays. Radiative weak decays are well understood in the same approach (Rauh, 1981; Gavela *et al.*, 1981b; Eeg, 1984). More complex phenomena like the neutron electric dipole moment (Gavela *et al.*, 1982a,b; Eeg, 1985) are amenable to the same treatments. A generally different approach is that of the Amherst group, based on the bag model (see e.g. Donoghue *et al.*, 1986). We must finally mention the contributions from the Zagreb group (see e.g. Milosević, Tadić and Trampetić, 1982).

The charmed weak decays are also more amenable to quark-model methods than kaon decays, although there are still important controversial issues. The study of exclusive D decays was initiated by Altarelli, Cabibbo and Maiani (1975), Gaillard, Lee and Rosner (1975) and Cabibbo and Maiani (1975), and in the explicit quark model by Stech and collaborators (Fakirov and Stech, 1978).

It is impossible to account for the immense literature on weak decays. Further references can be found for kaons and hyperons in Donoghue *et al.* (1986); for all recent aspects, we refer to the report of Buras at the Bari conference (Buras, 1985).

The subject of *exotic*-state decays is very interesting, and crucial for the identification of these states, in a situation where we lack firm experimental confirmation. But, of course, the other aspect of this situation is that we also lack experimental tests of the models. Only the dibaryons seem reasonably established, but the study of rearrangement processes along the lines proposed in the book would be rather complex. Diquonia, in addition to rearrangement decays into mesons described in this book (Gavela *et al.*, 1978), will have $N\bar{N}$ decays describable by the QPC model, the remarkable feature being a small partial width; radiative transitions are also interesting (on these subjects, see the papers of the Bordeaux group: Ader, 1980; Ader, Bonnier and Sood, 1981, 1982). At this point, it is adequate to note the results of Dover *et al.* (1986) on the complementary $N\bar{N}$ annihilation into mesons, *without* diquonium resonance, which seems satisfactorily described by pair annihilation and creation according to the QPC model. The hybrids, combining quarks and constituent gluons (Horn and Mandula, 1978; Hasenfratz, 1981; Ono, 1984), will have decays simply describable by the quark–gluon coupling, which are important to study for the elucidation of the states observed in the ψ photon decays (Tanimoto, 1982; Barnes, Close and De Viron, 1983; Le Yaouanc *et al.*, 1985b).

Proton decay is also a potentially interesting application of the quark model, in particular because of its prediction of branching ratios (Donoghue, 1980; Gavela, 1981a; Kane and Karl, 1980).

BEYOND THE NONRELATIVISTIC APPROXIMATION (INCLUDING PHENOMENOMENOLOGICAL APPLICATIONS) (CHAPTERS 3 AND 4)

The involvement of special relativity in the quark model is as old as the model itself. At the end of the 60s and the beginning of the 70s there was a lot of work based on the hypothesis of *very massive* quarks, with very strong binding; this was done in the framework of the Bethe–Salpeter equation (Bethe and Salpeter, 1951; Salpeter, 1951; Wick, 1954) and the Mandelstam (1955) formalism. Transitions (annihilation and strong decays) were studied by Kitazoe and Teshima (1968), Llewellyn-Smith (1969a) and Böhm *et al.* (1972, 1973a,b); there were also studies of the Bethe–Salpeter equation itself (Narayanaswamy and Pagnamenta, 1968; Sundaresan and Watson, 1970). At the same time, the desire for *covariance* led to numerous four-dimensional approaches; in particular four-dimensional oscillators (Fujimura *et al.*, 1967; Lipes, 1972; Feynman *et al.*, 1971; this work was continued by Kim and collaborators — for complete references see Kim and Noz, 1986). This was a rather different direction of work: although such four-dimensional equations could be deduced from very strong binding and very heavy quark masses, these approaches had *light* effective constituent-quark masses. In a still different approach, more akin to atomic physics, Compton-scattering *sum rules* stimulated a very careful study of relativistic effects in a systematic v/c expansion, including recoil effects. The most notable contributions were those of Brodsky and Primack (1969) and the important series of papers by Close and Osborn (of which the final one was Close and Osborn, 1971). This approach is also represented by more recent work on the

Adler–Weisberger sum rule (Donoghue, and Wyler, 1978; Gavela *et al.*, 1982c).

The general interest in the relativistic treatment of quark models faded away in the 70s, essentially because of the empirical success of the nonrelativistic quark model with light constituent-quark masses. The impressive successes of phenomenological applications led to the disregarding of the contradiction inherent in this "standard model" of hadronic physics: namely, with such light quark masses, the quark velocity is necessarily large.

It must be noted that some interest in the relativistic problems of light-quark physics continued throughout the 70s, in two main areas. The first was the MIT bag model (Chodos *et al.*, 1974a,b; De Grand *et al.*, 1975; De Grand, 1976; Hey *et al.*, 1978; and the work on weak nonleptonic decays by the Amherst group quoted above), which has been considered by many to be the only theoretically well-founded approach to light-quark phenomenology, and which has been the object of intense research; Close and Horgan (1980) have developed systematic methods for treating transitions and QCD perturbation theory in this model. The second area involves the isolated but persistent attempts to introduce relativistic improvements into the standard nonrelativistic model, especially by introducing a free-quark Dirac-spinor structure in the wave functions, and also Lorentz spin-boost and contraction effects. As examples of these, we note the contribution of Licht and Pagnamenta (1970), our own work on leptoproduction (Le Yaouanc *et al.*, 1973, 1977d; Andreadis *et al.*, 1974; Amer, 1978), the work of Cottingham and collaborators on electroproduction (Alcock and Cottingham, 1980), the work of Kellett (1974) extending that of Brodsky to baryons, and the detailed studies of the interplay of relativistic effects and SU(6) configuration mixing (Le Yaouanc *et al.*, 1977c). We must also note the systematic work of Bando and collaborators in the framework of the Dirac equation (see e.g. Abe *et al.*, 1980). Finally, there are the various attempts of Mitra and collaborators (Mitra, 1981; Mitra and Santhanam, 1981; and references therein).

After 1975, with the equally impressive development of charmonium phenomenology, a new prejudice appeared. Among certain circles, one began to think that one had to forget about "dirty" light-quark physics and do only "clean" charmonium physics, with a clearly justified Schrödinger equation and perturbative QCD. In this context, the first stage was the development of calculations of QCD *radiative corrections*, beginning with the work of Barbieri *et al.* (1975a,b). In the 80s, there has been increasing interest in the relativistic aspects of charmonium, in the context of the detailed study of *radiative* transitions. This has been done mainly using Hamiltonian methods. We note especially the work of Sucher and collaborators (see e.g. Hardekopf and Sucher, 1982) and of Byers and collaborators (see e.g. MacClary and Byers, 1983). We must not overlook, however, the possible interest of the BS equation with instantaneous interaction (Henriques *et al.*, 1976). Finally, the study of e^+e^-duality has stimulated the systematic study of the various corrections to e^+e^- annihilation of bound states (Durand, 1982).

An important area of activity concerning both light and heavy quarks has been devoted to the calculation of another type of corrections — the unitarity *coupled-channel corrections* (Eichten *et al.*, 1978; Ross and Törnqvist, 1980; Törnqvist and Zenczykowski, 1984; Heikkilä *et al.*, 1984; Zambetakis, 1986).

Finally, in recent years, a new subject of interest has arisen in the domain of relativistic effects in light-quark physics: this is the study of the dynamical breaking of chiral symmetry (see references in Chapter 1). However, the calculation of transitions in this approach is still in a very early stage.

GENERAL BOOKS ON RELATIVISTIC QUANTUM MECHANICS AND FIELDS

Good introductions to quantum field theory are the books by Bogoliubov and Shirkov (1960), Schweber (1961) and Bjorken and Drell (1964, 1965). A modern book on quantum field theories is that of Itzykson and Zuber (1980). On the particular subject of the Dirac equation, there are good treatments in Messiah (1959) and Bjorken and Drell (1964). On two-body equations, a classical book is that of Bethe and Salpeter (1957). On the basic problems of field theory, and on the connection between time-ordered perturbation theory and Feynman diagrams, see Schweber, Bethe and de Hoffmann (1956). On the Heisenberg picture and the Bethe–Salpeter formalism, see Schweber (1961).

References

Abe, Y., Fujii, K., Okazaki, T., Arisue, H., Bando, M. and Toya, M. (1980) *Prog. Theor. Phys.* **64**, 1363

Abers, E. S. and Lee, B. W. (1973) *Phys. Rep.* **9**, 1

Ader, J.-P. (1980) Bordeaux preprint PTB 111, Talk at the Bressanone Symposium

Ader, J.-P., Bonnier, B. and Sood, S. (1980) *Z. Phys.* **C5**, 85

Ader, J.-P., Bonnier, B. and Sood, S. (1981) *Phys. Lett.* **101B**, 427

Ader, J.-P., Bonnier, B. and Sood, S. (1982) *Nuovo Cim.* **68A**, 1

Adler, S. L. (1965) *Phys. Rev.* **140**, 736

Adler, S. L. and Dashen, R. F. (1968) *Current Algebra*. New York: Benjamin

Adler, S. L. (1969) *Phys. Rev.* **177**, 2426

Adler, S. L. (1970) In *Lectures on Elementary Particles and Quantum Field Theory: Proc. 1970, Brandeis University Summer Institute in Theoretical Physics*, edited by S. Deser, M. Grisaru and H. Pendleton. Cambridge, Mass.: MIT Press

Adler, S. L. and Davis, A. C. (1984) *Nucl. Phys.* **B224**, 469

Alcock, J. W. and Cottingham, W. N. (1984) *Z. Phys.* **C25**, 161

Alcock, J. W., Cottingham, W. N. and Dunbar, I. H. (1980) *Ann Phys. (NY)* **130**, 164

Altarelli, G. (1982) *Phys. Rep.* **81**, 1

Altarelli, G. (1985) In *Proc. Int. Europhysics Conf. on High Energy Physics, Bari*, edited by L. Nitti and G. Preparata. Bari: Laterza.

Altarelli, G. and Maiani, L. (1974) *Phys. Lett.* **52B**, 351

Altarelli, G., Cabibbo, N. and Maiani, L. (1975) *Phys. Lett.* **57B**, 277

Amer, A. (1978) *Phys. Rev.* **D18**, 2290

Amer, A., Le Yaouanc, A., Oliver, L., Pène, O., Raynal, J.-C. (1983) *Phys. Rev. Lett.* **50**, 87

Andreadis, P., Baltas, A., Le Yaouanc, A., Oliver, L., Pène, O., Raynal, J.-C. (1974) *Ann. Phys. (NY)* **88**, 242

Appelquist, T. and Politzer, H. D. (1975) *Phys. Rev. Lett.* **34**, 43

Appelquist, T., Barnett, R. M. and Lane, K. D. (1978) *Ann. Rev. Nucl. Particle Sci.* **28**, 387

Babcock, J., Rosner, J. L., Cashmore, R. J. and Hey, A. J. G. (1977) *Nucl. Phys.* **B126**, 87

Baker, M., Ball, J. S. and Zachariasen, F. (1985) *Phys. Rev.* **D31**, 2575

Barbieri, R., Gatto, R., Kögerler, R. and Kunszt, Z. (1975a) *Phys. Lett.* **57B**, 455

Barbieri, R., Gatto, R., Kögerler, R. and Kunszt, Z. (1975b) *Phys. Lett.* **56B**, 477

Barnes, T. (1986) Contribution to *7th Int. Workshop on Photon–Photon Collisions, Paris 86*, Report UTPT-86-05

Barnes, T., Close, F. E. and De Viron, F. (1983) *Nucl. Phys.* **B224**, 241

Becchi, C. and Morpurgo, G. (1965) *Phys. Lett.* **17**, 352

Becchi, C. and Morpurgo, G. (1966) *Phys. Rev.* **149**, 1284

Bell, J. S. (1965) In *Les Houches Lectures* 1965, edited by C. De Witt and M. Jacob. Paris: Gordon and Breach

Bell, J. S. and Jackiw, R. (1969) *Nuovo Cim.* **60A**, 47

Bernard, V., Brockmann, R. and Weise, W. (1985) *Nucl. Phys.* **A440**, 605

Bergström, L., Snellman, H. and Tengstrand, G. (1980) *Z. Phys.* **C4**, 215
Bertlmann, R. A. and Ono, S. (1981) *Z. Phys.* **C8**, 271; **C10**, 37; see also *Phys. Lett.* **96B**, 123 (1980)
Bethe, H. A. and Salpeter, E. E. (1951) *Phys. Rev.* **82**, 350
Bethe, H. A. and Salpeter, E. E. (1957) *Quantum Mechanics of One and Two Electron Atoms*. Berlin: Springer-Verlag
Bjorken, J. D. and Drell, S. D. (1964) *Relativistic Quantum Mechanics*. New York: McGraw-Hill
Bjorken, J. D. and Drell, S. D. (1965) *Relativistic Quantum Fields*. New York: McGraw Hill
Bjorken, J. D. and Paschos, E. A. (1969) *Phys. Rev.* **185**, 1975
Blatt, J. M. and Weisskopf, V. F. (1952) *Theoretical Nuclear Physics*. New York: Wiley
Bogolioubov, N. N. et Chirkov, D. V. (1960) *Introduction à la théorie quantique des champs*. Paris: Dunod
Bogolioubov, P. N. (1968) *Ann. Inst. Henri Poincaré* **8**, 163
Böhm, M., Joos, H. and Krammer, M. (1972) *Nuovo Cim.* **7A**, 21
Böhm, M., Joos, H. and Krammer, M. (1972a) *Nucl. Phys.* **B51**, 397
Böhm, M., Joos, H. and Krammer, M. (1973b) *Acta Physica Austriaca Suppl.* **11**, 3
Böhm, M., Joos, H. and Krammer, M. (1974) *Nucl. Phys.* **B69**, 349
Bonvin, M. (1984) *Nucl. Phys.* **B238**, 241
Bourquin, M. *et al.* (1983) *Z. Phys.* **C21**, 27
Bowler, K. C. (1970) *Phys. Rev.* **D1**, 926
Brodsky, S. and Primack, J. R. (1969) *Ann. Phys. (NY)* **52**, 315
Buccella, F., Kleinert, H., Savoy, C. A., Celeghini, E., and Sorace, E. (1970) *Nuovo Cim.* **69A**, 133
Buccella, F., Savoy, C. A., Sorba, P. (1974) *Lettere al Nuovo Cimento* **10**, 455
Buchmüller, W. (1984) Lectures at the International School of Physics of Exotic Atoms, Erice
Buras, A. J. (1985) *Proceedings of the International Europhysics Conference* edited by L. Nitti and G. Preparata. Bari: Laterza
Burkhardt, H. and Pulido, J. (1978) Nottingham–Birmingham Preprint UNAM-BUMP-78-31
Busetto, G. and Oliver, L. (1983) *Z. Phys.* **C20**, 247

Cabibbo, N. (1963) *Phys. Rev. Lett.* **10**, 531
Cabibbo, N. and Maiani, L. (1978) *Phys. Lett.* **73B**, 418
Carlitz, R. and Kislinger, M. (1970) *Phys. Rev.* **D2**, 336
Carlitz, R. and Tung, W. K. (1976) *Phys. Rev.* **D13**, 3466
Carruthers, P. (1966) *Introduction to Unitary Symmetry*. New York: Interscience
Chaichian, M. and Kögerler, R. (1980) *Ann. Phys. (NY)* **124**, 61
Chodos, A., Jaffe, R. L., Johnson, K., Thorn, C. B. and Weisskopf, V. F. (1974a) *Phys. Rev.* **D9**, 3471
Chodos, A., Jaffe, R. L., Johnson, K. and Thorn, C. (1974b) *Phys. Rev.* **D10**, 2599
Christ, N. H. and Lee, T. D. (1980) *Phys. Rev.* **D22**, 940
Close, F. E. (1979) *An Introduction to Quarks and Partons*. London: Academic Press
Close, F. E. and Osborn, H. (1971) *Phys. Lett.* **34B**, 400
Close, F. E. and Horgan, R. R. (1980) *Nucl. Phys.* **B164**, 413; see also *Nucl. Phys.* **B185**, 333 (1981)
Colglazier, E. W. and Rosner, J. (1971) *Nucl. Phys.* **B27**, 349
Copley, L. A., Karl, G. and Obryk, E. (1969) *Nucl. Phys.* **B13**, 303

Cortes, J. L., Morales, A., Nuñez-Lagos, R. and Sanchez-Guillén, J. (1978) Zaragoza Preprint IFNAE 1/78

Creutz, M. (1981) In *Lectures at the International School of Subnuclear Physics Ettore Majorana*, edited by A. Zichichi, p. 119. New York: Plenum Press

Dalitz, R. H. (1965) *High Energy Physics, Les Houches*, edited by C. De Witt and M. Jacob, New York: Gordon and Breach

Dalitz, R. H. (1966) *Proc. Oxford Int. Conf. on Elementary Particles, Rutherford High Energy Laboratory*

Dalitz, R. H. (1967a) In *Proc. Topical Conf. on Pion–Nucleon Scattering, Irvine, California*, edited by G. L. Shaw and D. Y. Wong. New York: Wiley-Interscience

Dalitz, R. H. (1967b) In *Proc. 13th Int. Conf. on High Energy Physics, Berkeley*. Berkeley: University of California Press

Dalitz, R. H. (1975) In *Proc. Summer Symp. on New Directions in Hadron Spectroscopy, Argonne National Laboratory*. Publication A.N.L.-HEP CP.75.58

Dalitz, R. H. (1976) In *Proc. 7th Int. Conf. on Few Body Problems in Nuclear and Particle Physics, University of Delhi*

Dalitz, R. H. (1977) In *Quarks and Hadronic Structure*, edited by G. Morpurgo. New York: Plenum

Dalitz, R. H. and Reinders, L. J. (1979) *Hadron structure as known from electromagnetic and strong interactions*. In *Proc. Hadron Structure Conf., High Tatras*, edited by S. Dubnicka. Bratislava: Veda Publishing House

Dalitz, R. H. and Sutherland, D. G. (1966) *Phys. Rev.* **146**, 1180

Dashen, R. and Weinstein, M. (1969) *Phys. Rev.* **183**, 1261

De Forcrand, P. and Stack, J. D. (1985) *Phys. Rev. Lett.* **55**, 1254

De Grand, T. A. (1976) *Ann. Phys. (NY)* **101**, 496

De Grand, T. A., Jaffe, R. L., Johnson, K. and Kiskis, J. (1975) *Phys. Rev.* **D12**, 2060

De Rújula, A., Georgi, H. and Glashow, S. L. (1975) *Phys. Rev.* **D12**, 147

De Rújula, A., Georgi, H., Glashow, S. L. (1976) *Phys. Rev. Letters* **37**, 398

De Rújula, A., Georgi, H., Glashow, S. L. (1977) *Phys. Rev. Letters* **38**, 317

De Swart, J. J. (1963) *Rev. Mod. Phys.* **35**, 916

Donoghue, J. F. (1980) *Phys. Lett.* **92B**, 99

Donoghue, J. F. and Holstein, (1982) *Phys. Rev.* **D25**, 206

Donoghue, J. F. and Wyler D. W. (1978) *Phys. Rev.* **D17**, 280

Donoghue, J. F., Golowich, E., Ponce, W. A., Holstein, B. R. (1980) *Phys. Rev.* **D21**, 186

Donoghue, J. F., Golowich, E., Holstein, B. R. and Ponce, W. A. (1981) *Phys. Rev.* **D23**, 1213

Donoghue, J. F., Golowich, E. and Hostein, B. (1986) *Phys. Rep.* **131**, 319

Dosch, H. and Gromes, D. (1986) *Phys. Rev.* **D33**, 1378

Dover, C. B., Fishbane, P. M. and Furui, S. (1986) *Phys. Rev. Lett.* **57**, 1538

Durand, B. L. (1982) *Phys. Rev.* **D25**, 2312

Eeg, J. O. (1984) *Z. Phys.* **21**, 253

Eeg, J. O. (1985) *Moriond Workshop on Flavour Mixing and CP violation*, edited par J. Tran Thanh Van. Editions Frontières

Eichten, E. and Feinberg, F. L. (1981) *Phys. Rev.* **D23**, 2724

Eichten, E., Gottfried, K., Kinoshita, T., Kogut, J., Lane, K. D. and Yan, T. M. (1975) *Phys. Rev. Lett.* **34**, 369

Eichten, E., Gottfried, K., Kinoshita, T., Lane, K. D. and Yan, T. M. (1976) *Rev. Lett.* **36**, 500

Eichten, E., Gottfried, K., Kinoshita, T., Lane, K. D. and Yan, T. M. (1978) *Phys. Rev.* **D17**, 3090

Eichten, E., Gottfried, K., Kinoshita, T., Lane, K. D. and Yan, T. M. (1980) *Phys. Rev.* **D21**, 203

Fadeev, L. D. and Slavnov, A. A. (1980) *Gauge Fields: Introduction to Quantum Theory*. Menlo Park, Ca.: Benjamin/Cummings

Faiman, D. and Hendry, A. W. (1968) *Phys. Rev.* **173**, 1720

Faiman, D. and Hendry, A. W. (1969) *Phys. Rev.* **180**, 1572

Faiman, D. and Plane, D. E. (1972) *Phys. Lett.* **39B**, 358; *Nucl. Phys.* **B50**, 379

Faiman, D. and Rosner, J. (1973) *Phys. Lett.* **45B**, 357

Fakirov, D. and Stech, B. (1978) *Nucl. Phys.* **B133**, 315

Feynman, R. P. (1969) *Phys. Rev. Lett.* **23**, 1415

Feynman, R. P. (1972) *Photon–Hadron Interactions*. Reading, Mass.: W. C. Benjamin

Feynman, R. P., Kislinger, M. and Ravndal, F. (1971) *Phys. Rev.* **D3**, 2706

Fierz, M. (1937) *Z. Phys.* **104**, 553

Finger, M., Horn, D. and Mandula, J. E. (1979) *Phys. Rev.* **D20**, 3253

Finjord, J. (1978) *Phys. Lett.* **76B**, 116

Flamm, D. and Schöberl, F. (1978) *Nuovo Cim.* **46A**, 61

Flamm, D. and Schöberl, F. (1982) *Introduction to the Quark Model of Elementary Particles*. New York: Gordon and Breach

Foldy, L. L. and Wouthuysen, S. A. (1950) *Phys. Rev.* **78**, 29

Franklin, J. (1979) *Phys. Rev.* **D20**, 1742

Friar, J. L. (1974) *Phys. Rev.* **C10**, 955

Friar, J. L. (1975) *Phys. Rev.* **C11**, 274

Friedberg, R. and Lee, T. D. (1978) *Phys. Rev.* **D18**, 2623

Fritzsch, H. and Gell-Mann, M. (1972) In *Proc. 16th Conf. on High Energy Physics*, Batavia, vol. 2, p. 133, published by National Accelerator Laboratory

Fritzsch, H., Gell-Mann, M. and Leutwyler, H. (1973) *Phys. Lett.* **47B**, 365

Fuchs, N. (1975) *Phys. Rev.* **D11**, 1569

Fujimura, K., Kobayashi, T., Kobayashi, T. and Namiki, M. (1967) *Prog. Theor. Phys.* **38**, 210

Gaillard, M. K. and Lee, B. W. (1974a) *Phys. Rev. Lett.* **33**, 108

Gaillard, M. K. and Lee, B. W. (1974b) *Phys. Rev.* **D10**, 897

Gaillard, M. K., Lee, B. W. and Rosner, J. L. (1975) *Rev. Mod. Phys.* **47**, 277

Gasiorowicz, S. (1966) *Elementary Particle Physics*. New York: Wiley

Gasser, J. and Leutwyler, H. (1982) *Phys. Rep.* **87**, 77

Gavela, M. B., Le Yaouanc, A., Oliver, L., Pène, O., Raynal, J.-C. and Sood, S. (1978) *Phys. Lett.* **79B**, 459

Gavela, M. B., Le Yaouanc, A., Oliver, L., Pène, O., Raynal, J.-C. and Sood, S. (1979) *Phys. Lett.* **82B**, 431

Gavela, M. B., Le Yaouanc, A., Oliver, L., Pène, O., Raynal, J.-C. and Sood, S. (1980a) *Phys. Rev.* **D21**, 182

Gavela, M. B., Le Yaouanc, A., Oliver, L., Pène, O. and Raynal, J.-C. (1980b) *Phys. Rev.* **D22**, 2906

Gavela, M. B., Le Yaouanc, A., Oliver, L., Pène, O. and Raynal, J.-C. (1981a) *Phys. Rev.* **D23**, 1580

Gavela, M. B., Le Yaouanc, A., Oliver, L., Pène, O., Raynal, J.-C. and Pham, T. N. (1981b) *Phys. Lett.* **101B**, 417

Gavela, M. B., Le Yaouanc, A., Oliver, L., Pène, O., Raynal, J.-C. and Pham, T. N. (1982a) *Phys. Lett.* **109B**, 83
Gavela, M. B., Le Yaouanc, A., Oliver, L., Pène, O., Raynal, J.-C. and Pham, T. N. (1982b) *Phys. Lett.* **109B**, 215
Gavela, M. B., Le Yaouanc, A., Oliver, L., Pène, O. and Raynal, J.-C. (1982c) *Phys. Rev.* **D25**, 1921, 1931
Gell-Mann, M. (1961) CALTECH Report CTSL-20 (1961)
Gell-Mann, M. (1964a) *Phys. Rev. Lett.* **8**, 214
Gell-Mann, M. (1964b) *Physics* **1**, 63
Gell-Mann, M. (1972) In *Proc. 11th Int. Universitätswochen für Kernphysik Schladming; Acta Physica Austriaca Suppl.* **9**, 733
Gell-Mann, M. and Levy, M. (1960a) *Nuovo Cim.* **16**, 53
Gell-Mann, M. and Levy, M. (1960b) *Nuovo Cim.* **16**, 705
Gell-Mann, M. and Ne'eman, Y. (1964c) *The Eightfold Way*. New York: Benjamin
Gell-Mann, M., Oakes, R. and Renner, B. (1968) *Phys. Rev.* **175**, 2195
Glashow, S. L. (1961) *Nucl. Phys.* **22**, 579
Glashow, S., Iliopoulos, J. and Maiani, L. (1970) *Phys. Rev.* **D2**, 1285
Goldstone, J. (1961) *Nuovo Cim.* **19**, 164
Gottfried, K. (1977) In *Proc. Symp. on Lepton and Photon Interactions at High Energies, Hamburg*, edited by F. Gutbrod, published by Deutsches Elektronen Synchrotron DESY
Greenberg, O. W. (1964) *Phys. Rev. Lett.* **13**, 598
Greenberg, O. W. and Resnikoff, M. (1967) *Phys. Rev.* **163**, 1844
Gromes, D. (1984a) *Z. Phys.* **C22**, 265
Gromes, D. (1984b) Lectures given at the *Yukon Advanced Study Institute*, Whitehorse, Yukon
Gromes, D. and Stamatescu, I. O. (1976) *Nucl. Phys.* **B112**, 213
Gromes, D. and Stamatescu, I. O. (1979) *Z. Phys.* **C3**, 43
Gronau, M. (1972) *Phys. Rev.* **D5**, 118
Gross, D. J. and Wilczek, F. (1973) *Phys. Rev. Lett.* **30**, 1343
Guberina, B., Tadić, D. and Trampetić, J. (1981) *Lett. Nuovo Cim.* **32**, 193
Gürsey, F. and Radicati, L. A. (1964) *Phys. Rev. Lett.* **13**, 173
Gürsey, F., Pais, A. and Radicati, L. A. (1964b) *Phys. Rev. Lett.* **13**, 299

Halprin, A., Lee, B. W. and Sorba, P. (1976) *Phys. Rev.* **D14**, 2343
Hansson, T. H., Johnson, K. and Peterson, C. (1982) *Phys. Rev.* **D26**, 2069
Harari, H. (1969) *Phys. Rev. Lett.* **22**, 562
Hardekopf, G. and Sucher, J. (1982) *Phys. Rev.* **D25**, 2938
Hasenfratz, P. (1983) *Lattice quantum chromodynamics*. CERN Report TH-3737
Hasenfratz P., Horgan, R. R., Kuti, J., and Richard, J.-M. (1981) *Phys. Lett.* **95B**, 299
Hasenfratz, P. and Kuti, J. (1978) *Phys. Rep.* **40**, 75
Henriques, A. B., Kellett, B. H. and Moorhouse, R. G. (1976) *Phys. Lett.* **64B**, 85
Heikkilä, K., Törnqvist, N. A. and Ono, S. (1984) *Phys. Rev.* **D29**, 110
Hey, A. J. G., Lichtfield, P. J. and Cashmore, R. J. (1975) *Nucl. Phys.* **B95**, 516
Hey, A. J. G., Holstein, B. R. and Sidhu, D. P. (1978) *Ann. Phys. (NY)* **117**, 5
Higgs, P. W. (1964) *Phys. Rev. Lett.* **12**, 132
Høgaasen, H. and Richard, J. M. (1983) *Phys. Lett.* **124B**, 520
Höhler, G. (1979) *Handbook of Pion–Nucleon Scattering; Physik Daten* **12–1**
Horgan, R. and Dalitz, R. H. (1973) *Nucl. Phys.* **B66**, 135
Horn, D. and Mandula, J. (1978) *Phys. Rev.* **D17**, 898

Iizuka, J. (1966) *Prog. Theor. Phys. Suppl.* **37/38**, 21
Iliopoulos, J. (1976) CERN Report 76–11
Isgur, N. (1975) *Phys. Rev.* **D12**, 3666
Isgur, N. and Karl, G. (1978) *Phys. Rev.* **D18**, 4187
Isgur, N. and Karl, G. (1979) *Phys. Rev.* **D20**, 1191
Isgur, N. and Wise, M. (1982) *Phys. Lett.* **117B**, 179
Isgur, N., Karl, G. and Koniuk, R. (1978) *Phys. Rev. Lett.* **41**, 1269
Itzykson, B. and Zuber, J.-B. (1980) *Quantum Field Theory*. New York: McGraw-Hill

Jackson, J. D. (1976) In Proc. SLAC Summer Inst. on Particle Physics. SLAC report 198
Jackson, J. D. (1977) In *Proc. 12th Rencontre de Moriond*, edited by J. Tran Thanh Van. Editions Frontières
Jaffe, R. L. and Low, F. E. (1979) *Phys. Rev.* **D19**, 2105
Jensen, T. *et al.* (1983) *Phys. Rev.* **D27**, 26

Kang, J. S. and Schnitzer, H. J. (1975) *Phys. Rev.* **D2**, 841
Kane, G. and Karl, G. (1980) *Phys. Rev.* **D22**, 2808
Kaufman, W. B. and Jacob, R. J. (1974) *Phys. Rev.* **D10**, 1051
Kellett, B. H. (1974) *Ann. Phys. (NY)* **87**, 60; *Phys. Rev.* **D10**, 2269
Kim, Y. S. and Noz, M. (1986) *Theory and Applications of the Poincaré Group*. Dordrecht: Reidel
Kitazoe, T. and Teshima, T. (1968) *Nuovo Cim.* **57A**, 497
Kobayashi, M. and Maskawa, T. (1972) *Prog. Theor. Phys.* **49**, 282
Kogut, J. (1983) *Rev. Mod. Phys. Rev.* **55**, 775
Kogut, J. and Susskind, L. (1975) *Phys. Rev.* **D11**, 395
Kokkedee, J. J. (1969) *The Quark Model*. Amsterdam: Benjamin
Kokoski, R. and Isgur, N. (1985) Toronto Report UTPT-85-05.
Königsmann, K. (1985) *Physics in Collisions* 5, edited by B. Aubert and L. Montanet. Éditions Frontières
Koniuk, R. and Isgur, N. (1980) *Phys. Rev. Lett.* **44**, 845; *Phys. Rev.* **D21**, 1868
Körner, J. (1970) *Nucl. Phys.* **B25**, 282
Kraseman, H. and Ono, S. (1979) *Nucl. Phys.* **B154**, 283
Krzywicki, A. and Le Yaouanc, A. (1969) *Nucl. Phys.* **B14**, 246
Kubota, T. and Ohta, K. (1976) *Phys. Lett.* **65B**, 374
Kühn, J. H. (1986) Talk at International Symposium on Production and Decay of Heavy Hadrons (Heidelberg) Report MPI-PAE/PTh44/86

Landau, L. and Lifshitz, E. (1972) *Théorie quantique relativiste, première partie*, édité par V. Berestetski, E. Lifshitz et L. Pitayevski. Moscow: Éditions MIR
Landau, L. and Lifshitz, E. (1973) *Théorie quantique relativiste, deuxième partie*, édité par E. Lifshitz et L. Pitayevski. Moscow: Éditions MIR
Lee, B. W. and Swift, A. (1964) *Phys. Rev.* **136B**, 228
Lee, T. D. (1979) *Phys. Rev.* **D19**, 1802
Lee, T. D. (1981) *Particle Physics and Introduction to Field Theory*. New York: Harwood Academic Publishers.
Leite Lopes, J. (1981) *Gauge Field Theories, An Introduction.* Oxford: Pergamon
Leutwyler, H. (1974) *Nucl. Phys.* **B76**, 413
Leutwyler, H. (1979) Lecture given at the *International Summer Institute on Theoretical* Physics, Kaiserlslautern
Le Yaouanc, A., Oliver, L., Pène, O. and Raynal, J.-C. (1973) *Phys. Rev.* **D8**, 2223

Le Yaouanc, A., Oliver, L., Pène, O. and Raynal, J.-C. (1975a) *Phys. Rev.* **D11**, , 680
Le Yaouanc, A., Oliver, L., Pène, O. and Raynal, J.-C. (1975b) *Phys. Rev.* **D11**, 1272
Le Yaouanc, A., Oliver, L., Pène, O. and Raynal, J.-C. (1977a) *Phys. Lett.* **71B**, 397; *Phys. Lett.* **72B**, 57
Le Yaouanc, A., Oliver, L., Pène, O. and Raynal, J.-C. (1977b) *Phys. Lett.* **72B**, 53
Le Yaouanc, A., Oliver, L., Pène, O. and Raynal, J.-C. (1977c) *Phys. Rev.* **D15**, 844
Le Yaouanc, A., Oliver, L., Pène, O., Raynal, J.-C. and Longuemare, C. (1977d) *Phys. Rev.* **D15**, 2447
Le Yaouanc, A., Oliver, L., Pène, O., Raynal, J.-C. and Sood, S. (1978a) *Phys. Lett.* **76B**, 484
Le Yaouanc, A., Oliver, L., Pène, O. and Raynal, J.-C. (1978b) *Phys. Rev.* **D18**, 1591
Le Yaouanc, A., Oliver, L., Pène, O. and Raynal, J.-C. (1979) *Nucl. Phys.* **B149**, 321
Le Yaouanc, A., Oliver, L., Pène, O. and Raynal, J.-C. (1984) *Phys. Rev.* **D29**, 1233
Le Yaouanc, A., Oliver, L., Ono, S., Pène, O. and Raynal, J.-C. (1985a) *Phys. Rev.* **D31**, 137
Le Yaouanc, A., Oliver, L., Ono, S., Pène, O. and Raynal, J.-C. (1985b) *Z. Phys.* **C28**, 309
Licht, A. L. and Pagnamenta, A. (1970) *Phys. Rev.* **D2**, 1150; *Phys. Rev.* **D2**, 1156
Lipes, R. G. (1972) *Phys. Rev.* **D5**, 2849
Lipkin, H. J. (1967) *Nucl. Phys.* **B1**, 597
Lipkin, H. J. (1969) *Phys. Rev.* **183**, 1221
Lipkin, H. J. (1973a) *Phys. Lett.* **45B**, 267
Lipkin, H. J. (1973b) Proc. SLAC Summer Inst. Particle Physics, p. 239
Lipkin, H. J. (1976) Fermilab Conf. 76/98-THY
Lipkin, H. J. and Meshkov, S. (1965) *Phys. Rev. Lett.* **14**, 670
Llewellyn-Smith, C. H. (1969a) *Ann. Phys. (NY)* **53**, 521
Llewellyn-Smith, C. H. (1969b) *Nuovo Cim.* **60A**, 348

MacClary, R. (1982) UCLA Thesis
MacClary, R. and Byers, N. (1983) *Phys. Rev.* **D28**, 1692
Mackenzie, P. B. and Lepage, G. P. (1981) In *Proc. Tallahassee Meeting; AIP Conf. Proc.* **74**, 176
Mandelstam, S. (1955) *Proc. R. Soc. Lond.* **A233**, 248
Mandelstam, S. (1979) *Phys. Rev.* **D19**, 2391
Manley, D. M. *et al.* (1984) Virginia Polytechnic Institute and State University Preprint
Marciano, W. and Pagels, H. (1978) *Phys. Rep.* **36**, 138
Marshak, R. E., Riazuddin and Ryan, C. P. (1968) *Theory of Weak Interactions in Particle Physics*. New York: Wiley-Interscience
Martin, A. (1977) Lectures given at the 15th Int. School on Subnuclear Physics Ettore Majorana, Erice, edited by A. Zichichi. New York: Plenum Press
Martin, A. (1980) *Phys. Lett.* **100B**, 338
Martin, A. (1981) *Phys. Lett.* **100B**, 511
Martin, A. (1982) In *21ème Conf. Int. de Physique des Hautes Energies*, édité par P. Petiau et M. Porneuf. Paris: Éditions de Physique
Martin, A. (1984) Lectures at 22nd Int. School of Subnuclear Physics Ettore Majorana, edited by A. Zichichi. New York: Plenum Press
Melosh, H. J. (1974) *Phys. Rev.* **D9**, 1095
Messiah, A. (1959) *Mécanique Quantique*, Vols. 1 and 2. Paris: Dunod
Michael, C. (1986) *Phys. Rev. Lett.* **56**, 1219
Micu, L. (1969) *Nucl. Phys.* **B10**, 521
Milošević, M., Tadić, D. and Trampetić, J. (1982) *Nucl. Phys.* **B207**, 461

Mitra, A. N. (1981) *Z. Phys.* **C8**, 25
Mitra, A. N. and Ross, M. (1967) *Phys. Rev.* **158**, 1630
Mitra, A. N. and Santhanam, I. (1981) *Z. Phys.* **C8**, 33
Mitra, A. N. and Sood, S. (1977) *Fortschr. Phys.* **25**, 649
Miura, K. and Minamikawa, T. (1967) *Prog. Theor. Phys.* **38**, 954
Mohapatra, R. N. and Lai, C. H. (1981) *Selected papers on gauge theories of fundamental interactions* (edited by). Singapore: World Scientific
Moorhouse, R. G. (1966) *Phys. Rev. Lett.* **16**, 771
Moorhouse, R. G. and Parsons (1973) *Nucl. Phys.* **B62**, 109
Morpurgo, G. (1965) *Physics* **2**, 95
Morpurgo, G. (1969) In *Theory and Phenomenology in Particle Physics*, Part A, edited by A. Zichichi. New York: Academic Press
Morpurgo, G. (1983) Intervention at meeting on "*The Frontiers of Physics*", Bologna
Mott, N. F. and Massey, H. S. W. (1949) *The Theory of Atomic Collisions*. Oxford: Clarendon Press

Nagels, M. M. *et al.* (1979) *Nucl. Phys.* **B147**, 189
Nambu, Y. (1966) In *Preludes in Theoretical Physics*, edited by A. de Shalit, H. Feshbach and L. Van Hove, p. 133. Amsterdam: North-Holland
Nambu, Y. and Jona-Lasinio, G. (1961a) *Phys. Rev.* **122**, 345
Nambu, Y. and Jona-Lasinio, G. (1961b) *Phys. Rev.* **124**, 246
Narayanaswamy, P. and Pagnamenta, A. (1968) *Nuovo Cim.* **53A**, 635
Nielsen, H. B. and Olesen, P. (1973) *Nucl. Phys.* **B61**, 45
Novikov, V. A., Okun, L. B., Shifman, M. A., Vainshtein, A. I., Voloshin, M. B. and Zakharov, V. I. (1978) *Phys. Rep.* **41**, 1
Nishijima, K. (1958) *Phys. Rev.* **111**, 995

Okubo, S. (1962) *Prog. Theor. Phys.* **27**, 949
Okubo, S. (1963) *Phys. Lett.* **5**, 165
Okubo, S. (1968) *Ann. Phys (NY)* **47**, 351
Ono, S. (1976) *Phys. Rev. Lett.* **37**, 655
Ono, S. (1979) *Phys. Rev.* **D20**, 2975
Ono, S. (1981a) *Phys. Rev.* **D23**, 1118
Ono, S. (1981b) Aachen University Preprint 81/31
Ono, S. (1983) Lectures at *23rd Cracow School of Theoretical Physics*, Zakopane
Ono, S. (1984) In *Proc. Hadronic Session of the Rencontre de Moriond*, p. 305, edited by J. Tran Thanh Van. Editions Frontières
Ono, S. (1985) *Z. Phys.* **C26**, 307
Ono, S. and Schöberl, F. (1982) *Phys. Lett.* **118B**, 419

Pagels, H. (1975) *Phys. Rep.* **16**, 219
Particle Data Group (1982) *Phys. Lett.* **111B**
Particle Data Group (1984) *Rev. Mod. Phys.* **56**, No. 2, Part II
Pati, J. C. and Woo, C. H. (1971) *Phys. Rev.* **D3**, 2920
Pauli, W. and Villars, F. (1949) *Rev. Mod. Phys.* **21**, 434
Petersen W. (1975) *Phys. Rev.* **D12**, 2700
Petersen, W. and Rosner, J. (1972) *Phys. Rev.* **D6**, 820
Pham, T. N. (1984) *Phys. Rev. Lett.* **53**, 326
Poggio, E. C. and Schnitzer, H. J. (1979) *Phys. Rev.* **D20**, 1175
Poggio, E. C. and Schnitzer, H. J. (1980) *Phys. Rev.* **D21**, 2034

Politzer, H. D. (1973) *Phys. Rev. Lett.* **30**, 1346
Politzer, H. D. (1974) *Phys. Rep.* **14**, 129
Porter, F. C. (1981) Talk presented at SLAC Summer Inst. SLAC Pub. 2796

Quigg, C. and Rosner, J. L. (1979) *Phys. Rep.* **56**, 167

Rauh, K. G. (1981) *Z. Phys.* **C10**, 81
Reid, J. H. and Troffimenkoff, N. N. (1972) *Nucl. Phys.* **B40**, 255
Reinders, L. J., Rubinstein, H. and Yazaki, S. (1985) *Phys. Rep.* **127**, 1
Richard, J. M. (1981) *Phys. Lett.* **100B**, 515
Richard, J. M. and Taxil, P. (1983) *Ann. Phys. (NY)* **150**, 267
Richardson, L. (1979) *Phys. Lett.* **82B**, 272
Roos, M. and Törnqvist, N. A. (1980) *Z. Phys.* **C5,** 205
Rosenzweig, C. (1976) *Phys. Rev. Lett.* **36**, 697
Rosner, J. L. (1969) *Phys. Rev. Lett.* **22**, 689
Rosner, J. L. (1980) In *High Energy Physics; AIP Conf. Proc.* **68**, 540

Sakurai, J. J. (1964) *Invariance Principles and Elementary Particles.* Princeton: Princeton University Press
Sakurai, J. J. (1969) *Currents and Mesons.* Chicago: University of Chicago Press
Salam, A. (1968) In *Elementary Particle Physics; Proc 8th Nobel Symp.*, edited by N. Svartholm, p. 367. Stockholm: Almquist and Wiksell
Salpeter, E. E. (1951) *Phys. Rev.* **84**, 1226
Schmid, C. (1977) *Phys. Lett.* **66B**, 353
Schindler, R. (1985) SLAC-PUB-3799
Schweber, S. S. (1961) *An Introduction to Relativistic Quantum Field Theory.* Evanston Illinois: Row, Peterson
Schweber, S. S., Bethe, H. A. and de Hoffmann, F. (1956) *Mesons and Fields*, Vol 1. Evanston Illinois: Row, Peterson
Schwinger, J. (1951) *Phys. Rev.* **82**, 664
Shifman, M. A. Vainshtein, A. I. and Zakharov, V. I. (1978) *Phys. Rev.* **41**, 1
Shifman, M. A. Vainshtein, A. I. and Zakharov, V. I. (1979) *Nucl. Phys.* **B147**, 385, 448, 519
Shifman, M. A. (1981) In *Proc. Int. Lepton–Photon Symp., Bonn*, edited by W. Pfeil. Bonn: Physikalisches Institut der Universitäts

Sucher, J. (1978) *Rep. Prog. Phys.* **41**, 1781
Sucher, J. and Kang, J. S. (1978) *Phys. Rev.* **D18**, 2698
Sundaresan, M. K. and Watson, P. J. (1970) *Ann. Phys. (NY)* **59**, 375
Susskind, L. (1977) *Phys. Rev.* **D16**, 3031

't Hooft, G. (1976) *Phys. Rev.* **D14**, 3432
't Hooft, G. (1979) In *Recent Developments in Gauge Theories, Cargèse Lectures*, edited by G. 't Hooft *et al.* New York: Plenum
't Hooft, G. (1980) *Acta Physica Austriaca Suppl.* **22**, 531
Tanimoto, M. (1982) *Phys. Lett.* **116B**, 198
Törnqvist, N. A. and Ono, S. (1984) *Z. Phys.* **C23**, 59
Törnqvist, N. A. and Zenczykowski, P. (1984) *Phys. Rev.* **D29**, 2139
Vainshtein, A. I., Zakharov, V. I. and Shifman, M. A. (1977) *Sov. Phys. JETP* **45**, 670

Van Royen, R. and Weisskoof, V. F. (1967) *Nuovo Cim.* **50A**, 617

Weinberg, S. (1967) *Phys. Rev. Lett.* **19**, 1264
Weinberg, S. (1969) *Phys. Rev.* **177**, 2604
Weinberg, S. (1974) *Rev. Mod. Phys.* **46**, 255
Weinberg, S. (1977) *Trans. NY Acad. Sci., Ser. II*, **38**, 185
Weisberger, W. I. (1966) *Phys. Rev.* **143**, 1302
Wick, G. C. (1954) *Phys. Rev.* **96**, 1124
Wilson, K. G. (1969) *Phys. Rev.* **179**, 1499
Wilson, K. G. (1974) *Phys. Rev.* **D10**, 2445
Witten, E. (1979) *Nucl. Phys.* **B160**, 57

Zambetakis, V. (1986) PhD thesis, UCLA; UCLA/86/TEP/2
Zimmermann, W. (1958) *Nuovo Cim.* **10**, 597.
Zweig, G. (1964) CERN Rep. 8419/TH-412; CERN Preprints TH-401, TH-412

Index

A_1 decay to $\rho\pi$, 105–7
Additivity, 82, 89
Adler-Weissberger sum rule, 193
Allowed transitions, 217
Angular momenta couplings, 116
Anti-SU(6)$_W$ signs, 119–21, 226–7
Antisymmetric representation, 40
α_S (QCD), 6, 49, 94, 247–9, 270–1
Asymptotic freedom (QCD), 4, 19, 49, 75, 93
Axial anomaly, 27, 163–70
Axial coupling, 194
  g_1/f_1, 198
  F/D ratio, 194, 198
  F and D coefficients, 197–8
Axial currents, 70, 90, 133, 164

B decay to $\omega\pi$, 105–7, 112, 226–7
Bag model, 12, 127–8, 188, 209, 263–4
Baryon, 19–20, 62
  number conservation, 27
  pole contribution, 258, 260, 266–7
  radius (see radius)
  SU(3)$_F$ matrix, 197
  wave function, 39–47
Basic processes, 183–4
Bethe-Salpeter
  equation, 176–9
  amplitude, 154–80
Bottonium, 12, 35
Bottom quark mass, 188
Breit equation, 151
Breit-Fermi interaction, 54–6

Cabibbo
  angle, 21
  suppressed interaction, 83
  theory, 194–6
Casimir operator, 31
Center of mass
  kinetic energy, 44
  momentum, 33, 44
  motion, 83
Central
  interquark interactions, 186
  region, 75
  wave function, 79, 86, 180, 242, 261–2
Centrifugal barrier, 119–21, 192, 235
Charge conjugation, 34
  of spinors, 76
Charge interaction, 64
Charmed mesons
  radiative transitions, 211
  semileptonic decay, 5
  spectrum, 37
Charm quark mass, 188
Charmonium
  decay, 109
  decay of the J/ψ, 92
  E_1 radiative transition, 124, 210–9
  $\eta'_c \to J/\psi\gamma$, 146, 211–2
  $\psi' \to \eta_c\gamma$, 146, 211–2
  paracharmonium, 96
  radial excitation, 36
  radial excitation decays, 233–8
  system, 12, 35
Chiral
  anomaly, (see axial anomaly)
  symmetry, 57, 191
  symmetry explicit breaking, 274
  symmetry spontaneous breaking, 6, 28, 273
Color, 2, 170
  electric and magnetic fields, 47
  electric flux, 50, 110
  saturation, 32, 114
  singlet states (see hadrons), 19–20
  wave function, 43
Commutator contribution, 258
Confinement, 2, 6, 19
Confining
  Lorentz structure of the confining force, 56
  potential, 10

Constituent quarks (see valence quarks), 11, 20
Constituent masses, 10–1, 128
Contact interaction, 86
Convective current, 64, 69, 207, 212–5, 229
Cornell model (and Cornell potential), 108–10, 150, 153, 213–6, 221, 232, 234, 244, 248
Coulomb (see potential)
Coupled channel corrections, 110, 212, 218, 248
Current algebra 193, 200, 257
Current masses, 11, 26, 128
Current-current interaction, 23

$\Delta(1236)$, 4
  coupling to $N\pi$, 143
  decay to $p\gamma$, 132, 190, 203–4, 210
Decay width
  partial waves, 122
  into three body, 72, 74
  into two body, 67
  from cross section, 94–6
$\Delta I=1/2$ rule, (see non leptonic enhancement)
Dibaryon, 2
Diquonium 2, 17, 113–6
Dirac
  like free part, 62, 149–50
  equation, 13, 126–48, 151, 153–4, 198, 208–9, 216
  spinors normalization, 61, 72
Dispersion relation, 184, 200
Distance
  intermediate, 243
  large distance potential, 49
  short distance QCD potential (see potential-Coulomb), 49, 52, 242
  short distance wave function, (see central wave function)

E_1 transition, 212, 218–9
Electromagnetic
  interactions, 1–2
  current, 22, 60, 64–5, 70, 75, 129
Electroproduction, 64
  of baryon resonances, 203, 241
Electroweak interactions, 2, 20–3
Energy gap, 28
η decay into $\gamma\gamma$, 170
Even operators, (see Foldy-Wouthuysen)
Exotic states, 17

Factorizable contribution (see weak non leptonic decay), 252
Fermi coupling, 22–3
  formula, 67
Feynman diagrams, 154, 165, 256
Fierz transformation, 148
First quantization operators, 64, 83, 130
Flavor, 18
  conservation, 3, 26
  wave function, 41–2
Flux tube (see color electric), 110
Foldy-Wouthuysen
  transformation, 135–47
  odd, even operators, 135–40
Forbidden transitions, 212–3, 217
Form factors, 16, 185–6
F_π and f_π, 164, 252–3

Gauge bosons, 18
Gell-Mann matrices, 18, 196–7
Gell-Mann-Okubo mass formula, 25
Glueballs, 2
Gluon, 2
  condensate (QCD), 6, 50
Gluonic field (hadronic medium), 110
Goldberger-Treiman relation, 267
Goldstone bosons, 10–1, 26–9, 169, 191
  quasi-Goldstone bosons, 28, 245, 273
Green functions, 8, 155–74
Group of permutations, 40

Hadron (see color singlet states), 2
  charge in Mandelstam formalism, 162–3
  mass matrix formalism, 152
  usual hadrons, 2
Hard processes (see asymptotic freedom), 6
Harmonic oscillator
  potential, 115, 185, 209
  spectrum, 186, 215
  wave functions, 44, 185–6, 204, 222–31, 234–5, 240–1, 276
Heavyside-Lorentz units, 61
Heisenberg fields, 155
Helicity
  amplitudes, 107, 203–4
  of antiquarks, 148
Higgs bosons, 20
Hybrids, 2
Hyperons, (see strange baryons)

Intermediate region (see distance, intermediate)
Internal baryon conversion (see weak internal baryon conversion)
Internal wave function, 66, 77–8
  in Bethe-Salpeter, 179–80
Irreducible kernels, 157–8
Irreducible representations (SU)(3)), 24
Isospin, 25

j-j coupling, 130

Kaon (K), 24–6
  semi-leptonic decay, 74, 199–200
  to $\pi\pi$ decay, 17, 80, 264
Kinetic energy (bag model), 128
Kobayashi-Maskawa matrix, 23
Λ baryon, 27
  decay to Nπ, 80, 250–3, 262–8, 273–5
Λ parameter (QCD), 19, 242–3, 247

Lagrangian mass term (see current mass), 10
Lattice QCD, 7, 50–1, 193
Left-handed fermions, 20–1, 27
Lepton, 18
  currents, 71
  number conservation, 3
Lorentz
  boost, 159, 179–80
  contraction, 180
L_z representations, 142–4

Magnetic interaction, 64, 207, 229
Matter particles, 18
Meson, 19–20
  charge radius, 186
  wave function, 24–5, 33–9
  wave function radius (see radius)
Mixed symmetric representation, 40
Moorhouse selection rule, 206

N* decays, 4
  to Nγ, 201–10
  to Nπ, 104, 116–21
  to $\Delta\pi$, 225–30
  to Nρ, 228–30
N_c ($1/N_c$ expansion), 92–3
Nearby singularity, 277
Neutrino-production, 70
Nodes (wave functions), 231–38
Non Abelian group, 2, 19
Non-leptonic enhancement, 253, 261–2, 264–5, 271
Non relativistic
  approximation, 30, 63, 177–9, 189, 211, 243
  limit, 56, 125, 153, 175
  wave function, 62, 149, 175–9
Normal order, 83, 272
Normalization
  of hadron states, 61, 78–9, 163
  of spinors (see Dirac spinors)
  convention related to phase space, 189–91

Nucleon
  β decay, 74, 143, 194–9
  coupling to Nπ, 104–6, 143
  magnetic moment, 132, 201–2

Odd operators (see Foldy-Wouthuysen)
Okubo-Zweig-Ilzuka (OZI) rule, 4, 92, 186, 219–242, 248–50
ω decay
  to $\pi\gamma$, 124, 189, 210–1
  to e^+e^-, see ρ decay to e^+e^-
Ω decay to $\Xi\pi$, 280
Operator product expansion (Wilson), 7, 268
Orbital
  rotation, 33
  excitation of baryons (see N*), 45, 223
  excitations of mesons (see A_1, B), 37

Parity, 34
Parity doublets, 27
Partial conservation of axial current (PCAC), 11, 89, 168, 254, 273, 275
Pauli approximation (spinors), 63, 133
Pauli-Villars regularization, 167
Penguin diagrams, 270, 275
Peripheral
  regions, 75
  quantities, 187
Perturbative QCD
  (see hard processes and asymptotic freedom), 9, 93–4, 248–50,
  corrections (see radiative corrections), 180–1, 246–7, 249–50, 259, 268–75
Phenomenological potentials (see potential), 51
Photon (see radiative transitions), 201–19
  absorption and emission, 65, 66
  absorption and emission (Hamiltonian formalism), 152
Photoproduction, 183–4, 202
Pion (π)
  absorption, 140
  axial vector coupling, 90–1, 134, 140, 220
  coupling (direct term), 91, 104, 221, 223, 224, 225–8
  coupling (recoil term), 91, 104, 221, 223, 225–8
  decay to $\gamma\gamma$, 94, 163–70
  elementary π emission model, 88–91, 140–3, 220–8
  pole contribution, 168, 255
  pseudoscalar coupling, 91, 134, 220
  semi leptonic decay, 199
  leptonic decay ($\rightarrow\mu\nu$), 75, 79–80, 94
  Weak direct π emission (see weak non-leptonic decay)

φ decay
  to $K\bar{K}$ and 3π, 92
  to e^+e^-, see ρ decay to e^+e^-
Polarization
  hadrons, 77
  leptons, 73
  photon, 203
Pole contribution (see π, baryon)
Potential
  Coulomb, 5, 10, 47–8, 61, 242
  instantaneous, 150
  Lorentz scalar, 32, 56–8, 127
  Lorentz vector, 32, 56–8, 127, 178–9
  Martin, 244
  relativistic, 54
  Richardson, 244
Proton
  charge radius, 186–7
  magnetic moment, 132, 187
Pseudoscalar density (see quark pseudoscalar density)

Quantum Chromodynamics (QCD), 2
  Lagrangian, 18, 47
  lattice (see lattice QCD)
  perturbation (see perturbative QCD)
  duality sum rules, 7, 193
Quark, 18
  annihilation into lepton pairs, 74–9, 95–6, 159–60, 244–8
  annihilation into *n* gluons, 93, 248–50
  axial current (see axial currents)
  condensate, 6, 272
  electromagnetic annihilation (see annihilation into lepton pairs),
  electromagnetic current (see electromagnetic current),
  heavy (c,b,t), 12, 247
  internal velocity, 11, 124
  light (u,d,s), 10–1, 14
  masses (see constituent masses and current masses), 7, 188, 243–5, 274–5
  orbital angular momentum, 33
  pair corrections (see Z graphs)
  pair creation constant γ, 16, 100, 185, 230
  pseudoscalar density, 164
  relativistic quark models, 11
  structure, 187
  total spin, 33
  valence quarks (see constituent quarks), 150
  weak current (see weak current)
Quarkonia
  heavy, 6, 14, 87, 248
Quasi-Goldstone bosons (see Goldstone bosons)
ρ
  coupling to ππ, 105–6
  decay to e^+e^-, 75–8, 245
ρ′(1600) decay, 232–3

Radial excitation
  of baryons (see Roper resonance), 45
  of mesons (see charmonium, ρ′(1600)), 37
Radial quantum number, 33
Radiative corrections (see perturbative QCD), 180–1
  of axial currents, 143, 198–9
Radius of the wave function, 186–8
Reduction formula
  photon, 169
  general, 170, 254
Reference frame dependence, 189–91
Regge trajectory of Δ, 223–5
Relative coordinates, 44, 68
Relativistic corrections, 123, 126, 212, 216–8, 243, 275–7
  to Schrödinger energies, 57
Renormalization point (QCD), 6
Resonance, 4, 184
Right handed fermions, 20–1, 27
Roper resonance, 209–10, 238–42

Σ decay to Λγ, 132
S-matrix, 66, 114
Salpeter equation, 178
Schrödinger
  equation, 5, 10, 29, 62, 125
  wave function, 141, 175
Second quantization, 14, 130
Second class current, 195
Short distance QCD potential, (see distance) and Potential (Coulomb)
Short distance QCD corrections, (see perturbative QCD)
Soft pion theorems, 29
Spatial wave functions, 37, 43–4
Spectator quarks, 99
Spin
  rotation, 33
  wave-function, 37, 41
  dependent forces, 14, 52–8
Spin-flavor wave function, 42–3
Spin-flavor-space wave function, 45–6
Spin-orbit
  current, 144–5, 208
  forces (see spin dependent forces), 53

Spin-spin forces (see spin dependent forces), 53, 266
Stable particles, 3
Standard model, 2
Strange baryons
  strong decays, 16
  weak decays, (see weak non leptonic decay)
String tension, 50, 112
Strong interactions, 2
Strong binding, 124
Strong coupling calculations, 7
$SU(3)_C$ gauge group, 2
$SU(3)_F$
  symmetry, 24
  breaking, 25
SU(6) and SU(6)×O(3)
  symmetry 35–6, 191–2
  analysis, 191–2, 207–9
$SU(6)_W$(anti-$SU(6)_W$), 192
  signs, 119–21, 226
Superconductivity, 50
Symmetric representation, 40

Transition rate (see decay width), 67

Vacuum
  of QCD, 27
  expectation value, 28, 272–5
Valence quark approximation (see constituent quarks), 62
Van der Waals force, 114
Vector current (see electromagnetic current and quark weak current)
  conserved vector current (CVC), 194, 199
Vector meson dominance (VMD), 89, 242
Velocity
  internal quark velocity, 124–5, 132

Ward identity, 164–8
Weak
  charged interaction, 3, 147
  current, 22–3, 70–1
  direct pion emission, 250–4
  internal baryon conversion, 250–1, 254–68
  mixing angle, 22
  parity conserving (violating) Hamiltonian, 81–2
Weak binding of quarks, 124
Wick's theorem, 272
Wilson
  loop, 57
  expansion, (see operator product expansion)

Yukawa coupling, (see pion pseudoscalar coupling), 267

Z graphs, 127, 130, 153, 179, 253–4, 256–7